Protected Areas, Sustainable Tourism and Community Livelihood Linkages

The book uses a multi-disciplinary approach to address lessons learned and challenges encountered over the years in different ecological, economic, political and cultural contexts.

Protected areas were originally established as recreational spaces and to protect some components of nature; however, today they are also expected to provide an increasing range of benefits to an array of people. Protected areas no longer simply "protect" but they also provide ecosystem services and facilitate poverty reduction via local development, ecotourism and sustainable resource use. Integrating tourism and conservation with existing local historical, socio-economic and institutional landscapes is associated with the promotion of local community participation in resource management. The book adopts an interdisciplinary approach to understand social-ecological systems that explain the relationship between protected areas, tourism and community livelihoods linkages. The book provides a platform for dialogue to develop a better understanding of the complex relationships between protected areas, tourism, and community livelihoods linkages. Due to the role tourism plays in poverty alleviation, conservation, empowerment and addressing other environmental and social challenges, the book also connects tourism with the UN Sustainable Development Goals.

This book will be of great interest to students, scholars and policymakers of tourism, conservation, natural resource management, sustainable development as well as professionals and policymakers involved in conservation policy. The chapters in this book were originally published as a special issue of *Journal of Sustainable Tourism*.

Moren Tibabo Stone is Associate Professor of environmental science and tourism studies at the University of Botswana, Department of Environmental Science. His research interests include sustainable tourism development and management, ecotourism, community-based tourism, protected areas conservation and community livelihoods dynamics.

Lesego Senyana Stone is Associate Professor in tourism management at the University of Botswana, Department of Tourism and Hospitality Management. Her research interests are in sustainable tourism development with specific reference to nature-based tourism, community-based tourism and community participation in tourism.

Gyan P. Nyaupane is Professor in the School of Community Resources & Development at Arizona State University. He has research experience in the fields of protected areas, public lands and sustainable tourism planning and policy. His research interests include understanding human–environment interactions, sustainable communities, resilience, indigeneity, nature-based tourism and policy and planning.

Protected Areas, Sustainable Tourism and Community Livelihood Linkages

Edited by
Moren Tibabo Stone, Lesego Senyana Stone and Gyan P. Nyaupane

Routledge
Taylor & Francis Group
LONDON AND NEW YORK

First published 2025
by Routledge
4 Park Square, Milton Park, Abingdon, Oxon OX14 4RN

and by Routledge
605 Third Avenue, New York, NY 10158

Routledge is an imprint of the Taylor & Francis Group, an informa business

British Library Cataloguing in Publication Data
A catalogue record for this book is available from the British Library

ISBN13: 978-1-032-83175-6 (hbk)
ISBN13: 978-1-032-83188-6 (pbk)
ISBN13: 978-1-003-50818-2 (ebk)

DOI: 10.4324/9781003508182

Typeset in Myriad Pro
by Newgen Publishing UK

Publisher's Note
The publisher accepts responsibility for any inconsistencies that may have arisen during the conversion of this book from journal articles to book chapters, namely the inclusion of journal terminology.

Disclaimer
Every effort has been made to contact copyright holders for their permission to reprint material in this book. The publishers would be grateful to hear from any copyright holder who is not here acknowledged and will undertake to rectify any errors or omissions in future editions of this book.

Contents

Citation Information

The chapters in this book were originally published in the *Journal of Sustainable Tourism*, volume 30, issue 11 (2022). When citing this material, please use the original page numbering for each article, as follows:

Chapter 1

Theorizing and contextualizing protected areas, tourism and community livelihoods linkages
Moren Tibabo Stone, Lesego Senyana Stone and Gyan P. Nyaupane
Journal of Sustainable Tourism, volume 30, issue 11 (2022), pp. 2495–2509

Chapter 2

Resident support of community-based tourism development: Evidence from Gunung Ciremai National Park, Indonesia
Prasetyo Nugroho and Shinya Numata
Journal of Sustainable Tourism, volume 30, issue 11 (2022), pp. 2510–2525

Chapter 3

Impacts of tourism on support for conservation, local livelihoods, and community resilience around Maasai Mara National Reserve, Kenya
Kathleen Krafte Holland, Lincoln R. Larson, Robert B. Powell, W. Hunter Holland, Lawrence Allen, Moriaso Nabaala, Salaton Tome, Simon Seno and James Nampushi
Journal of Sustainable Tourism, volume 30, issue 11 (2022), pp. 2526–2548

Chapter 4

Community-Based ecotourism and bushmeat consumption dynamics: Implications for conservation and community development
Moren Tibabo Stone and Lesego Senyana Stone
Journal of Sustainable Tourism, volume 30, issue 11 (2022), pp. 2549–2573

Chapter 5

Tourism development discourse dynamics in a context of conflicts between mining and nature conservation in the Brazilian Cerrado Hotspot
Nayara Marques, Mozart Fazito and André Cunha
Journal of Sustainable Tourism, volume 30, issue 11 (2022), pp. 2574–2594

Chapter 6

Connecting landscape-scale ecological restoration and tourism: stakeholder perspectives in the great plains of North America
Connor Clark and Gyan P. Nyaupane
Journal of Sustainable Tourism, volume 30, issue 11 (2022), pp. 2595–2613

Chapter 7

Environmentality, green grabbing, and neoliberal conservation: The ambiguous role of ecotourism in the Green Life privatised nature reserve, Sumatra, Indonesia
Chantal Elizabeth Wieckardt, Stasja Koot and Nadya Karimasari
Journal of Sustainable Tourism, volume 30, issue 11 (2022), pp. 2614–2630

Chapter 8

Privately protected areas in increasingly turbulent social contexts: strategic roles, extent, and governance
William Thomas Borrie, Trace Gale and Keith Bosak
Journal of Sustainable Tourism, volume 30, issue 11 (2022), pp. 2631–2648

Chapter 9

The use of tourism as a social intervention in indigenous communities to support the conservation of natural protected areas in Mexico
Gerda Warnholtz, Neil Ormerod and Chris Cooper
Journal of Sustainable Tourism, volume 30, issue 11 (2022), pp. 2649–2664

Chapter 10

Social and cultural capitals in tourism resource governance: the essential lenses for community focussed co-management
Muhammad Shoeb-Ur-Rahman, David Simmons, Michael C. Shone and Nazmun N. Ratna
Journal of Sustainable Tourism, volume 30, issue 11 (2022), pp. 2665–2685

Chapter 11

Governance of protected areas: an institutional analysis of conservation, community livelihood, and tourism outcomes
Gyan P. Nyaupane, Surya Poudel and Abigail York
Journal of Sustainable Tourism, volume 30, issue 11 (2022), pp. 2686–2705

Notes on Contributors

Lawrence Allen, Department of Parks, Recreation, & Tourism Management, Clemson University, Clemson, SC, USA.

William Thomas Borrie, School of Life and Environmental Sciences, Deakin University, Burwood, Victoria, Australia.

Keith Bosak, Department of Society and Conservation, University of Montana, Missoula, MT, USA.

Connor Clark, School of Community Resources and Development, Arizona State University, Phoenix, AZ, USA,

Chris Cooper, School of Events, Tourism and Hospitality Management, Leeds Beckett University, Leeds, UK.

André Cunha, Lab for Biodiversity and Protected Areas, Department of Ecology, University of Brasilia, Brasilia, Brazil; Research Centre UnB Cerrado, University of Brasilia, Brasilia, Brazil.

Mozart Fazito, Tourism Department, Federal University of Rio Grande do Norte, Natal, Brazil.

Trace Gale, Human-Environmental Interactions Group, Center for Patagonia Ecosystem Investigation, Coyhaique, Chile.

W. Hunter Holland, College of Health and Human Services, University of North Carolina Wilmington, Wilmington, NC, USA.

Kathleen Krafte Holland, Department of Environmental Sciences, University of North Carolina Wilmington, Wilmington, NC, USA.

Nadya Karimasari, Sociology of Development and Change, Wageningen University, Wageningen, The Netherlands.

Stasja Koot, Sociology of Development and Change, Wageningen University, Wageningen, The Netherlands; Department of Geography, Environmental Management & Energy Studies, University of Johannesburg, Auckland Park, South Africa.

Lincoln R. Larson, Department of Parks, Recreation, & Tourism Management, North Carolina State University, Raleigh, NC, USA.

Nayara Marques, Centre for Excellence in Tourism, University of Brasilia, Brasilia, Brazil.

Moriaso Nabaala, Koiyaki Guiding School, Sekenani, Kenya.

James Nampushi, School of Tourism, Hospitality and Leisure Studies, Maasai Mara University, Narok, Kenya.

Prasetyo Nugroho, Vocational College of Universitas Gadjah Mada, Yogyakarta, Indonesia; Department of Tourism Science, Tokyo Metropolitan University, Tokyo, Japan.

Shinya Numata, Department of Tourism Science, Tokyo Metropolitan University, Tokyo, Japan.

Gyan P. Nyaupane, School of Community Resources & Development, Arizona State University, Phoenix, AZ, USA; College of Business & Economics, University of Johannesburg, Johannesburg, South Africa.

Neil Ormerod, School of Events, Tourism and Hospitality Management, Leeds Beckett University, Leeds, UK.

Surya Poudel, School of Community Resources & Development, Arizona State University, Phoenix, AZ, USA.

Robert B. Powell, Department of Parks, Recreation, & Tourism Management, Clemson University, Clemson, SC, USA.

Nazmun N. Ratna, Department of Global Value Chains and Trade. Lincoln University, Lincoln, New Zealand.

Simon Seno, School of Tourism, Hospitality and Leisure Studies, Maasai Mara University, Narok, Kenya.

Muhammad Shoeb-Ur-Rahman, Department of Tourism and Hospitality Management, Faculty of Business Studies, University of Dhaka, Dhaka, Bangladesh.

Michael C. Shone, Ara Institute of Canterbury, Christchurch, New Zealand.

David Simmons, Department of Tourism, Sport and Society, Lincoln University, Lincoln, New Zealand.

Lesego Senyana Stone, Department of Hospitality & Tourism Management, University of Botswana, Gaborone, Botswana.

Moren Tibabo Stone, Department of Environmental Science, University of Botswana, Gaborone, Botswana.

Salaton Tome, School of Tourism, Hospitality and Leisure Studies, Maasai Mara University, Narok, Kenya.

Gerda Warnholtz, School of Events, Tourism and Hospitality Management, Leeds Beckett University, Leeds, UK.

Chantal Elizabeth Wieckardt, Sociology of Development and Change, Wageningen University, Wageningen, The Netherlands; LANDac –The Netherlands Land Academy, Utrecht University, Utrecht, The Netherlands.

Abigail York, School of Human Evolution and Social Change, Arizona State University, Tempe, AZ, USA.

Theorizing and contextualizing protected areas, tourism and community livelihoods linkages

Moren Tibabo Stone, Lesego Senyana Stone and Gyan P. Nyaupane

ABSTRACT

Although efforts have been made to connect the conservation of protected areas (PAs) to the livelihoods of people living in adjacent localities through sustainable tourism development, the debate is still evolving and inconclusive. This special issue contributes to the discourse by endeavouring to advance discussions on relationships that exist between protected areas, tourism and community livelihoods. This paper synthesizes accomplishments and challenges that exist in studying linkages between PAs, tourism and community livelihoods, which to date have been sporadic and mostly atheoretical. Empirical evidence from different researchers are highlighted to guide interested stakeholders, policy planners and practitioners to glean and adapt their strategies to achieve integrated conservation and community development. The linkages between PAs, tourism and community livelihoods can best be understood by considering different theories, conceptual frameworks and methodologies used to understand the relationship dynamics with a view to resolve the demands of livelihoods and conservation that have been advanced over the last fifty years.

Introduction

For about 50 years, tourism has been touted as having the potential to improve the connection between biodiversity conservation and livelihood (Sene-Harper et al., 2019). Since then, the conservation and development nexus debate has been endemic in tourism and conservation literature. Local communities, through tourism participation, often depend on adjacent protected areas for their livelihoods, while the state and other outside groups have other conservation goals (Madden & McQuinn, 2014). This has resulted in conflicting goals that form part of important conservation and development discussions and the development of different efforts designed to manage and/or reconcile resultant outcomes over the last decades.

Important to note is that globally protected areas are not neutral landscapes but highly contested (Eagles et al., 2002). Originally established as conservation and recreation spaces, the protected area domain today is expected to provide an increasing range of benefits to an array of stakeholders (Hossu et al., 2017). Protected areas no longer simply "protect" but they also provide ecosystem services and facilitate poverty reduction through local development, ecotourism, and sustainable resource use. Consequently, people living in and around these protected areas often have high expectations of what PAs and tourism can offer them.

As protected area tourism occurs mainly in isolated and remote rural regions, it is often assumed that such regions will experience the stimulation of economic activities induced by tourism from which local people will be able to derive tangible benefits (UNEP-WCMC & IUCN, 2016). Evidence from different parts of the world suggest that this is rarely the case (see Snyman, 2012; Stone & Stone, 2011). Indeed, in most protected areas, benefits hardly reach the local community that bears the heaviest burden of protected area co-existence and management (Fisher, 2006). Others (e.g. Eagles et al., 2002) point that when a protected area is established and opened for tourism, it is often outsiders who rush in, to siphon-off a major portion of the tourism income generated locally.

While some scholars (e.g. Jamal & Stronza, 2009; Mbaiwa, 2008) call for the participation of local communities in protected area management to help benefit from such resources, others are cautious (e.g. Fisher, 2006) as they reason that protected areas are not necessarily bound in any predetermined relationship to poverty or wealth. Nevertheless, there is an understanding that protected areas must be integrated into the surrounding land and with the neighbouring human populations (e.g. Bennett et al., 2012; Nyaupane & Poudel, 2011; Stone & Nyaupane, 2018). Integrating tourism and conservation with existing local historical, socio-economic, and institutional landscapes is associated with the promotion of local community participation in resource management. This integrated approach is an appealing concept, often quoted as essential in promoting conservation and rural development (Tumusiime & Vedeld, 2012).

In studying relationships that exist between protected areas and visitors, a number of indicator-based frameworks have been proposed to conceptualize, frame, predict and manage visitor impacts on protected areas tourism. To mention but a few, these include Limits of Acceptable Change, Visitor Impact Management, Visitor Activity Management Process, Recreational Opportunity Spectrum, Visitor Carrying Capacity and the Visitor Impact Management Model (Kuss et al., 1990; Newsome, Moore & Dowling, 2002; Salafsky & Wollenberg, 2000; Cater, 1994).

To explain the relationship between conservation and tourism, Budowski (1976) developed a framework that explains three relationship scenarios of conflict, coexistence, and symbiosis. Salafsky and Wollenberg (2000) conceptual framework comprises three scenarios of no linkage, indirect linkage, and direct linkage. Cater (1994) conceptualizes four possible links: win/win, win/lose, lose/win and lose/lose. Scheyvens (1999) developed an ecotourism empowerment framework, which comprises four dimensions: social, economic, political, and psychological empowerment. The focus of these frameworks seems to be on the current state of the system without considering complex interactions and interdependencies between resources and stakeholders (Strickland-Munro et al., 2010). More recently, Nyaupane and Poudel (2011) developed and tested a framework to show the two-way relationships between biodiversity conservation, tourism development and livelihood improvement.

Emerging thinking conceptualizes tourism as a complex system, consisting of multiple interacting components (Lacitignola et al., 2007; Strickland-Munro et al., 2010), others call for systems thinking in the conceptualization of the relationship (e.g. Poudel & Nyaupane, 2018; Stone & Nyaupane, 2017). Therefore, any PAs and tourism study conducted without the explicit recognition of interacting variables e.g. political, social, cultural, physical, economic, historic, and ecological, will reveal an incomplete and possibly confusing picture as the complex interactions between system components will not be apparent. It is on this background that this special issue and paper are premised.

However, this approach does not strive for a simple answer to solve complex problems (Berkes et al., 2009), but tries to build on the work of scholars who have undertaken carefully, well documented and theoretically sound studies on socio-economic systems of tourism and ecosystems (Berkes et al., 2009; Budowski, 1976; Cater, 1994; Salafsky & Wollenberg, 2000). Protected areas tourism has problems which tend to be systems problems, where aspects of behaviour are complex and unpredictable and where causes, while at times simple, are always multi-faceted (Ostrom, 2007).

This special issue aims to critically analyse existing frameworks, methods, and state of knowledge, and proposes future directions on linkages research. Inter and multi-disciplinary approaches are adopted to try to understand social-ecological systems that explain the relationship between protected areas, tourism and community livelihoods. The special issue volume endeavours to provide a platform for dialogue, contributing further to the enhancement of understanding of the complex relationships between protected areas, tourism and community livelihoods linkages. Due to the role tourism plays in poverty alleviation, conservation, and empowerment and addressing other environmental and social challenges, this special issue connects protected areas, tourism and livelihoods with the UN Sustainable Development Goals.

Having highlighted on the focus, scope of the special issue and aim of this introductory paper, the paper proceeds with five other sections, namely; contextualizing protected areas, tourism and livelihoods relationships; community-based tourism and livelihoods improvement debate; synthesis of accomplishments and challenges of the debate (state of the art), followed by the section on the way forward and, lastly the conclusion.

Contextualizing protected areas, tourism and livelihoods relationships

It is clear that protected areas (PAs) have contributed enormously in the development of tourism, particularly nature-based tourism (e.g. Nyaupane & Poudel, 2011; Park et al., 2019; Sebele, 2010) as tourist attractions, and as contributors to the world's socio-economic growth (e.g. Dlamini & Masuku, 2013; Zhang et al., 2019). Unlike other forms of tourism that may be dependent on man-made infrastructural development for their growth and hence favour only those regions which have developed infrastructure, protected area tourism is not selective and may favour those areas that are not relatively developed as it depends on pristine and relatively undisturbed natural environments. Paradoxically, while the developed world is endowed with infrastructure-related development, the developing world is endowed with natural and cultural resources. Developing countries have comparatively realised success in protected areas tourism development as one of its economic drivers (Jones et al., 2013). Protected area tourism has garnered support for environmental conservation from non-traditional actors in public and private sector institutions (Backman & Munanura, 2017), and has led to wildlife conservation and the promotion of human-wildlife coexistence, community participation and improvements in community livelihoods (Stone & Nyaupane, 2017) through numerous community tourism enterprises and conservation initiatives.

To provide the context of this conceptualization, extant studies (e.g. Nyaupane & Poudel, 2011; Stone & Nyaupane, 2017, 2018) have explored relationships between biodiversity conservation, tourism development and livelihood improvements. Taking a leaf from such studies, it has emerged that PAs no longer serve the sole objective of conserving biodiversity just for its existence, but have accommodated other land uses such as tourism, that in turn serve the interests of tourists and destination communities (Gezon, 2014). The resultant outcomes of the relationship between biodiversity conservation, tourism development and livelihood improvements have produced mixed results (Stone, 2013).

There is no winning formula on which these relationships can co-exist without adverse consequences for both conservation and development (Stone & Nyaupane, 2016). As a result, Nyaupane and Poudel (2011) report that the distribution of benefits and costs of conservation are disproportionate, while Balmford and Whitten (2003) contend that the greatest benefits of protected areas and their bequest and existence values largely benefit global citizens, whereas Matiku (2008) argues that the costs incurred are absorbed by local communities living near protected areas.

As far as we know, there is no single study that reports only positive relationships between protected areas, tourism and community livelihoods without any adverse consequence emanating from such engagements. For example, many case studies indicate that the relationships between conservation and tourism (Lowe et al., 2019), tourism and community livelihood improvement (Aazami & Shanazi, 2020); and community livelihoods and conservation (Pozo et al., 2021) are dynamic, complex, conflicting, multifaceted and contradictory such that there is no single framework to examine the complex interactions and yield a conclusive debate.

Nevertheless, emerging literature (e.g. Stone & Nyaupane, 2018) indicates that protected areas and the community-based tourism model still promises local economic growth, inclusive development processes, and flows of benefits to local communities. While this model has delivered on economic opportunities in some cases, the viability of its impacts in some other areas is questionable as the inequitable distribution of benefits and lack of broad participation by the community remain critical debilitating factors (Stone & Stone, 2020). Yet, for conservation efforts to be effective there needs to be backing from and commitment from local communities (Doyle-Capitman et al., 2018) marking a departure from initial conservation approaches that largely mirrored fortress and protectionist dogmas that militantly evicted inhabitants violently (Adams, 2013), regardless of the historical relationship inhabitants have had with their environment.

By and large, ideals of fortress conservation have long been abandoned in favour of integrated conservation and development (ICD) approaches that emerged with a goal to improve and reconcile co-existence among protected areas and community development (Franks & Blomley, 2004). The ICD approach envisaged the notion that community support for conservation goals can be viable only when community benefits outweigh the costs of living with such resources (Swemmer et al., 2017). This paradigm shift in conservation and development witnessed the call for tourism as a potential development vehicle that can connect conservation landscapes and community development. While tourism was perceived as the right development tool, it was received with mixed feelings because communities were not empowered to adopt and run enterprises right away (see Büscher, 2013; Stone & Stone, 2020).

Literature indicates a myriad of tourism problems such as lack of community participation (Wondirad & Ewnetu, 2019), lack of business skills and acumen (Summers et al., 2019), development of elitism in community-based tourism projects (Li et al., 2020), lack of benefits distribution plans (Andrade, 2018), misappropriation and embezzlement of revenue generated (Nelson, 2012).

Community-based tourism and livelihoods improvement debate

Community-based tourism links with community livelihoods improvement have been presented as a remedy for amalgamating conservation and development goals and objectives (Mathis & Rose, 2016) to forge conservation and development together (Stone & Stone, 2017). Community-based tourism is promoted to encourage community political and psychological empowerment (Butler, 2017); accrual of financial benefits (Naidoo et al., 2016); employment creation (Mishra & Verma, 2017); social trust and cohesion (Nunkoo, 2017); human-wildlife coexistence (Broekhuis et al., 2020; König et al. 2020); and ultimately building value towards natural resources (Wang et al., 2016).

Though tourism has delivered in terms of conservation and community development in many parts of the world (Sebele, 2010), it still experiences some problems due to many challenges that often lead to the disillusionment of conservation goals. For instance, the benefits of revenue generated from tourism may not be equitably distributed within the community (Balmford & Whitten, 2003). Therefore, sustainable tourism should be the guiding principle in linking conservation and improvements in community livelihoods. Sustainable tourism development should be aligned to community lifestyles, goals, values, and practices and carefully coordinated to meet desired standards of livelihoods and minimize unintended consequences (Walker, Micthell & Wismer, 2001).

Careful tourism development planning is key as tourism is generally adopted into a portfolio of economic activities as a diversification option, reinforcing synergies or trade-offs by either complementing or clashing with the seasonality of traditional livelihoods like agriculture, livestock management or fishing (Sene-Harper et al., 2019). Therefore, it is vital to appreciate and probe how tourism matches the community livelihood portfolio. Focusing on the relationship between protected areas, tourism and community livelihoods broadens the context of investigation and attracts multi-disciplinary approaches that have a propensity to yield rich multifaceted results that can provide better insights on the relationship. Hence, Tao and Wall (2009) argue that if the links between tourism and other sectors are not adequately appreciated, and if the context in which tourism occurs is overlooked, then understanding is likely to be a fraction of the whole.

Synthesis of accomplishments and challenges – state of the art debate

In this section, the conservation and development dialogue centred on protected areas and tourism, is synthesized, largely informed by useful insights generated by papers from this volume. Results from articles making up this volume indicate that the relationships between PAs, tourism and livelihoods are complex, multi-faceted, multidimensional, contradictory, divergent, and nested in many factors such as the level of tourism development, agents facilitating conservation and development efforts, implementation processes, diverse stakeholders' participation and institutional arrangements that provide guidance in planning and implementation processes. The key findings derived from contributing papers point to theory and practice, policy frameworks, institutional landscapes and diverse stakeholders as the "*conservation and development mix*" that account for ultimate outcomes (i.e. intended and unintended), henceforth, determining the performance and relationships of protected areas, tourism and livelihoods interplay dynamics. These conservation and development mix are discussed in detail below.

From theory to practice

There is a general consensus that protected areas provide tourism resources and in turn tourism serves communities living within and around PAs by providing livelihood opportunities. The contest is how best PAs and tourism can serve both conservation and development objectives sustainably, without yielding detrimental outcomes that undermine the same efforts. We argue here that the contest emanates from the transition from theory to practice, as this can be challenging.

Theory and practice should provide guidance to practitioners or stakeholders leading PAs and tourism organisational change processes. Protected areas based on current scientific knowledge are rarely mono-functionally oriented, rather they often fulfil a multitude of different tasks simultaneously (Mose & Weixlbaumer, 2007). The lack of comprehensive theory approaches or mismatch of theory and practice (implementation) to inform the fulfilment of multitude functions of PAs contribute to conservation and development knowledge gap. Questions directed to theory and practice are still unanswered. For instance, do established PAs fulfil multifunctional expectations of different stakeholders as models of conservation and development? Many studies that examine this discourse thus far convey mixed and to a certain extent doubtful images (see Sebele, 2010; Stone & Stone, 2020).

Despite the fact that PAs have only partially met previously established standards and expectations, at least three important advances seem to have been achieved – a heightened societal awareness of the meaning of area protection, significant reservation of areas with endangered natural and cultural landscapes, and the adoption of tourism as an integral tool that can reinforce conservation and development (Mose & Weixlbaumer, 2007).

As outlined by some papers in this volume, the disparity between theory and practice is imminent. For instance, Clark and Nyaupane (2020) took an exploratory approach and adopted the grounded theory approach to understand stakeholder attitudes towards tourism development and ecological restoration. Their results revealed four overarching themes inclined towards theory and practice; *contrasting world views, landscape-scale land management, ecological paradigm, and lack of shared community vision.* The contrasting world views validate the argument that the relationship between PAs, tourism and livelihoods is not perceived homogeneously throughout the world. The *landscape-scale land management* reveals the degree of linkages between PAs, tourism and livelihoods, and varies based on the stage of tourism development. In the sites where tourism is highly developed, people receive more economic benefits, are more empowered, and take more pride in PAs than in underdeveloped sites, which is consistent with other studies (Nyaupane & Poudel, 2011). The challenge is how we move from theory to practice.

We argue that for PAs, tourism and community livelihoods to yield intended outcomes, guidance should be grounded in both theory and practice. Nevertheless, efforts have been geared to this juncture as demonstrated by the call and use of such integrated conservation and development projects (ICDP) methodologies such as collaborative, public private, joint venture management and community-based partnerships in an endeavour to find the most working formula for PAs, tourism and livelihoods. The theory and practice knowledge gap interpretation and application of the management regimes for PAs in Europe differs from those in other continents, for example, North America (Bixler et al., 2015). The transition from "protection without use" (with no tourism – no integration) to "protection through use" (with tourism – with integration) with little understanding for integration without seriously thinking beyond the boundaries of the protected areas (Mose & Weixlbaumer, 2007) is a reflection of the deficit between theory and practice. There is a need for fundamental frameworks that have underpinnings that can yield better understanding. This explains the resultant conflict caused by opposing interests between PAs and their surroundings or even due to the clashing interests between PAs constituencies. Therefore, the spatial overlapping of nature protection and tourism and to some extent agriculture exemplifies this theory and practice short-fall.

Many case studies have suggested that the relationship between livelihoods and conservation (Stone & Nyaupane, 2017; Stone & Stone, 2020); tourism and livelihood improvement (Croes & Vanegas, 2008; Harrison, 2008); and conservation and tourism/development (Nyaupane & Thapa, 2004; Salafsky & Wollenberg, 2000) are dynamic, complex, and locally specific such that there is no single framework to examine the complex relationships between these global concerns, and the implementation challenges. Clear frameworks to steer tourism in protected areas as a green initiative, to promote sustainability and continue to support biodiversity conservation and socio-economic development are needed. More empirical research is also needed to navigate, direct and yield science-informed materials that can inform both theory/conceptual underpinnings and sustainable practice.

Policy frameworks

Policy frameworks have contributed to heightened social awareness of the meaning of "area protection" and the reservation of areas with endangered natural and cultural landscapes (Mose & Weixlbaumer, 2007). Several international policy papers have helped PAs change their roles to find their way into conceptions of PAs worldwide. This was informed by regional land use planning that is oriented towards sustainable development as well as "reasonable use" of natural resources through tourism; however, there was no guidance on how to attain sustainability and how reasonable use should look like (Cortes-Vazquez, 2020). As is evident from some of the papers in this volume, it is clear that in response to perceived shortcomings of the dominant protected areas approach of the world, the 1980s and 1990s witnessed an upsurge in people-centred conservation approaches to bridge the conservation-development gap (Holland et al., in this volume). Such policy instruments yielded integrated conservation development programs (ICDP). Despite this growing tendency toward integrated approaches at the landscape level, the inherent complexity of landscapes always presents persistent and significant challenges such as balancing multiple objectives, equitably including all relevant stakeholders, dealing with power and gender asymmetries, and moving beyond existing administrative, jurisdictional, and sectorial silos (Ros-Tonen et al., 2018).

Though policy instruments have spelt out how to transit from exclusive to inclusive conservation and development approaches, to do this remains unclear given the lack of common systemic practices and understanding of these approaches (Reed et al., 2015). Moreover, in the development of conservation and development policies, power has often been neglected as a useful concept in analysing and understanding evolving conservation and development governance processes. In fact, governance and partnerships tend to be presented as depoliticized and consensual policymaking by independent actors in seemingly power-free processes (Kuindersma et al., 2012). The adoption of tourism as a development tool in a non-depoliticized environment has brought more uncertainties such as how best to involve diverse stakeholder participants (Czernek, 2013), equitable distribution of revenue generated from tourism (Stone & Stone, 2020), the role of community participation (Park & Kim, 2016), balancing of resources use for conservation and development (Mathis & Rose, 2016) and empowerment and resident support (Strzelecka et al., 2017), to highlight but a few.

Pellis et al. (2018) further argue that in the implementation of conservation and development projects, what is considered as "solved" in one location and moment in time, may form the stage for yet another problem if we move out from one landscape to another. Hence, policy instruments in conservation and development agenda need to be informed by in situ than ex situ contextual settings as each landscape is different from another. Besides, inclusive policies and programs are criticized elsewhere as promoting elitism (Dubbink, 2013) and being exclusionary and favouring the interests of specific partners (Rhodes, 1997). This lead to power imbalances (Visseren-Hamakers, 2009) and the favouring of "capable" partners and the exclusion of others in governance processes (Bitzer & Glasbergen, 2015). Nevertheless, all is not lost, while it is still rare that land use and nature protection legislation are harmonized, there is a trend towards increasing integration (Dewi et al., 2013).

Institutional arrangements

In view of promoting harmony in PAs, tourism and livelihoods linkages, institutional mechanisms have been created to co-ordinate and integrate the actions of various stakeholders. One of the vital subjects in the debate on sustainable development addresses the question of what institutional arrangements are the most promising in advancing the process of progressive transformation (Glasbergen, 2007). Organisational scholars have hyped collaborative partnerships of institution as a useful and necessary instrument for problem-solving within a problem realm (Selin & Beason, 1991). Institutions are considered important and effective when people realise their benefits, when they are ready to listen to, discuss, and make decisions in consultation with people and when they are capable of providing the right solutions to their problems (Shoo & Songorwa, 2013).

Collaboration is used to describe interactions between institutions based on at least some agreed rules or norms, intended to address a common issue (Bramwell & Lane, 2000). The domination of and over-reliance on certain individual institutions in most cases affects the degree of all participating institutions. For instance, donors and NGOs who sometimes sponsor conservation and development projects are found to dominate collaborative partnership arrangements (Stone, 2015). Although various authors have examined the effectiveness of institutional partnerships, especially at the global level (Visseren-Hamakers, 2013), there is limited understanding on how partnerships contribute to governing integrated landscapes across regions (Nyaupane, Poudel, and York (2020)). It is not clear what could be done by institutions to discharge their functions to help PAs to achieve their many functions and roles. An array of recommendations exist as far as institutions' involvement in guiding PAs. For instance, Badola (2018) posits that weak institutions and governance systems may lead to mass tourism development in PAs controlled by powerful stakeholders. The author also argues that local institutions with strong intrinsic spatial linkages are most adaptive to challenges of PAs, and that, traditional local institutions assisted by governments and NGOs are best suited to achieve PAs mandate. In view of Badola's (2018) assessment, it is important to appreciate that some papers in this volume have highlighted the importance of involving all relevant institutions, but also underline how difficult it is to manage power relations that emerge due to the involvement of many institutions.

Diverse stakeholders

The multitude of diverse expectations from PAs has great potential to create conflict about the very purpose of the beneficial use of the resources. In a diversified community setting, each stakeholder controls resources such as knowledge, expertise, and capital, but on its own it is unlikely to possess all the resources necessary to achieve its objectives and to plan effectively for its future in relation to a significant development project (Jamal & Stronza, 2009). The stakeholders' interdependence means that there are potential mutual or collective benefits from stakeholders collaborating with each other (Selin & Beason, 1991). Stakeholders may work together if they consider that the probability of realising their goals and creating new opportunities in a problem domain are greater by performing jointly rather than acting alone (Czernek, 2013).

A variety of terms are used to describe different collaborative and partnership arrangements in the tourism development literature. These include coalitions, forums, alliances, task forces, community – private partnerships, community – private agreements, joint venture partnerships, and public – private partnerships (Bramwell & Lane, 2000; Jamal & Stronza, 2009; Jones, 2002; Rozemeijer, 2001).

As demonstrated by articles in this volume and existing literature, all these arrangements have been adopted and put to the test; however, the resultant outcomes have been mixed. Nevertheless, Bramwell and Lane (2000) caution that the definition of collaborative partnerships makes no assumptions about which stakeholders will participate, how much power they may have, how representative they may be, or about the total number of stakeholders involved. Furthermore, Stone and Stone (2020) posit that projects with many stakeholders often have issues with power relations as some stakeholders by virtue of their existence possess more power than others, hence they always influence conservation and tourism stakeholders collaborative partnership projects.

Bramwell and Lane (2000) caution that the complexity of engaging diverse stakeholders in collaborative partnerships makes it difficult to involve them all equally, the power of some partnerships may be too great, leading to the creation of cartels. It is important to note the role power plays in collaborative efforts. Gerschick et al. (1990) also challenge the assumption that power is a zero-sum commodity. That is, increasing the power of one community or individual or partner implies decreasing the power of another (Bartunek & Keys, 1982). Due to the involvement of diverse stakeholders, the structural nature of stakeholders in PAs and tourism development projects, what is to be developed and conserved, the inclusiveness of stakeholders' opinions, and the array of goals become very wide-ranging and difficult to accomplish equitably. Based on articles in this special issue, we can fairly remark that the current structural design of PAs and tourism development may not in itself guarantee the achievement of environmental conservation and development.

Hall (2011) argues that collaborative partnership methodologies in tourism planning have the potential to take in a wide set of stakeholders, escalate political participation and social equality, and contribute to more sustainable forms of tourism. Hall (2011) also cautions that theoretically informed collaborative arrangements may not be satisfactorily inclusive in nature. Thus far, an inclusive approach to PAs management remains a key focus that still faces many challenges. Important to note is that the complexities of managing change in practice reflect distinctive organizational environments and cultures (Haynes, 2015).

Moving forward: implications for protected areas, tourism and community livelihoods

In the late 1980s, concerns over social issues of poverty and vulnerability in rural communities outside protected areas propelled a shift in the dominant conservation paradigm towards a focus on local livelihoods and the engagement of local communities in resource management (Brandon & Wells, 1992; West et al., 2006). This paradigm shift brought a growing recognition that development and conservation were not mutually exclusive but rather complementary goals (Alpert, 1996). This new development in conservation and development, therefore fuelled donors, non-governmental organizations (NGOs), government agencies and local communities to join forces in the design and implementation of different forms of participatory interventions to enhance local livelihoods while continuing to conserve the biodiversity of protected areas. Indications are that considerable institutional arrangement changes in the field of protected area governance are evident and reflect evolutionary management styles (Stone, 2013).

As is evident from this special issue, emerging thinking conceptualizes protected areas, tourism and community livelihoods as a complex system consisting of multiple interacting components. Therefore, any tourism study conducted without explicit recognition of interacting variables (e.g. political, social, cultural, physical, economical, historic, and ecological), will reveal incomplete and possibly contradictory outcomes, as the complex interactions between system components will not be apparent. What is evident from most of the articles constituting this special issue is the call for holistic, diverse, partnerships, collaborations, malleable thinking approaches and adaptive management regimes in studying and managing this tripartite (i.e. PAs, tourism and livelihoods) relationship. By accepting the tourism industry, protected areas require adaptation to new situations and challenges that come with this conservation and development evolution nexus (Stone & Nyaupane, 2017).

This special issue extends the debate about how to conceptualise PAs, tourism, and community livelihoods giving an upturn to several models that attempt to represent the relationship. While there are a number of models that attempt to contribute to the understanding of the relationship between protected areas, tourism and community livelihoods, it is also critical to take cognisance of what other scholars highlight as the root cause of the failure to reconcile these relationships and move towards a more sustainable conservation, tourism and livelihoods improvement. For instance, Gossling et al. (2012) are sceptical and convinced that moving to more sustainable tourism is a very difficult task as it requires transformations of well-established and interlocking systems and social practices that are not easy to revolutionize. While that might be the case, Saarinen and Lenao (2014) argue that comprehensive sustainable tourism needs to go beyond focusing on the economic, social, and environmental benefits but needs to also include cultural, human, and political planning processes, challenging the robustness of the conceptual frameworks used as underpinnings to understand this relationship.

This special issue volume is in line with the current thinking and its contribution to knowledge is premised on the use of different approaches that have the propensity to yield new dispositions in advancing this tripartite relationship discourse. The debate is still open and needs bridging blocks to fill the knowledge vacuum that still exists. Until such a time the knowledge gap is bridged, PAs, tourism and livelihoods relationship discourse debate will persist and fuel researchers to put more resources in an endeavour to find a solution to the problem as it is inevitable that tourism as a new land use will need to be hosted by protected areas, while the resources custodians in the form of adjacent communities need to tap from tourism benefits that are accommodated by protected areas.

It is vital to recognize the role played by PAs in conservation and development interconnections. It is expected that present-day biodiversity predicaments would be much starker had the creation of PAs in the last century not ensued. PAs have offered alternative means of support to many societies that today rely on them for numerous means of sustenance. As a result, amplified support for protected areas worldwide continues to be a requirement if they are to deliver their potential and capabilities. In practice, it seems protected areas are known to make much wider ecological, social and economic contributions to human society, nonetheless, it is not yet clear what the general impact of these cumulative demands on PAs will be, or whether extra demands will arise in future, adding to more complications and pressure for PAs.

PAs tourism is expected to grow, so, if it is not cognisant of sustainable tourism development principles, it can certainly become one of the human-led threats to the existence of PAs. Efforts are needed to promote reconciliation between the multiple roles played by PAs and their delivery of conservation and development goals. This requires the adoption of an inclusive planning and development frameworks that take into account all relevant stakeholders and advances comprehensive sustainable tourism development as espoused by Saarinen and Lenao (2014). This will ensure that PA resources are conserved and used in a sustainable manner while also improving the lives of communities and all who benefit from their existence. Protected area tourism cannot be dealt with in the same way as a single product or service. It therefore, should involve a continual process of assessing actions and outputs, obtaining feedback from customers and making improvements or realignment adaptively. It should be concerned with the effective use of resources, and the level of participation and satisfaction of all PA stakeholders.

Nonetheless, the accommodation of tourism as an additional land use by PAs sounds like a noble idea; others view it as an illustration of a community development "imposter" driven by economic imperatives and a neo-liberal agenda maintained to further exploit local communities (Blackstock, 2005). All the same, the loss of biodiversity resources accommodated by PAs will similarly lead to loss of revenue from tourism, thereby adversely affecting communities who live around PAs. Many of these communities live in extreme poverty and rely on benefits from tourism, these exposes them to food insecurity, human welfare challenges, and human–wildlife conflicts. Supporting the contribution of PAs to conservation and tourism development and reconciling resultant unintended outcomes is imperative, otherwise the contribution of PAs to regional conservation and development will remain a contentious issue.

To this end, we do not have any reservations that the overall findings of the articles in this volume have lessons for researchers, protected areas' managers, policy architects and tourism and community planners whose goal is to design programs or projects that are geared towards assisting protected areas to sustain their conservation and community development goals. There is little doubt either that the success of PA tourism requires effective implementation guidance from theory to practice, informed policy frameworks, strong institutional arrangements and cognizant diverse stakeholders. Furthermore, the findings enhance more knowledge to extant literature on the linkages between protected areas, tourism and community livelihoods. Contributions in this special issue have provided progress in the field and illuminated some areas where more work is still needed for sustainable PA tourism development to occur.

Conclusion

Tourism has been seen as a tool that communities can use to provide a positive link between PAs and their livelihoods; therefore, the relationship between tourism, conservation and community livelihoods plays an important role on the sustainability of protected areas. However, how to integrate tourism into the PA-community nexus is still evolving. This special issue is a significant step towards our understanding of the dynamic and complex relationship by drawing on the literature and individual cases that are often sporadic, disconnected, and atheoretical. In essence, this paper offer a few directions to move forward the debate in the following ways. First, PAs are often treated as a localized entity to serve surrounding communities and tourists. However, the ecological services PAs provide transcend local and national borders, and the distribution of costs and benefits of PAs are unevenly distributed to local, national, and global communities. This is a bigger justice and equity issue that has to be broadened at the global scale. Second, PAs are managed for multiple, often competing goals, including biodiversity conservation, community livelihoods, and tourism. To meet these competing goals, there is a need for an adaptive governance through meaningful, just and equitable collaborations among government, nonprofit, and private entities involved in conservation, tourism, and community development.

Third, navigating linkages within the bigger political-economic framework is daunting. The role of power and politics needs to be better integrated in understanding and developing linkages. Many argue PAs are a mechanism for the introduction of neoliberalism and not just a product of it (Duffy, 2015; Ives, 2015). The outcome of these capitalist processes is a loss in the social embeddedness of nature as locals become supplanted with global markets and conservation agendas. This is evidenced by the external pressures and priorities imposed on communities, often by external stakeholders with differing values leading to asymmetrical power relations with potential conflicts in the offing. Fourth, creating a positive linkage is a dynamic and complex process that requires multi-disciplinary approaches in assessing PAs, tourism and community livelihoods relationships to discern well with complex social–ecological relationships and advance our thinking to the level at which we devise mitigation measures directed to difficult situations marked by complex and multi-interactions and the absence of instantaneous apparent solutions. Lastly, COVID-19 has brought a new set of challenges and opportunities to the existing linkages. Although there is a growing appreciation of nature, the global pandemic has added additional pressure to the already crowded natural spaces for both recreation and community livelihoods. Despite some early research on the impacts of COVID-19 on PAs and tourism, the long-term impacts on the PAs and their relationship with community and tourism needs further research.

Disclosure statement

No potential conflict of interest was reported by the authors.

ORCID

Moren Tibabo Stone http://orcid.org/0000-0003-4379-8123
Lesego Senyana Stone http://orcid.org/0000-0002-6851-3545
Gyan P. Nyaupane http://orcid.org/0000-0002-0814-8507

References

Aazami, M., & Shanazi, K. (2020). Tourism wetlands and rural sustainable livelihood: The case from Iran. *Journal of Outdoor Recreation and Tourism*, *30*, 100284. https://doi.org/10.1016/j.jort.2020.100284

Adams, W. B. (2013). *Against extinction: The story of conservation*. Routledge.

Alpert, P. (1996). Integrated conservation and development projects. *BioScience*, *46*(11), 845–855. https://doi.org/10.2307/1312970

Andrade, A. (2018). Capacity management and tourism distribution channels: The case of the Brazilian National Park of Fernando de Noronha, a World Heritage Site. *Journal of Park and Recreation Administration*, *36*(3), 63–81. https://doi.org/10.18666/JPRA-2018-V36-I3-8341

Backman, K., & Munanura, I. E. (2017). Introduction: Ecotourism in Africa over the past 30 years. In K. Backman & I. E. Munanura (Eds.), *Ecotourism in sub-saharan Africa: Thirty years of practice* (pp. 7–22) Taylor & Francis.

Balmford, A., & Whitten, T. (2003). Who should pay for tropical conservation, and how could the costs be met? *Oryx*, *37*(2), 238–250. https://doi.org/10.1017/S0030605303000413

Bartunek, J. M., & Keys, C. B. (1982). Power equalization in schools through organization development. *Journal of Applied Behavioral Science*, *18*(2), 171–183. https://doi.org/10.1177/002188638201800205

Bennett, N., Lemelin, R. H., Koster, R., & Budke, I. (2012). A capital assets framework for appraising and building capacity for tourism development in aboriginal protected area gateway communities. *Tourism Management*, *33*(4), 752–766. https://doi.org/10.1016/j.tourman.2011.08.009

Berkes, F., Kofinas, G. P., & Chapin, F. S. (2009). Conservation, community, and livelihoods: Sustaining, renewing, and adapting cultural connections to the land. In Folke, C, Kofinas, G.P and Chapin, F., S. (Eds.), *Principles of ecosystem stewardship* (pp. 129–147). Springer.

Bitzer, V., & Glasbergen, P. (2015). Business–NGO partnerships in global value chains: Part of the solution or part of the problem of sustainable change? *Current Opinion in Environmental Sustainability*, *12*, 35–40. https://doi.org/10.1016/j.cosust.2014.08.012

Bixler, R. P., Dell'Angelo, J., Mfune, O., & Roba, H. (2015). The political ecology of participatory conservation: Institutions and discourse. *Journal of Political Ecology*, *22*(1), 164–182. https://doi.org/10.2458/v22i1.21083

Bramwell, B., & Lane, B. (2000). *Tourism collaboration and partnerships: Politics, practice and sustainability* (Vol. 2, pp 1–19). Channel View Publications.

Brandon, K. E., & Wells, M. (1992). Planning for people and parks: Design dilemmas. *World Development*, *20*(4), 557–570. https://doi.org/10.1016/0305-750X(92)90044-V

Broekhuis, F., Kaelo, M., Sakat, D. K., & Elliot, N. B. (2020). Human–wildlife coexistence: Attitudes and behavioural intentions towards predators in the Maasai Mara, Kenya. *Oryx*, *54*(3), 366–374. https://doi.org/10.1017/S0030605318000091

Budowski, G. (1976). Tourism and environmental conservation: Conflict, coexistence, or symbiosis? *Environmental Conservation*, *3*(1), 27–31. https://doi.org/10.1017/S0376892900017707

Büscher, B. (2013). *Transforming the frontier: Peace parks and the politics of neoliberal conservation in Southern Africa*. Duke University Press.

Cater, E. (1994). Introduction. In E. Cater & G. Lowman (Eds.), *Ecotourism: A sustainable option?* (pp. 3–16). John Wiley & Sons.

Cortes-Vazquez, J. A. (2020). In the name of the people: The populist redefinition of nature conservation in post-crisis Spain. *Geoforum*, *108*, 110–118. https://doi.org/10.1016/j.geoforum.2019.12.004

Croes, R., & Vanegas, M. (2008). Cointegration and causality between tourism and poverty reduction. *Journal of Travel Research*, *47*(1), 94–103. https://doi.org/10.1177/0047287507312429

Czernek, K. (2013). Determinants of cooperation in a tourist region. *Annals of Tourism Research*, *40*, 83–104. https://doi.org/10.1016/j.annals.2012.09.003

Dearden, P., Bennet, M., & Johnston, J. (2005). Trends in global protected area governance, 1992-2002. *Environmental Management*, *36*(1), 89–100. https://doi.org/10.1007/s00267-004-0131-9

Dewi, S., Van Noordwijk, M., Ekadinata, A., & Pfund, J. L. (2013). Protected areas within multifunctional landscapes: Squeezing out intermediate land use intensities in the tropics? *Land Use Policy*, *30*(1), 38–56. https://doi.org/10.1016/j.landusepol.2012.02.006

Dlamini, C. S., & Masuku, M. (2013). Towards sustainable financing of protected areas: A brief overview of pertinent issues. *International Journal of Biodiversity and Conservation*, *5*(8), 436–445.

Doyle-Capitman, C. E., Decker, D. J., & Jacobson, C. A. (2018). Toward a model for local stakeholder participation in landscape-level wildlife conservation. *Human Dimensions of Wildlife*, *23*(4), 375–390. https://doi.org/10.1080/10871209.2018.1444215

Dubbink, W. (2013). Assisting the invisible hand. *Contested relations between market, state and civil society* (Vol. 18): Springer.

Duffy, R. (2015). Nature-based tourism and neoliberalism: Concealing contradictions. *Tourism Geographies*, *17*(4), 529–543. https://doi.org/10.1080/14616688.2015.1053972

Eagles, P. F., & McCool, S. F. (2002). Tourism in national parks and protected areas: planning and management. Cabi.

Fisher, D. G. (2006). The potential for rural heritage tourism in the Clarence Valley of Northern New South Wales. *Australian Geographer*, *37*(3), 411–424. https://doi.org/10.1080/00049180600954815

Franks, P., & Blomley, T. (2004). Fitting ICD into a project framework: A CARE perspective. In T. O. McShane & M. P. Wells (Eds.), *Getting biodiversity projects to work: Towards more effective conservation and development* (pp. 77–97). Columbia University.

Gerschick, T., Israel, B., & Checkoway, B. (1990). *Means of empowerment in individuals, organizations, and communities*. Program on Conflict Management Alternatives. University of Michigan.

Gezon, L. L. (2014). Who wins and who loses? Unpacking the "local people" concept in ecotourism: A longitudinal study of community equity in Ankarana, Madagascar. *Journal of Sustainable Tourism*, *22*(5), 821–838. https://doi.org/10.1080/09669582.2013.847942

Glasbergen, P. (2007). Setting the scene: The partnership paradigm in the making. In: Glasbergen, P., Biermann, F., and Mol, P. J. A. (Eds.), *Partnerships, governance and sustainable development reflections on theory and practice* (pp. 1–26). Edward Elgar Publishing.

Gossling, S., Hall, C., Ekstrom, F., Engeset, A., & Aall, C. (2012). Transition management: A tool for implementing sustainable tourism scenarios? *Journal of Sustainable Tourism*, *20*(6), 899–916. https://doi.org/10.1080/09669582.2012.699062

Hall, M. (2011). Policy learning and policy failure in sustainable tourism governance: From first- and second-order to third-order change? *Journal of Sustainable Tourism*, *19*(4-5), 649–671. https://doi.org/10.1080/09669582.2011.555555

Harrison, D. (2008). Pro-poor tourism: A critique. *Third World Quarterly*, *29*(5), 851–868. https://doi.org/10.1080/01436590802105983

Haynes, P. (2015). *Managing complexity in the public services*. Routledge.

Hossu, C. A., Ioja, I. C., Nita, M. R., Hartel, T., Badiu, D. L., & Hersperger, A. M. (2017). Need for a cross-sector approach in protected area management. *Land Use Policy*, *69*, 586–597. https://doi.org/10.1016/j.landusepol.2017.10.012

IUCN (International Union for the Conservation of Nature). (2015a). *IUCN red list: Trends in the changes in status of biodiversity*. http://www.iucnredlist.org/about/summary-statistics

Ives, A. (2015). Neoliberalism and the concept of governance: Renewing with an older liberal tradition to legitimate the power of capital. *Mémoire(s), Identité(s), marginalité(s) dans le monde occidental contemporain*, (14). https://doi.org/10.4000/mimmoc.2263

Jamal, T., & Stronza, A. (2009). Collaboration theory and tourism practice in protected areas: Stakeholders, structuring and sustainability. *Journal of Sustainable Tourism*, *17*(2), 169–189. https://doi.org/10.1080/09669580802495741

Jones, B. (2002). *Chobe Enclave*. Botswana – lessons learnt from a CBNRM project (1993). IUCN/SNV CBNRM Support Programme.

Jones, P. J. S., Qiu, W., & De Santo, E. M. (2013). Governing marine protected areas: Social–ecological resilience through institutional diversity. *Marine Policy*, *41*, 5–13. https://doi.org/10.1016/j.marpol.2012.12.026

Kamble, Z., & Bouchon, F. (2016). Developing a framework for assessing social cohesion via tourism. *Tourism Review*, *71*(4), 272–286. https://doi.org/10.1108/TR-08-2015-0035

König, H. J., Kiffner, C., Kramer-Schadt, S., Fürst, C., Keuling, O., & Ford, A. T. (2020). Human–wildlife coexistence in a changing world. *Conservation Biology*, *34*(4), 786–794. https://doi.org/10.1111/cobi.13513

Kuindersma, W., Arts, B., & van der Zouwen, M. (2012). Power faces in regional governance. *Journal of Political Power*, *5*(3), 411–429. https://doi.org/10.1080/2158379X.2012.735116

Kuss, F. R., Graefe, A. R., & Vaske, J. J. (1990). *Visitor impact management: A review of research* (Vol. 1). National Parks and Conservation Association.

Lacitignola, D., Petrosillo, I., Cataldi, M., & Zurlini, G. (2007). Modelling socio-ecological tourism-based systems for sustainability. *Ecological Modelling*, *206*(1-2), 191–204. https://doi.org/10.1016/j.ecolmodel.2007.03.034

Lark, C., & Nyaupane, G. P. (2020). Connecting landscape-scale ecological restoration and tourism: stakeholder perspectives in the great plains of North America. *Journal of Sustainable Tourism*, 1–19. https://doi.org/10.1080/09669582.2020.1801698

Li, Y., Knight, D. W., Luo, W., & Hu, J. (2020). Elite circulation in Chinese ethnic tourism. *Annals of Tourism Research*, *85*, 103030. https://doi.org/10.1016/j.annals.2020.103030

Lowe, J., Tejada, J. F. C., & Meekan, M. G. (2019). Linking livelihoods to improved biodiversity conservation through sustainable integrated coastal management and community based dive tourism: Oslob Whale Sharks. *Marine Policy*, *108*, 103630. https://doi.org/10.1016/j.marpol.2019.103630

Madden, F., & McQuinn, B. (2014). Conservation's blind spot: The case for conflict transformation in wildlife conservation. *Biological Conservation*, *178*, 97–106. https://doi.org/10.1016/j.biocon.2014.07.015

Mathis, A., & Rose, J. (2016). Balancing tourism, conservation, and development: A political ecology of ecotourism on the Galapagos Islands. *Journal of Ecotourism*, *15*(1), 64–77. https://doi.org/10.1080/14724049.2015.1131283

Matiku, P. (2008). Poverty and mortality indicators: Data for the povertyconservation debate. *Oryx*, *42*(01), 1. https://doi.org/10.1017/S0030605308042154

Mbaiwa, J. E. (2008). The realities of ecotourism development in Botswana. In A. Spenceley (Ed.), *Responsible tourism: Critical issues for conservation and development* (pp. 205–223). Earthscan.

Mishra, P. K., & Verma, J. K. (2017). Tourism and peace in economic development perspective of India. *Journal of Environmental Management & Tourism*, *8*(20), 927–934.

Mose, I., & Weixlbaumer, N. (2007). A new paradigm for protected areas in Europe. In I. Mose (Ed.), *Protected areas and regional development in Europe. Towards a new model for the 21st century* (pp. 3–20). Routledge.

Naidoo, R., Weaver, L. C., Diggle, R. W., Matongo, G., Stuart-Hill, G., & Thouless, C. (2016). Complementary benefits of tourism and hunting to communal conservancies in Namibia. *Conservation Biology*, *30*(3), 628–638.

Nelson, F. (2012). Blessing or curse? The political economy of tourism development in Tanzania. *Journal of Sustainable Tourism*, *20*(3), 359–375. https://doi.org/10.1080/09669582.2011.630079

Nunkoo, R. (2017). Governance and sustainable tourism: What is the role of trust, power and social capital? *Journal of Destination Marketing & Management*, *6*(4), 277–285. https://doi.org/10.1016/j.jdmm.2017.10.003

Nyaupane, G. P., & Poudel, S. (2011). Linkages among biodiversity, livelihood, and tourism. *Annals of Tourism Research*, *38* (4), 1344– 1366. https://doi.org/10.1016/j.annals.2011.03.006

Nyaupane, G. P., Poudel, S., & Timothy, D. J. (2018). Assessing the sustainability of tourism systems: A social–ecological approach. *Tourism Review International*, *22*(1), 49–66. https://doi.org/10.3727/154427218X15202734130440

Nyaupane, G. P., & Thapa, B. (2004). Evaluation of ecotourism: A comparative assessment in the Annapurna Conservation Area Project. *Journal of Ecotourism*, *3*(1), 20–45. https://doi.org/10.1080/14724040408668148

Nyaupane, G. P., Poudel, S., & York, A. (2020). Governance of protected areas: an institutional analysis of conservation, community livelihood, and tourism outcomes. *Journal of Sustainable Tourism*, 1–20. https://doi.org/10.1080/09669582.2020.1858089

Ostrom, E. (2007). Challenges and growth: the development of the interdisciplinary field of institutional analysis. *Journal of Institutional Economics*, *3*(3), 239–264. https://doi.org/10.1017/S1744137407000719

Park, E., Choi, B. K., & Lee, T. J. (2019). The role and dimensions of authenticity in heritage tourism. *Tourism Management*, *74*, 99–109. https://doi.org/10.1016/j.tourman.2019.03.001

Park, E., & Kim, S. (2016). The potential of Cittaslow for sustainable tourism development: Enhancing local community's empowerment. *Tourism Planning & Development*, *13*(3), 351–369. https://doi.org/10.1080/21568316.2015.1114015

Pellis, A., Pas, A., & Duineveld, M. (2018). The persistence of tightly coupled conflicts. The case of Loisaba. *Conservation and Society*, *16*(4), 387–396. https://doi.org/10.4103/cs.cs_17_38

Poudel, S., & Nyaupane, G. P. (2018). Understanding environmentally responsible behaviour of ecotourists: The reasoned action approach. *Tourism Planning & Development*, *14*(3), 337–352. https://doi.org/10.1080/21568316.2016.1221851

Pozo, R. A., LeFlore, E. G., Duthie, A. B., Bunnefeld, N., Jones, I. L., Minderman, J., Rakotonarivo, O. S., & Cusack, J. J. (2021). A multispecies assessment of wildlife impacts on local community livelihoods. *Conservation Biology*, *35*(1), 297–306. https://doi.org/10.1111/cobi.13565

Reed, M. G., Godmaire, H., Guertin, M. A., Potvin, D., & Abernethy, P. (2015). Engaged scholarship: reflections from a multi-talented, national partnership seeking to strengthen capacity for sustainability. *Engaged Scholar Journal: Community-Engaged Research, Teaching and Learning*, *1*(1), 167–183. https://doi.org/10.15402/esj.2015.1.A010

Ros-Tonen, M. A., Reed, J., & Sunderland, T. (2018). From synergy to complexity: the trend toward integrated value chain and landscape governance. *Environmental Management*, *62*(1), 1–14. https://doi.org/10.1007/s00267-018-1055-0

Rozemeijer, N. (2001, November). Investment in partnership. Review of community based natural resource management in Botswana. In *Proceedings of the Second National CBNRM Conference in Botswana* (pp. 14–16).

Saarinen, J., & Lenao, M. (2014). Integrating tourism to rural development and planning in the developing world. *Development Southern Africa*, *31*(3), 363–372. https://doi.org/10.1080/0376835X.2014.888334

Salafsky, N., & Wollenberg, E. (2000). Linking livelihoods and conservation: A conceptual framework and scale for assessing the integration of human needs and biodiversity. *World Development*, *28*(8), 1421–1438. https://doi.org/10.1016/S0305-750X(00)00031-0

Scheyvens, R. (1999). Ecotourism and the empowerment of local communities. *Tourism Management*, *20*(2), 245–249. https://doi.org/10.1016/S0261-5177(98)00069-7

Sebele, L. S. (2010). Community-based tourism ventures, benefits, and challenges: Khama Rhino Sanctuary Trust, Central District. *Tourism Management*, *31*(1), 136–146. https://doi.org/10.1016/j.tourman.2009.01.005

Selin, S., & Beason, K. (1991). Interorganizational relations in tourism. *Annals of Tourism Research*, *18*(4), 639–652. https://doi.org/10.1016/0160-7383(91)90079-Q

Sene-Harper, A., Matarrita-Cascante, D., & Larson, L. R. (2019). Leveraging local livelihood strategies to support conservation and development in West Africa. *Environmental Development*, *29*, 16–28. https://doi.org/10.1016/j.envdev.2018.11.002

Shoo, R., & Songorwa, A. (2013). Contribution of ecotourism to nature conservation and improvement of livelihoods around Amani nature reserve. *Journal of Ecotourism*, *12*(2), 75–89. https://doi.org/10.1080/14724049.2013.818679

Snyman, S. L. (2012). The role of tourism employment in poverty reduction and community perceptions of conservation and tourism in southern Africa. *Journal of Sustainable Tourism*, *20*(3), 395–416. https://doi.org/10.1080/09669582.2012.657202

Stone, L. S., & Stone, T. M. (2011). Community-based tourism enterprises: Challenges and prospects for community participation; Khama Rhino Sanctuary Trust. *Journal of Sustainable Tourism*, *19*(1), 97–114. https://doi.org/10.1080/09669582.2010.508527

Stone, M. T. (2015). Community-based ecotourism: a collaborative partnerships perspective. *Journal of Ecotourism*, *14*(2-3), 166–184. https://doi.org/10.1080/14724049.2015.1023309

Stone, L. S., & Stone, T. M. (2017). Domestic tourism in Botswana: An examination of nature-based tourism constraints. *Journal of Sustainable Tourism*, *10*(1), 97–116.

Stone, L. S., Stone, M. T., & Mbaiwa, J. E. (2017). Tourism in Botswana in the Last 50 Years: A review. *Botswana Notes and Records, 49*, 57–72.

Stone, M. T. (2013). *Protected areas, tourism and rural community livelihoods in Botswana* [Unpublished PhD dissertation]. Arizona State University.

Stone, M. T., & Nyaupane, G. P. (2016). Protected areas, tourism and community livelihoods linkages: A comprehensive analysis approach. *Journal of Sustainable Tourism, 24*(5), 673–693. https://doi.org/10.1080/09669582.2015.1072207

Stone, M. T., & Nyaupane, G. P. (2017). Ecotourism influence on community needs and the functions of protected areas: A systems thinking approach. *Journal of Ecotourism, 16*(3), 222–246. https://doi.org/10.1080/14724049.2016.1221959

Stone, M. T., & Nyaupane, G. P. (2018). Protected areas, wildlife-based community tourism and community livelihoods dynamics: Spiraling up and down of community capitals. *Journal of Sustainable Tourism, 26*(2), 307–324. https://doi.org/10.1080/09669582.2017.1349774

Stone, M. T., & Stone, L. S. (2020). Challenges of community-based tourism in Botswana: A review of literature. *Transactions of the Royal Society of South Africa, 75*(2), 181–193. https://doi.org/10.1080/0035919X.2020.1715510

Strickland-Munro, J., Allison, H., & Moore, A. (2010). Using resilience concepts to investigate the impacts of protected area tourism on communities. *Annals of Tourism Research, 37*(2), 499–519. https://doi.org/10.1016/j.annals.2009.11.001

Strzelecka, M., Boley, B. B., & Strzelecka, C. (2017). Empowerment and resident support for tourism in rural Central and Eastern Europe (CEE): The case of Pomerania, Poland. *Journal of Sustainable Tourism, 25*(4), 554–572. https://doi.org/10.1080/09669582.2016.1224891

Summers, J., Cavaye, J., & Woolcock, G. (2019). Enablers and barriers of tourism as a driver of economic and social-cultural growth in remote Queensland. *Economic Papers: A Journal of Applied Economics and Policy, 38*(2), 77–94.

Swemmer, L., Mmethi, H., & Twine, W. (2017). Tracing the cost/benefit pathway of protected areas: A case study of the Kruger National Park. *Ecosystem Services, 28*, 162–172. https://doi.org/10.1016/j.ecoser.2017.09.002

Tao, T. C., & Wall, G. (2009). A livelihood approach to sustainability. *Asia Pacific Journal of Tourism Research, 14*(2), 137–152. https://doi.org/10.1080/10941660902847187

Tumusiime, D. M., & Vedeld, P. (2012). False promise or false premise? Using tourism revenue sharing to promote conservation and poverty reduction in Uganda. *Conservation and Society, 10*(1), 15–28. https://doi.org/10.4103/0972-4923.92189

Visseren-Hamakers, I. J. (2009). *Partnerships in biodiversity governance: An assessment of their contributions to halting biodiversity loss* (pp. 1–177). Nederlandse Geografische Studies.

Visseren-Hamakers, I. J. (2013). Partnerships and sustainable development: the lessons learned from international biodiversity governance. *Environmental Policy and Governance, 23*(3), 145–160. https://doi.org/10.1002/eet.1612

Wang, S. H., Lee, M. T., Château, P. A., & Chang, Y. C. (2016). Performance indicator framework for evaluation of sustainable tourism in the Taiwan coastal zone. *Sustainability, 8*(7), 652. https://doi.org/10.3390/su8070652

West, P., Igoe, J., & Brockington, D. (2006). Parks and peoples: The social impact of protected areas. *Annual Review of Anthropology, 35*(1), 251–277. https://doi.org/10.1146/annurev.anthro.35.081705.123308

Wondirad, A., & Ewnetu, B. (2019). Community participation in tourism development as a tool to foster sustainable land and resource use practices in a national park milieu. *Land Use Policy, 88*, 104155. https://doi.org/10.1016/j.landusepol.2019.104155

Zhang, Y., Lee, T. J., & Xiong, Y. (2019). A conflict resolution model for sustainable heritage tourism. *International Journal of Tourism Research, 21*(4), 478–492. https://doi.org/10.1002/jtr.2276

Resident support of community-based tourism development: evidence from Gunung Ciremai National Park, Indonesia

Prasetyo Nugroho and Shinya Numata

ABSTRACT

While many studies have examined community-based tourism (CBT) in the national parks of developing countries, studies that focused on residents' mechanism of cost and benefit perception for tourism development support are rare. To understand mechanism of resident support of CBT developments in the early stage, an extended social exchange theory (SET) was used to examine the relationships among the following latent variables of support of tourism development: community involvement, community attachment, perceived benefits, and perceived costs in 12 villages adjacent to Gunung Ciremai National Park (GCNP), Indonesia. In total, 934 usable questionnaires were collected. CBT provides residents access to and benefits from the GCNP area. Perceived economic benefits and community involvement exerted the greatest influence on resident support of tourism development. Perceived benefits and costs were significant mediators of the relationship between community attachment and involvement in, and support of, tourism development. These findings confirm the interrelationships among determinant variables that explain resident support of tourism development based on SET, albeit with different mechanisms, effects, and values. The findings also imply that park managers should focus on not only allocation of the benefits of tourism among local residents but also minimizing the level of disturbance of the communities surrounding GCNP.

Introduction

National parks (NPs) are important nature-based tourist destinations. However, the sustainable development of tourism in Indonesian NPs is hampered by negative issues (Campbell et al., 2013; Myers & Muhajir, 2015; Shivakoti & Shivakoti, 2008; Yusran et al., 2017). Traditional exclusionary and repressive approaches to NP management have frequently affected the livelihoods of adjacent communities (Soliku & Schraml, 2018; West et al., 2006), which disproportionately bear the costs of conservation (Mackenzie, 2012; Ninan & Sathyapalan, 2005). Displacement and restricted access to natural resources are the most prevalent socioeconomic costs of NPs (Agrawal & Redford, 2009; West et al., 2006), and these can reduce the incomes of local residents and lead to impoverished communities (Brockington & Wilkie, 2015; Eneji et al., 2009; Ma et al., 2019).

For sustainable NP management, formulating and implementing participatory NP management are important for overcoming the socioeconomic costs of NPs (Clifton, 2013; Dearden et al., 2005; Soliku & Schraml, 2018). Redistributing spaces and opportunities for locals to benefit from NPs is strongly advocated (Clifton, 2013; MacKenzie et al., 2017; Sirivongs & Tsuchiya, 2012). The support and involvement of residents in natural resource protection in NPs are crucial (Campbell et al., 2013; Rastegar, 2019). At this point, tourism is promoted as a sustainable way by which local people may benefit from nature conservation by regenerating rural communities and local economies (Eagles et al., 2002; Ghoddousi et al., 2018; Kim et al., 2013; Mackenzie, 2012; Udaya Sekhar, 2003).

In Indonesia, improved NP management has increased tourism in recent years (KLHK, 2017). Gunung Ciremai National Park (GCNP), an area of high tropical forest biodiversity in West Java (Rozak & Gunawan, 2015), has been the pilot of the development of community-based tourism (CBT) (Wiratno, 2018). CBT implementation is predicated on the active involvement and empowerment of adjacent local communities in NP management (Okazaki, 2008; Russell, 2000). National regulations have affirmed the importance of community involvement in tourism management in the utilization zone of GCNP and strengthened the identity of local residents (Wiratno, 2018). Hence, the number of CBT sites in GCNP has increased since 2015 (BTNGC, 2018). Simultaneously, the residents of adjacent communities are experiencing the benefits and costs of tourism due to the increasing numbers of visitors. Therefore, within the high uncertainties in the early stage of exchange relationships (Whitener et al., 1998), understanding residents' perceptions and how they determine their support of CBT initiatives are critical when designing appropriate strategies for improving management and policy decisions.

Social exchange theory (SET) is the most frequently applied theoretical basis for understanding the perceptions and attitudes of residents toward tourism worldwide (Sharpley, 2014). SET argues that rational humans base their behavioral tendencies on maximizing benefits and minimizing costs (Ap, 1992; Cook & Rice, 2003; Homans, 1961), implying that if residents maximize their benefits from a NP, they will subsequently come to support the conservation efforts and existence of the NP (Dearden et al., 2005; Ghoddousi et al., 2018; Mackenzie, 2012; Sirivongs & Tsuchiya, 2012). Findings indicate that the socioeconomic effects have been the most discernible impact of tourism activities on adjacent communities in protected areas worldwide (Lee, 2013; Lepp, 2007; Mbaiwa & Stronza, 2011; Molina-Murillo et al., 2016; Sebele, 2010; Udaya Sekhar, 2003). Tourism provides a variety of employment sources, improving living standards and increasing village infrastructure. Conversely, studies have also found evidence of the shortcomings of tourism. Akyeampong (2011) indicated that the socioeconomic benefits of tourism in Kakum National Park in Ghana were only modest. Tourism development can harm aspects of society, culture, and the environment by, for example, generating noise, traffic congestion, and overcrowding (Huong & Lee, 2017; Nunkoo et al., 2013; Šegota et al., 2017). Tourism development can also alter traditional cultures, disrupt community life, and promote conflicts among residents (Choi & Sirakaya, 2005; Huong & Lee, 2017; Ko & Stewart, 2002). Furthermore, Sene-Harper et al. (2019) found that in Senegal, the absence of legal framework for communal land tenure and tourism concession rights have reduced the channels through which communities can benefit from tourism around national parks. Conversely, Scheyvens and Russell (2012) found that in Fiji, a stronger communal land policy supports the interests of landowners against private investors to ensure that communities can maximize tourism benefits.

The original SET framework is robust to understand perceived benefits and costs in determining residents' attitudes (Sharpley, 2014), but may not examine the mechanism on how resident perceives costs and benefits of tourism development in particular social circumstances. Therefore, an additional framework to extend original SET may help to understand the mechanism of costs and benefits perception of locals. There are several possible factors influencing perceived costs and benefits have been separately discussed (Gursoy & Rutherford, 2004; Nicholas et al., 2009; Nunkoo, 2016; Presenza et al., 2013; Rasoolimanesh et al., 2015; Sirivongs & Tsuchiya,

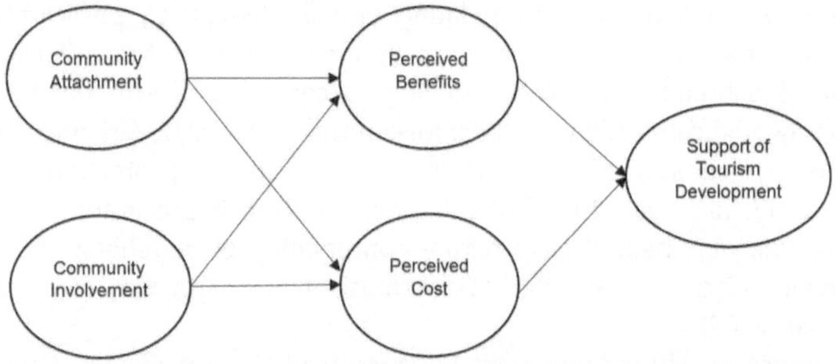

Figure 1. Basic structural model of resident support of tourism development proposed by Lee (2013).

2012; Udaya Sekhar, 2003), and Lee (2013) provided an extended SET model to examine mechanism of tourism development support. Lee's model explains how perceived benefits/costs are built by community attachment and community involvement. Figure 1 shows the structural framework proposed by Lee (2013). Nevertheless, Lee (2013) also stresses the need for more empirical research on the influence of his framework across different contexts and social circumstances.

In the context of sustainable development of tourism in NPs in Indonesia, acknowledging the sociocultural circumstances of neighboring communities is inevitable. Most of the adjacent local communities are highly-attached to the forest area, it comprises residents with strong socioeconomic and cultural links to the NP area (Clifton, 2013; Myers & Muhajir, 2015; Pudyatmoko et al., 2018). Their livelihoods rely on the natural resources of the NP. Moreover, encouraging community involvement in the governance of Indonesian NPs is also thought to be important for sustainable natural resource management and biodiversity conservation (Campbell et al., 2013; Pudyatmoko et al., 2018). Indonesian national regulations on NP management have shifted toward encouraging adjacent local communities to be actively involved in planning, decision-making, establishing, and managing tourism sites within the utilization zone of NPs (Wiratno, 2018). Also, community involvement may empower residents, enhance their awareness of the potential impacts of tourism, and boost their respect for their culture and values (Látková & Vogt, 2012; Molina-Murillo et al., 2016; Tosun, 2006). The rationales above indicate that community attachment and involvement should play important roles in CBT at an early stage of development in Indonesian NPs.

Therefore, by using the case of CBT in GCNP, which is the fastest-growing CBT in a NP in Indonesia (KLHK, 2017), we used Lee's framework to extend the original SET to understand how perceived benefits and costs are built by community attachment and community involvement. We adopted Lee's structural framework and implementing it in the context of the inception stage of CBT development in the national parks of developing countries, which to our knowledge has not been done in the past. Indeed, incorporating community attachment and community involvement into SET to explore the perceptions of residents toward newly emerged CBT development in Indonesian NPs may provide a better understanding of the mechanism on how residents support tourism development. This study may contribute to enhance the robustness of the relation amongst construct beyond SET in shaping residents' attitudes toward tourism development and simultaneously will enable the establishment of guidelines for making policy and management decisions to promote further CBT development in NPs.

Research methods

This study focused on the perception of residents of communities neighboring GCNP concerning tourism development in the utilization zone of GCNP, and the determinants of their perceptions. We adopted the structural framework of Lee (Lee, 2013) and evaluated the model in a NP in

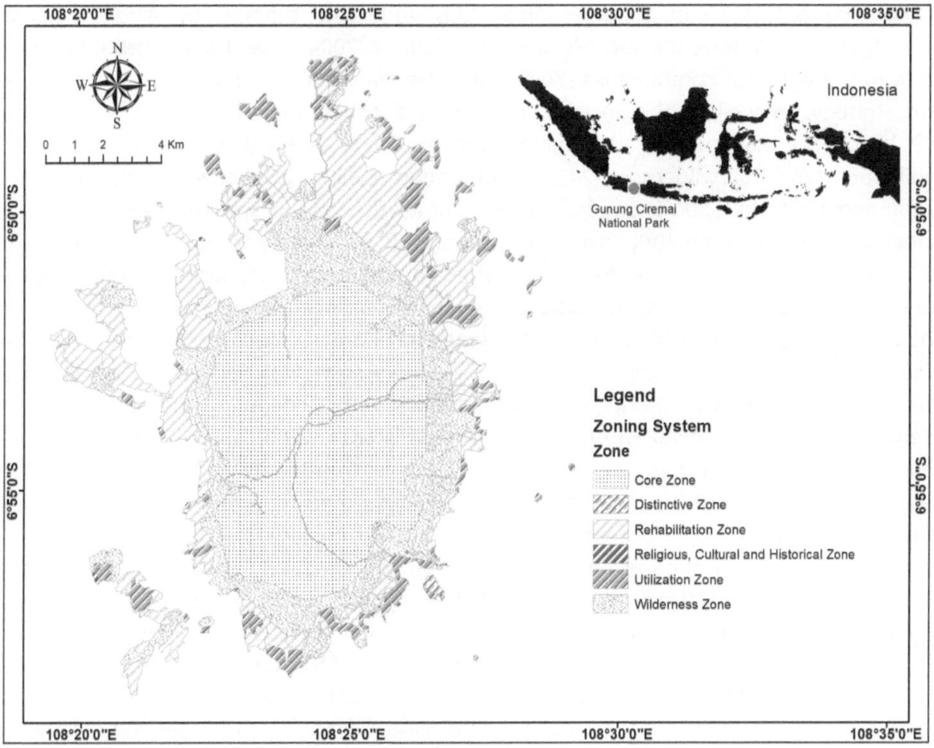

Figure 2. Location and map of Gunung Ciremai National Park.

Indonesia, which is at an early stage of development and local residents have little or limited experience of tourism.

Study site

Gunung Ciremai National Park covers 14,841.30 ha in Kuningan, Majalengka, and Cirebon Regencies, West Java Province, Indonesia. It is surrounded by 51 villages with a total population in 2016 of 149,522 (BPS Kabupaten Majalengka, 2017; BPS Kabupaten Kuningan, 2017). In total, 31 of 51 villages have tourism attractions. Most of the local residents are ethnic Sundanese who have strong cultural links to the forest (Nijman & Nekaris, 2014; Noerdjito & Royyani, 2009). Most households stated that working in farming-related activities on their own land was their primary source of income, while a small fraction of residents are employed as government officers, entrepreneurs, and laborers. Farming, mainly carrot, cabbage, shallot, collard, and potato cultivation, is a prominent livelihood activity and the primary source of income for many households (BPS Kabupaten Majalengka, 2017; BPS Kabupaten Kuningan, 2017).

Gunung in the local language means mountain, and Gunung Ciremai is the highest mountain in West Java at 3,078 m above sea level (asl). Gunung Ciremai National Park, established in 2004, is a high conservation priority due to its high level of tropical forest biodiversity and the presence of three ecosystems: lowland rain forest (200–1,000 m asl), montane rainforest (1,000–2,400 m asl), and subalpine forest (> 2.400 m asl) (BTNGC, 2018; Rozak & Gunawan, 2015).

Before its establishment in 2004, the GCNP area was long managed by a state-owned company as a protected and production forest. The only legal access by adjacent communities was for pine-sap harvesting and short-term agricultural cropping for 2 years during the reforestation period of the production forest (Bratamihardja et al., 2005). However, the establishment of GCNP changed the access of adjacent communities to the area. Land-based activities, including short-term crop cultivation by planting seasonal crops under the canopy and pine-sap harvesting, were restricted. Therefore, the establishment of GCNP in 2004 caused many protests and resistance from the adjacent communities (Yuningsih & Herdiansyah, 2019). Concerns over the conservation strategy, as well as the social, economic, and environmental impacts on locals, led to GCNP becoming managed by the zoning system. GCNP is divided into the core (40.35%), rehabilitation (26.37%), wilderness (24.90%), utilization (8.05%), distinctive (0.27%), and religious, cultural, and historical (0.06%) zones. Figure 2 shows the zoning system in GCNP. Moreover, a national park–local community partnership through CBT development was officially implemented in 2015. This change was critical for moving from top-down protectionism to bottom-up participatory management. According to national regulations, tourism development is permitted only in the utilization zone of GCNP, and this development prioritizes the adjacent local communities.

The goals of the CBT initiative were to provide more sustainable economic activities by redirecting tourism benefits to the adjacent communities and encouraging the communities to be actively involved in conservation efforts (Mbaiwa & Stronza, 2011; Russell, 2000; Sebele, 2010). Within this scheme, GCNP issues licenses for registered residents or groups of residents to develop and manage tourist sites. The license is evaluated and renewed every 2 years. Under the supervision of GCNP officials, licensed residents build and manage tourism attractions and facilities at the tourist sites. Residents are given opportunities to actively engage in the planning and decision making of tourist site development. Employment opportunities include tour guides, shopkeepers, lodge staff, and selling souvenirs, traditional food, and beverages. In addition to providing employment income, the profits generated from CBT were to be deposited as government and village funds and reinvested to improve and maintain social services, such as charities for orphans and elderly residents, improvement of village facilities, and other cultural events. The number of CBT sites in GCNP has increased since 2015 (BTNGC, 2018), and GCNP now hosts the most successful CBT in an Indonesian NP (KLHK, 2017), with increasing numbers of visitors. By 2017, the total amount of money received by local communities exceeded 196,000 USD (BTNGC, 2018).

Questionnaire design

A questionnaire survey was developed to evaluate the determinants of residents' attitudes toward tourism development and sociodemographic characteristics. The framework and selected latent variables were adapted from Lee (Lee, 2013), and the measurement variables were obtained from previous research on perceived economic, social, and environmental benefits and costs (Andereck & Vogt, 2000; Gursoy & Rutherford, 2004; Huong & Lee, 2017; Lee, 2013; Rasoolimanesh et al., 2015), community attachment (Lee, 2013; Nicholas et al., 2009; Rasoolimanesh et al., 2015), community involvement (Nicholas et al., 2009; Rasoolimanesh et al., 2015), and support of tourism development (Andereck & Vogt, 2000; Lee, 2013; Nunkoo & So, 2016). The initial questionnaire comprised 57 items. Responses were on a five-point Likert scale, with 1 meaning 'strongly disagree' and 5 'strongly agree'. The final part of the questionnaire inquired about age, sex, education level, occupation, involvement in tourist activities, and monthly household income.

Sampling and survey procedures

A non-probability sampling was used to determine the targeted villages and respondents. Nonprobability sampling uses procedures in which the sampling selection is not based on a predetermined probability but rather on the research purpose, availability of subjects, subjective judgment, or a variety of other non-statistical criteria (Guo & Hussey, 2004). Purposive sampling technique was employed to determine targeted villages. Purposive sampling allows for selection for a sampling frame that may be most affected by the specific issues, thus it was useful in its ability to test theoretical frameworks and identify the phenomena of broader population (Etikan et al., 2016; Rivera, 2019). To provide the best information to suffice the objectives study, targeted villages should be adjacent to GCNP, availability of tourist destinations in the village, data availability, and accessibility. A series of discussions with GCNP managers, tourism site managers, and field observation were conducted to obtain the data of tourism in GCNP and the candidates of sampling villages. As a result, 12 of the 31 villages with tourist sites in the Kuningan and Majalengka Regency were selected as survey sites. Data were collected from July to August 2018.

The questionnaires were administered in a direct face-to-face manner. Within a certain time and resources limitations, direct face-to-face interview allows interviewers to locate and persuade intercepted respondents to participate, improving contact and cooperation. Also, this approach enables respondents to complete the questionnaire on site. Likewise, face-to-face surveys tend to have high response rates (Heerwegh & Loosveldt, 2008). Ten undergraduate university students majoring in forestry sciences were trained to perform the interviews, administer the questionnaire, encourage participation, and collect the survey data. However, total residents in the 12 targeted villages were 38,190 (BPS Kabupaten Majalengka, 2017; BPS Kabupaten Kuningan, 2017) (see Appendix). Hence, randomization, as well as testing the entire population, are practically almost impossible due to the limitation of resources (e.g., time, human resources, financial). At this point, a convenience sampling method is recommended (Etikan et al., 2016). A convenience sampling method was used to determine the targeted respondents based on residence in the selected villages, availability at certain times, accessibility, and willingness to participate. Also, limited financial and time resources available for the investigation were taken into consideration in targeting respondents. To capture the response from those who involved and not involved in tourism activities, on-site interviews were conducted at the tourist site and door-to-door in settlement areas nearby. The intercepted residents were first asked whether they were residents of the selected villages and were aged > 18 years. During the door-to-door surveys, individual respondents from the same household were avoided. Next, the residents were informed about the aim of the study and asked if they were willing to participate. The interviewers then asked all the question items to the intercepted residents. Once the interview was completed, the resident was given a small gift.

Data analysis

Descriptive statistics, confirmatory factor analysis (CFA), and structural equation modelling (SEM) were performed in this study. Descriptive statistics were used to outline respondent characteristics (*e.g.* age, sex, educational level). First, the data were screened for suitability and applicability before performing the analyses. Questionnaires with an unanswered item were eliminated, and accordingly, only completed questionnaires will be used to subsequent data analysis. Second, CFA was used to specify the posited relations of the measured variables to the underlying latent variables. Before testing the overall measurement model, the unidimensionality of each latent variable was assessed to assure that each set of other indicators had only one underlying trait or latent variable in common (Sethi & King, 1994). The latent variables with unacceptable fits were respecified by deleting the measured variables that failed to preserve the unidimensionality of the measurement (Anderson & Gerbing, 1988; Iacobucci, 2009) as well as suspected redundant items (Ahmad et al., 2016; Fuchs & Diamantopoulos, 2009). This decreased the number of measured variables in each latent variable. Third, we assessed the model fits (absolute, incremental and parsimonious fits), indicator reliability, acquired convergence, internal consistency and discriminant validity. Fourth, the structural relationships among latent variables were tested by SEM. A maximum likelihood (ML) method was used to estimate all of the SEM parameters. A robust

Table 1. Overall value of latent variables of the questionnaire.

Latent variables	Overall value		
	Average	SD	Cronbach's alpha
Community attachment (*n* = *5*)	3.90	0.77	0.711
Perceived benefits (*n* = *24*)	3.34	0.96	0.938
Perceived costs (*n* = *14*)	2.49	0.99	0.819
Community involvement (*n* = *6*)	2.82	1.15	0.770
Support of tourism development (*n* = *8*)	4.00	0.74	0.873

ML method was used for the CFA and SEM. The ML method provides high efficiency and consistency in estimating parameters under the condition of multivariate normality of the observed variables (Bollen, 1989), and it is reasonably robust to violations of the multivariate normality assumption (Bentler, 2010; Iacobucci, 2009). Fifth, the directions and significant relationships were determined. The measurement and structural model fits were assessed using the lavaan package, ver. 0.63, for Windows (Rosseel, 2012) in R studio, ver. 1.1.463 (RStudio Team, 2015).

Results

General results

In total, 934 usable questionnaires were collected. Multivariate normality is assumed in SEM. Mardia's standardized coefficient was 69.5, indicating violation of multivariate normality (Bentler, 2010; Zhang & Xu, 2019). Most of the respondents were male (50.3%), and 48.7% were 30–50 years of age; 2.7% were illiterate, 65.5% had completed primary or secondary education, and 31.8% had completed higher education. Most worked in the agricultural sector farmers/farm workers (61.5%) and 34.5% were involved in tourism activities. Of the respondents, 60.0% had lived in the community for > 30 years. Finally, 82.2% of the respondents had a monthly income of < IDR 3,000,000 ($210) and 17.8% of > IDR 3,000,000 (1 US$ = 14,245 IDR as of 2 May 2019).

The initial questionnaire consisted of 5 latent variables and 57 measurement variables (see Appendix). The overall average values of community attachment, perceived benefits, perceived costs, community involvement, and support of tourism development were 3.90, 3.34, 2.49, 2.82, and 4.00, respectively.

The standard deviations of community attachment, perceived benefits, perceived costs, community involvement, and support of tourism development were 0.77, 0.96, 0.99, 1.15, and 0.74, respectively. Regarding the overall reliability, Cronbach's alpha values for community attachment, perceived benefits, perceived costs, community involvement, and support of tourism development were 0.711, 0.938, 0.819, 0.770, and 0.873, respectively. Table 1 shows the overall values of the latent variables in the questionnaire.

Assessment of the measurement model

The measurement model comprised five latent variables. The initial CFA result of measurement model analysis failed to provide acceptable fit model indices. Thus, a total of 38 items (i.e. 2 items for community attachment, 19 for perceived benefits, 11 for perceived cost, 2 for community involvement, and 4 items for support of tourism development) were removed to create an acceptable measurement model. The chi-squared value of the model was 653.563 ($df = 143$). The results of the revised CFA indicated that the model did not fit the data well, possibly because the chi-squared statistic assesses the absolute fit of the model to the data but is susceptible to sample size (Bollen, 1989; McDonald & Ho, 2002) and can be ignored if the sample size is > 200 (Rajalahti & Kvalheim, 2011). Thus, the parsimonious fit index for this study was chi-squared/df (653.563/143) = 4.57) and an acceptable value was < 5 (Schumacker & Lomax, 2004). The other

Table 2. Factor loadings, average variance extracted, and composite reliability of the measurement model.

Variable	Factor loading	AVE	CR
Community Attachment		0.548	0.852
I think of myself as being from my village.	0.752		
I have an emotional attachment to this place - it has meaning to me.	0.766		
I am willing to invest my talent or time to make this an even better place.	0.808		
Perceived Benefits		0.604	0.805
Increasing the number of tourists to a community improves the local economy	0.649		
Tourism encourages more business for local people.	0.691		
Tourism contributes to improving incomes and standard of living.	0.739		
Tourism generates substantial tax revenues for local government	0.757		
Tourism enhances residents' quality of life	0.797		
Perceived Costs		0.526	0.811
Community life has become disrupted as a result of the development of tourism	0.630		
Tourism increases noise	0.790		
Tourism causes overcrowding in our community	0.777		
Community Involvement		0.592	0.845
I am involved in decision-making about tourism planning and development	0.651		
The residents of this community have been involved in the management of tourism	0.779		
The residents of this village have been involved in decision-making about tourism development and preservation of GCNP	0.853		
I have the opportunity to participate in tourism planning and development in the future	0.795		
Support of Tourism Development		0.548	0.852
I support community-based tourism in GCNP	0.609		
I would like to see more tourists in GCNP	0.752		
I support tourism having a vital role in this community	0.766		
I believe that tourism should be actively encouraged in my community.	0.808		

model fit indices included a normed fit index of 0.917, Gunning Fog Index of 0.939, adjusted goodness of fit index of 0.907, comparative fit index of 0.934, Tucker and Lewis index of 0.921, and root mean square error of approximation of 0.062. Based on the preceding model fit indices, the measurement model fit the sample data well. Furthermore, the convergent validity and discriminant validity of the scale were assessed for each latent variable; the CRs were above 0.7 and the AVEs were > 0.5. Overall, the convergent validity of the latent variables was supported (Ahmad et al., 2016; Bagozzi & Yi, 1988), demonstrating a high level of internal consistency. Factor loadings, average variance extracted, and composite reliability of the measurement model are provided in Table 2.

Table 3 shows that all of the intercorrelations between pairs of latent variables were less than the square root of the AVE estimates of the two variables, indicating discriminant validity.

Structural model

The structural model achieved the required level of fitness. The structural relationship and its impact are shown in Figure 3. The perceived benefits directly, positively, and significantly affected resident support of tourism development ($\beta = 0.352$, $p < 0.001$). The perceived costs directly, negatively, and significantly affected support of tourism development ($\beta = -0.419$, $p < 0.001$).

Community attachment directly, positively and significantly affected the perceived benefits and indirectly, positively, and significantly affected support of tourism development ($\beta = 0.207$, $p < 0.001$). Community attachment directly, negatively, and significantly affected the perceived

Table 3. Discriminant validity index for latent variables.

Latent variable	Support of tourism development	Community attachment	Perceived benefits	Perceived costs	Community involvement
Support of tourism development	**0.769**				
Community attachment	0.159	**0.741**			
Perceived benefits	0.455	0.294	**0.740**		
Perceived costs	−0.293	−0.087	−0.003	**0.777**	
Community involvement	0.247	0.257	0.558	0.022	**0.725**

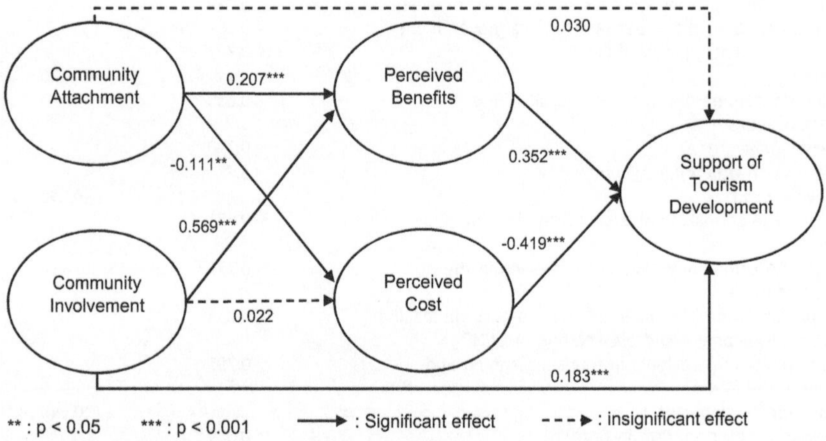

Figure 3. Structural model of resident support of tourism development.

costs (β = –0.111, $p < 0.005$) and indirectly, negatively, and significantly affected support of tourism development. Community attachment directly, positively, and non-significantly affected support of tourism development (β = 0.030, $p < 0.001$).

Community involvement directly, positively and significantly affected the perceived benefits (β = 0.569, $p < 0.001$) and indirectly, positively, and significantly affected support of tourism development. Community involvement indirectly, positively and non-significantly affected the perceived costs (β = 0.022, $p < 0.001$) and indirectly, negatively, and significantly affected support of tourism development. Community involvement directly, positively, and significantly affected resident support of tourism development (β = 0.183, $p < 0.001$).

Discussion

Our results implied that CBT development in the utilization zone of GCNP provides opportunities for residents to regain access to GCNP and poses both positive and negative impacts. The economic, environmental, and sociocultural impacts of tourism in NPs are critical for the sustainable protection of natural resources (Mackenzie, 2012), and a balance among all three impact types may enable achievement of conservation goals (MacKenzie et al., 2017; Walpole & Goodwin, 2001). Our analysis indicated that the perceived benefits directly, positively, and significantly influenced the support of CBT development by residents. Conversely, perceived costs directly, negatively, and significantly influenced residents support of tourism development. These results generally confirm the robustness of SET postulates.

However, the negative impacts of tourism had stronger effects on resident support of CBT development. This is inconsistent with the findings of Lee (2013), who conducted research in a tourism development area during the consolidation stage in Taiwan. These contradictory findings in the social exchange relationship might be related to the stage of development. The adjacent communities of GCNP have traditionally depended on the agricultural sector, and thus tourism sector is relatively new for them. Whitener et al. (1998) suggests that social exchanges in the early stages of relationships involve uncertainties. Positive perceptions are less frequent due to the relative newness of the tourist destinations in GCNP, the positive impacts of which may be limited. Nevertheless, the perceptions of residents regarding the economic benefits had a greater influence on their support of tourism development than did perceptions regarding sociocultural or environmental benefits. The adjacent communities view tourism as an important alternative economic generator for their community. Hence, improving and distributing equitable tourism benefits would increase residents' support for tourism development. Government intervention and regulation are pivotal. Formulating policies to distribute wider tourism benefits to the adjacent communities may improve residents' awareness of the importance of CBT in GCNP. Accordingly, Sebele (2010) argued that acquiring sufficient economic benefits from tourism is essential for stimulating residents' feelings of obligation to conserve natural resources.

Our findings also indicate that while residents saw the economic benefits, they also realized that the rapid growth of tourism would have an adverse impact on their community, such as disruption of community life, increased noise, and overcrowding. Accordingly, during the early stage of tourism development, residents might hold temporary, rather than long-term, negative perceptions due to rapid changes in their community life. Based on CBT implementation in communities adjacent to Etosha National Park in Namibia, Saarinen (2010) indicated that prior knowledge pertaining to tourism and its potential impacts was substantial in determining the attitudes of residents toward tourism. Our finding was corroborated by Lepp (2008), who reported that when residents have little or no prior knowledge about tourism, the initial development of tourism will be met with suspicion, anxiety, and fear. Residents may have negative reactions toward the rapid changes initially but later adjust to those changes as their negative perceptions become less intense (e.g., Brown et al., 2005; England & Albrecht, 1984). Thus, the perceptions of residents regarding perceived costs may change with their ability to adjust and react.

Community involvement in planning, decision making, and participation in tourism-related activities was found to be related to resident support of tourism development. Compared with the report by Lee (2013), the effect of community involvement on the perceived benefits was much stronger in GCNP. This may be due to the availability of supporting regulations, as well as policies, issued by GCNP managers to encourage the involvement of residents in the adjacent communities in the planning, decision-making, and management of tourist sites in the utilization zone of GCNP. Our findings support the research of Sene-Harper et al. (2019), who advocate the importance of the legal frameworks in providing locals access and benefits from tourism around national parks in Senegal. Active involvement in tourism management may lead to their positive perceptions of GCNP management and eventually boost their willingness to support the conservation efforts and existence of the national park (Ghoddousi et al., 2018; Mackenzie, 2012). Therefore, policy support is important for ensuring community involvement in all aspects of tourism management in GCNP. Residents who support tourism development will be more involved in tourism activities. Hence, resident participation that provides extra-economic benefits could influence resident perception and support of CBT development.

During the consolidation stage in Cigu, a positive non-significant relationship was found between community attachment and perceived costs (Lee, 2013). However, we found that community attachment exerted a significant effect on the perceived benefits and perceived costs. The effect of community attachment on perceived benefits was positive, and that on perceived costs was negative. The findings were contrary to Harrill (2004) reports that the higher the attached residents are to their community, the more negative they are about tourism development. This contradictory finding may be because, during the early stage of tourism development, highly attached residents have high expectations of receiving economic benefits. Residents with a positive feeling toward their community, and who are willing to invest their talent and time, will have a more positive perception of the impact of tourism development. We concluded that residents who express a high level of attachment to their community are more likely to expect greater economic benefits from, and perceive lower costs of, tourism development.

In this study, community attachment did not directly, positively, or significantly affect the support of residents for CBT development. This could have been because the higher level of attachment to community at an early stage of tourism development enhances indirect effects. Resident support of tourism development was linked to the perceived benefits and costs. Our results confirm that community attachment in a social exchange relationship is dependent on the perceived benefits and perceived costs (Gursoy & Kendall, 2006; Lee, 2013). Therefore, the direct relationship between community attachment and support of tourism development remains unclear. Moreover, community involvement did not directly, negatively, or significantly affect the perceived costs or support of residents. This may be because a high level of community involvement in tourism activities increases the opportunities of residents to benefit from tourism. Highly involved residents will receive more benefits and so increase their support of tourism development; hence, community involvement was not significantly associated with perceived costs.

Conclusions and implications

Perceived benefits, perceived costs, community involvement, and community attachment play important roles in determining resident support of tourism development in the CBT initiative in GCNP. Our findings indicate that resident support of tourism development is based on the perceived benefits and perceived costs. Furthermore, it is affected by community attachment and community involvement, which are related to the perceived benefits and costs.

Theoretical implications

This study contributes to knowledge of behavioral models based on the SET framework, particularly in CBT in a NP. While the findings generally confirm the robustness SET postulates, it also provides some new theoretical evidence in terms of relations and strength of its effects. This can be summarized as follows. First, recent existing literature conceptualizes that if residents perceive that the benefits of tourism outweigh the costs, they will subsequently support for tourism development (Gursoy & Rutherford, 2004; Lee, 2013; Sharpley, 2014). Interestingly, our results indicated that residents' perceived costs in tourism had stronger effects on residents' support for CBT development than did perceived benefits. It is probably the reason that Whitener et al. (1998) considers the early stage of a relationship in social exchanges characterized by high uncertainties. The relative newness of tourism in the communities may limit residents' perceived benefits and subsequently affect their level of support for tourism development. Therefore, we suggest that the specific context in which tourism occurs and its level of development should be strongly considered when assessing residents' level of support. Second, we found that community attachment directly, negatively, and significantly affected perceived costs. The higher attached residents toward their community will have a more positive perception of perceived benefits and lessen in tourism perceived costs. These results were contradicted with those who conducted investigation in mature tourist' destinations (Gursoy & Kendall, 2006; Gursoy & Rutherford, 2004; Lee, 2013). The difference may be explained by differences pertaining to the different stages of tourism development. It is probably related to Hummon (1992) considers community attachment as a reflection of "community sentiment". He argued that the built environment may also contribute to such emotional ties. The rapid development of tourism may disrupt their community life circumstances. Thus, residents' feeling attached to the communities is not only as an attribution of a particular social structure, but it has consequences on their community development trend. Third, the perception of negative impacts is affected by contradictory forces as tourism development progresses. During the early stage of tourism development, when residents have no or little experience of tourism, they may hold strong temporary, rather than long-term, negative perceptions due to an inability to adapt to the rapid changes in their community life.

Practical implications

The empirical findings demonstrate that residents' perceptions are affected by a variety of factors. Sustainable CBT management could be achieved by strategies tailored to each stage of tourism development. Our results, although empirical, have practical implications for GCNP managers.

First, in the context of perceived benefits, in the early stages of tourism development economic benefits are more recognized by residents than are environmental and social benefits. Given the legal status of the NP and concern for its preservation, residents adjacent to GCNP had little or no experience of tourism prior to 2015. This situation changed due to the enactment of new inclusive policies to promote tourism in the NP. Consequently, tourism increased rapidly, generating economic benefits for the residents, who formerly focused on agriculture. However, any economic euphoria during the early stages of tourism development may not be sustained and may lead to conflicts among residents (Lee & Jan, 2019) due to unequal distribution of tourism revenue (Blake et al., 2008; Hunt & Stronza, 2014). Based on our findings, GCNP managers must formulate policies to distribute the benefits of tourism to residents.

Second, our findings indicate that despite its benefits, residents consider tourism development to exert a negative effect on their community, such as by disrupting community life and increasing the levels of noise and overcrowding. The pace of tourism growth, especially in the early stages of tourism development, may lead to social disruption. Therefore, we suggest that GCNP and managers of tourist sites provide guidelines to reduce noise and avoid disturbance of local communities. For example, formulating policies on specific types of attractions and arranging the temporal and spatial distribution of visitors.

Third, community involvement is significantly correlated with the perceived benefit and support of tourism development. Because of the importance of community involvement in resident support of tourism development, this significant correlation indicates increasing resident participation in the planning, decision-making, and management of tourist destinations. Tourism studies suggest that community involvement improves the awareness of the potential impacts of tourism, strengthens respect for culture and values (Látková & Vogt, 2012; Lee, 2013; Nunkoo & So, 2016), and leads to stronger conservation management and governance (Dearden et al., 2005; Mackenzie, 2012). Furthermore, expanding collaborations involved in tourism management to village-owned enterprises, community empowerment institutions, and other existing organizations may provide opportunities for residents to be involved in all aspects of development. This would increase the benefits to village residents and subsequently the support of sustainable tourism management.

Fourth, despite the relatively homogeneous initial response to tourism development due to its economic benefits, previous studies have indicated that the perception of residents regarding the impact of tourism can change from positive to negative (Butler, 1980; Hunt & Stronza, 2014). Accordingly, GCNP should monitor and evaluate the program to ensure its sustainability.

Limitations and future research

The direct role of community attachment in resident support of tourism development remains unclear, and thus further research is needed. Next, the use of nonprobability sampling allows us to use predetermined criteria to suffice the study objectives. This study was not covering all of the villages, but we focused on those adjacent villages of GCNP with tourist' destinations. Residents on those villages without tourist' destinations may hold different opinions toward CBT development in GCNP. Hence, the survey over a broader spectrum of residents is promising for future studies. Moreover, a limitation of our study is that the study was conducted in the context of the early stage of CBT development in NPs. The different types of communities as well as different ethnicities may have different perceptions of CBT development. Future research should aim to overcome this limitation by assessing the reliability of this behavioral model in other cases.

Acknowledgments

The authors would like to express appreciation to the Tokyo Metropolitan Government for their support for this research through the Tokyo Human Resources Fund for City Diplomacy. The authors thank Gunung Ciremai national park staff and members of faculty of forestry of Universitas Kuningan for their assistance during data collection. We are also thankful to Mr. Etsuro Takagi for his valuable and constructive advice. We thank the reviewers very much for valuable comments on the manuscript.

ORCID

Prasetyo Nugroho http://orcid.org/0000-0002-7275-9708

References

Agrawal, A., & Redford, K. (2009). Place, conservation, and displacement. *Conservation and Society*, *7*(1), 56. doi:10.4103/0972-4923.54797

Ahmad, S., Zulkurnain, N., & Khairushalimi, F. (2016). Assessing the validity and reliability of a measurement model in Structural Equation Modeling (SEM). *British Journal of Mathematics & Computer Science*, *15*(3), 1–8. doi:10.9734/BJMCS/2016/25183

Akyeampong, O. A. (2011). Pro-poor tourism: residents' expectations, experiences and perceptions in the Kakum National Park Area of Ghana. *Journal of Sustainable Tourism*, *19*(2), 197–213. doi:10.1080/09669582.2010.509508

Andereck, K. L., & Vogt, C. A. (2000). The relationship between residents' attitudes toward tourism and tourism development options. *Journal of Travel Research*, *39*(1), 27–36. doi:10.1177/004728750003900104

Anderson, J. C., & Gerbing, D. W. (1988). Structural equation modeling in practice: A review and recommended two-step approach. *Psychological Bulletin*, *103*(3), 411–423. doi:10.1037/0033-2909.103.3.411

Ap, J. (1992). Residents' perceptions on tourism impacts. *Annals of Tourism Research*, *19*(4), 665–690.(92)90060-3 doi:10.1016/0160-7383(92)90060-3

Bagozzi, R. P., & Yi, Y. (1988). On the evaluation of structural equation models. *Journal of the Academy of Marketing Science*, *16*(1), 74–94. doi:10.1007/BF02723327

Bentler, P. M. (2010). *NIH Public Access*, *75*(2), 243–248. 10.1016/j.jcps.2010.03.002.SEM

Blake, A., Arbache, J. S., Sinclair, M. T., & Teles, V. (2008). Tourism and poverty relief. *Annals of Tourism Research*, *35*(1), 107–126. doi:10.1016/j.annals.2007.06.013

Bollen, K. A. (1989). A new incremental fit index for general structural equation models. *Sociological Methods & Research*, *17*(3), 303–316. doi:10.1177/0049124189017003004

BPS Kabupaten Majalengka. (2017). *Kabupaten Majalengka dalam Angka (Majalengka Regency in Figures)*. Retrieved from https://majalengkakab.bps.go.id/publication

Bratamihardja, M., Satyawan, & Kartasubrata. (2005). Forest management in Java 1975–1999: towards collaborative management. *ICRAFT Southeast Asia Working Paper, No. 2005-1. ICRAFT Southeast Asia Regional Office, Bogor, Indonesia*. Retrieved from http://www.worldagroforestry.org/sea/Publications/files/workingpaper/WP0055-05.pdf

Brockington, D., & Wilkie, D. (2015). Protected areas and poverty. *Philosophical Transactions of the Royal Society B: Biological Sciences*, *370*(1681), 20140271. doi:10.1098/rstb.2014.0271

Brown, R. B., Dorins, S. F., & Krannich, R. S. (2005). The Boom-Bust-Recovery Cycle: Dynamics of Change in Community Satisfaction and Social Integration in Delta, Utah*. *Rural Sociology*, *70*(1), 28–49. https://doi.org/10.1526/0036011053294673

Brown, R. B., Dorins, S. F., & Krannich, R. S. (2005). Community satisfaction and social integration in a Boomtown: A longitudinal analysis. *Rural Sociology*, *70*(1), 28–49. doi:10.1526/0036011053294673

BTNGC. (2018). *Statistik Balai Taman Nasional Gunung Ciremai Tahun 2017*. Balai Taman Nasional Gunung Ciremai.

Butler, R. W. (1980). The concept of a tourist area cycle of evolution: implications for management of resources. *The Canadian Geographer/Le Géographe Canadien*, *24*(1), 5–12. doi:10.1111/j.1541-0064.1980.tb00970.x

Campbell, S. J., Kartawijaya, T., Yulianto, I., Prasetia, R., & Clifton, J. (2013). Co-management approaches and incentives improve management effectiveness in the Karimunjawa National Park, Indonesia. *Marine Policy*, *41*, 72–79. doi:10.1016/j.marpol.2012.12.022

Choi, H.-S. C., & Sirakaya, E. (2005). Measuring residents' attitude toward sustainable tourism: Development of Sustainable Tourism Attitude Scale. *Journal of Travel Research*, *43*(4), 380–394. doi:10.1177/0047287505274651

Clifton, J. (2013). Refocusing conservation through a cultural lens: Improving governance in the Wakatobi National Park, Indonesia. *Marine Policy*, *41*, 80–86. doi:10.1016/j.marpol.2012.12.015

Cook, K. S., & Rice, E. (2003). Social exchange theory. In K. S. Cook (Ed.), *Handbooks of sociology and social research. Handbook of social psychology* (pp. 53–76). Stanford U, Dept of Sociology, Kluwer Academic/Plenum Publishers.

Dearden, P., Bennett, M., & Johnston, J. (2005). Trends in global protected area governance, 1992–2002. *Environmental Management*, *36*(1), 89–100. doi:10.1007/s00267-004-0131-9

Eagles, P. F. J., Mccool, S. F., & Haynes, C. D. (2002). Sustainable tourism in protected areas. In A. Phillips (Ed.), *Best practice protected area guidelines*. 10.2305/IUCN.CH.2002.PAG.8.en

Eneji, V. C. O., Gubo, Q., Okpiliya, F. I., Aniah, E. J., Eni, D. D., & Afangide, D. (2009). Problems of public participation in biodiversity conservation: the Nigerian scenario. *Impact Assessment and Project Appraisal*, *27*(4), 301–307. doi:10.3152/146155109X479431

England, J. L., & Albrecht, S. L. (1984). Boomtowns and Social Disruption. In *Rural Sociology*.

Etikan, I., Musa, S. A., & Alkassim, R. S. (2016). Comparison of convenience sampling and purposive sampling. *American Journal of Theoretical and Applied Statistics*, *5*(1), 1. doi:10.11648/j.ajtas.20160501.11

Fuchs, C., & Diamantopoulos, A. (2009). Using single-item measures for construct measurement in management research: Conceptual issues and application guidelines. *Die Betriebswirtschaft*, *69*(2), 195–211. 10.1021/ja01488a024

Ghoddousi, S., Pintassilgo, P., Mendes, J., Ghoddousi, A., & Sequeira, B. (2018). Tourism and nature conservation: A case study in Golestan National Park, Iran. *Tourism Management Perspectives*, *26*, 20–27. (December 2017). doi:10.1016/j.tmp.2017.12.006

Guo, S., & Hussey, D. L. (2004). Nonprobability Sampling in Social Work Research. *Journal of Social Service Research*, *30*(3), 1–18. doi:10.1300/J079v30n03_01

Gursoy, D., & Kendall, K. W. (2006). Hosting mega events. Modeling Locals' Support. *Annals of Tourism Research*, *33*(3), 603–623. doi:10.1016/j.annals.2006.01.005

Gursoy, D., & Rutherford, D. G. (2004). Host attitudes toward tourism. *Annals of Tourism Research*, *31*(3), 495–516. doi:10.1016/j.annals.2003.08.008

Harrill, R. (2004). Residents' attitudes toward tourism development: a literature review with implications for tourism planning. *Journal of Planning Literature*, *18*(3), 251–266. doi:10.1177/0885412203260306

Heerwegh, D., & Loosveldt, G. (2008). Face-to-face versus web surveying in a high-internet-coverage population: Differences in response quality. *Public Opinion Quarterly* , *72*(5), 836–846. doi:10.1093/poq/nfn045

Homans, G. C. (1961). *Social behaviour and its elementary forms*. Harcourt, Brace and World.

Hummon, D. M. (1992). Community attachment. In *Place attachment* (pp. 253–278). doi:10.1007/978-1-4684-8753-4_12

Hunt, C., & Stronza, A. (2014). Stage-based tourism models and resident attitudes towards tourism in an emerging destination in the developing world. *Journal of Sustainable Tourism*, *22*(2), 279–298. doi:10.1080/09669582.2013.815761

Huong, P. M., & Lee, J. H. (2017). Finding important factors affecting local residents' support for tourism development in Ba Be National Park. Vietnam. *Forest Science and Technology*, *13*(3), 126–132. doi:10.1080/21580103.2017.1354337

Iacobucci, D. (2009). Everything you always wanted to know about SEM (structural equations modeling) but were afraid to ask. *Journal of Consumer Psychology*, *19*(4), 673–680. doi:10.1016/j.jcps.2009.09.002

Kim, K., Uysal, M., & Sirgy, M. J. (2013). How does tourism in a community impact the quality of life of community residents? *Tourism Management*, *36*, 527–540. doi:10.1016/j.tourman.2012.09.005

KLHK. (2017). *Statistik Lingkungan Hidup dan Kehutanan Tahun 2016*. Kementerian Lingkungan Hidup dan Kehutanan.

Ko, D. W., & Stewart, W. P. (2002). A structural equation model of residents' attitudes for tourism development. *Tourism Management*, *23*(5), 521–530.(02)00006-7 doi:10.1016/S0261-5177(02)00006-7

BPS Kabupaten Kuningan . (2017). *Kabupaten Kuningan dalam Angka (Kuningan Regency in Figures)*. Retrieved from https://kuningankab.bps.go.id/publication

Látková, P., & Vogt, C. A. (2012). Residents' attitudes toward existing and future tourism development in rural communities. *Journal of Travel Research, 51*(1), 50–67. doi:10.1177/0047287510394193

Lee, T. H. (2013). Influence analysis of community resident support for sustainable tourism development. *Tourism Management, 34*, 37–46. https://doi.org/10.1016/j.tourman.2012.03.007

Lee, T. H. (2013). Influence analysis of community resident support for sustainable tourism development. *Tourism Management, 34*, 37–46. doi:10.1016/j.tourman.2012.03.007

Lee, T. H., & Jan, F.-H. (2019). Can community-based tourism contribute to sustainable development? Evidence from residents' perceptions of the sustainability. *Tourism Management, 70*, 368–380. (September 2018),doi:10.1016/j.tourman.2018.09.003

Lepp, A. (2007). Residents' attitudes towards tourism in Bigodi village. Uganda. *Tourism Management, 28*(3), 876–885. doi:10.1016/j.tourman.2006.03.004

Lepp, A. (2008). Attitudes towards initial tourism development in a community with no prior tourism experience: The case of Bigodi. Uganda. *Journal of Sustainable Tourism, 16*(1), 5–22. doi:10.2167/jost630.0

Ma, B., Cai, Z., Zheng, J., & Wen, Y. (2019). Conservation, ecotourism, poverty, and income inequality – A case study of nature reserves in Qinling. China. *World Development, 115*, 236–244. doi:10.1016/j.worlddev.2018.11.017

Mackenzie, C. A. (2012). Accruing benefit or loss from a protected area: Location matters. *Ecological Economics, 76*, 119–129. doi:10.1016/j.ecolecon.2012.02.013

MacKenzie, C. A., Salerno, J., Hartter, J., Chapman, C. A., Reyna, R., Tumusiime, D. M., & Drake, M. (2017). Changing perceptions of protected area benefits and problems around Kibale National Park. Uganda. *Journal of Environmental Management, 200*, 217–228. doi:10.1016/j.jenvman.2017.05.078

Mbaiwa, J. E., & Stronza, A. L. (2011). Changes in resident attitudes towards tourism development and conservation in the Okavango Delta, Botswana. *Journal of Environmental Management, 92*(8), 1950–1959. doi:10.1016/j.jenvman.2011.03.009

McDonald, R. P., & Ho, M.-H R. (2002). Principles and practice in reporting structural equation analyses. *Psychological Methods, 7*(1), 64–82. doi:10.1037/1082-989X.7.1.64

Molina-Murillo, S. A., Otárola, M. F., & Shreeve, K. N. (2016). Understanding the link between benefits from protected areas and their relationship with surrounding communities: An exploration in Costa Rica. *Parks, 22*(1), 79–88. doi:10.2305/IUCN.CH.2016.PARKS-22-1SAM-M.en

Myers, R., & Muhajir, M. (2015). Searching for Justice: Rights vs "Benefits" in Bukit Baka Bukit Raya National Park, Indonesia. *Conservation and Society, 13*(4), 370. doi:10.4103/0972-4923.179886

Nicholas, L. N., Thapa, B., & Ko, Y. J. (2009). Residents' perspectives of a world heritage site. The Pitons Management Area, St. Lucia. *Annals of Tourism Research, 36*(3), 390–412. doi:10.1016/j.annals.2009.03.005

Nijman, V., Nekaris, K. A. (2014). *Traditions, taboos and trade in slow lorises in Sundanese communities in southern Java, Indonesia. 25*(Redford 2011), 79–88. doi:10.3354/esr00610

Ninan, K. N., & Sathyapalan, J. (2005). The economics of biodiversity conservation: a study of a coffee growing region in the Western Ghats of India. *Ecological Economics, 55*(1), 61–72. doi:10.1016/j.ecolecon.2004.10.005

Noerdjito, M., & Royyani, M. F. (2009). Peran Adat dan Pensakralan Mata Air Terhadap Konservasi Air di Lereng Ciremai. Jurnal Biologi Indonesia *5*(3), 363–376. Retrieved from http://e-journal.biologi.lipi.go.id/index.php/jurnal_biologi_indonesia/article/view/3188/2772

Nunkoo, R. (2016). Toward a more comprehensive use of social exchange theory to study residents' attitudes to tourism. *Procedia Economics and Finance, 39*, 588–596.(16)30303-3 doi:10.1016/S2212-5671(16)30303-3

Nunkoo, R., Smith, S. L. J., & Ramkissoon, H. (2013). Residents' attitudes to tourism: a longitudinal study of 140 articles from 1984 to 2010. *Journal of Sustainable Tourism, 21*(1), 5–25. doi:10.1080/09669582.2012.673621

Nunkoo, R., & So, K. K. F. (2016). Residents' support for tourism. *Journal of Travel Research, 55*(7), 847–861. doi:10.1177/0047287515592972

Okazaki, E. (2008). A community-based tourism model: Its conception and use. *Journal of Sustainable Tourism, 16*(5), 511–529. doi:10.1080/09669580802159594

Presenza, A., Del Chiappa, G., & Sheehan, L. (2013). Residents' engagement and local tourism governance in maturing beach destinations. Evidence from an Italian case study. *Journal of Destination Marketing & Management, 2*(1), 22–30. doi:10.1016/j.jdmm.2013.01.001

Pudyatmoko, S., Budiman, A., & Kristiansen, S. (2018). Towards sustainable coexistence: People and wild mammals in Baluran National Park. Indonesia. *Forest Policy and Economics, 90*, 151–159. (April 2017),doi:10.1016/j.forpol.2018.02.006

Rajalahti, T., & Kvalheim, O. M. (2011). Multivariate data analysis in pharmaceutics: A tutorial review. *International Journal of Pharmaceutics, 417*(1–2), 280–290. doi:10.1016/j.ijpharm.2011.02.019

Rasoolimanesh, S. M., Jaafar, M., Kock, N., & Ramayah, T. (2015). A revised framework of social exchange theory to investigate the factors influencing residents' perceptions. *Tourism Management Perspectives, 16*, 335–345. doi:10.1016/j.tmp.2015.10.001

Rastegar, R. (2019). Tourism development and conservation, do local resident attitudes matter? *International Journal of Tourism Sciences, 19*(3), 181–191. doi:10.1080/15980634.2019.1663998

Rivera, J. D. (2019). When attaining the best sample is out of reach: Nonprobability alternatives when engaging in public administration research. *Journal of Public Affairs Education, 25*(3), 314–342. doi:10.1080/15236803.2018.1429821

Rosseel, Y. (2012). lavaan : An R Package for Structural Equation. *Journal of Statistical Software, 48*(2), 1–36. doi:10.18637/jss.v048.i02

Rozak, A. H., & Gunawan, H. (2015). Altitudinal gradient affects on trees and stand attributes in Mount Ciremai National Park, West Java, Indonesia. *Jurnal Penelitian Kehutanan Wallacea, 4*(2), 93. doi:10.18330/jwallacea.2015.vol4iss2pp93-99

RStudio Team. (2015). RStudio: Integrated Development for R. *[Online] RStudio, Inc., Boston, MA.* http://www. rstudio.com. 10.1007/978-81-322-2340-5

Russell, P. (2000). Community-based tourism. *Travel & Tourism Analyst,* (5), 89–116.

Saarinen, J. (2010). Local tourism awareness: Community views in Katutura and King Nehale Conservancy, Namibia. *Development Southern Africa, 27*(5), 713–724. doi:10.1080/0376835X.2010.522833

Scheyvens, R., & Russell, M. (2012). Tourism, land tenure and poverty alleviation in Fiji. *Tourism Geographies, 14*(1), 1–25. doi:10.1080/14616688.2011.593188

Schumacker, R. E., & Lomax, R. G. (2004). A beginner's guide to structural equation modeling. *Technometrics, 47,* 522. 10.1198/tech.2005.s328

Sebele, L. S. (2010). Community-based tourism ventures, benefits and challenges: Khama Rhino Sanctuary Trust, Central District, Botswana. *Tourism Management, 31*(1), 136–146. doi:10.1016/j.tourman.2009.01.005

Šegota, T., Mihalič, T., & Kuščer, K. (2017). The impact of residents' informedness and involvement on their perceptions of tourism impacts: The case of Bled. *Journal of Destination Marketing & Management, 6*(3), 196–206. doi:10.1016/j.jdmm.2016.03.007

Sene-Harper, A., Matarrita-Cascante, D., & Larson, L. R. (2019). Leveraging local livelihood strategies to support conservation and development in West Africa. *Environmental Development, 29,* 16–28. (November 2017),doi:10.1016/j.envdev.2018.11.002

Sethi, V., & King, W. R. (1994). Development of measures to assess the extent to which an information technology application provides competitive advantage. *Management Science, 40*(12), 1601–1627. doi:10.1287/mnsc.40.12.1601

Sharpley, R. (2014). Host perceptions of tourism: A review of the research. *Tourism Management, 42,* 37–49. doi:10.1016/j.tourman.2013.10.007

Shivakoti, Y., & Shivakoti, G. P. (2008). Decentralization and co-management of protected areas in Indonesia. *The Journal of Legal Pluralism and Unofficial Law, 40*(57), 141–165. doi:10.1080/07329113.2008.10756620

Sirivongs, K., & Tsuchiya, T. (2012). Relationship between local residents' perceptions, attitudes and participation towards national protected areas: A case study of Phou Khao Khouay National Protected Area, central Lao PDR. *Forest Policy and Economics, 21,* 92–100. doi:10.1016/j.forpol.2012.04.003

Soliku, O., & Schraml, U. (2018). Making sense of protected area conflicts and management approaches: A review of causes, contexts and conflict management strategies. *Biological Conservation, 222,* 136–145. (August 2017),doi:10.1016/j.biocon.2018.04.011

Tosun, C. (2006). Expected nature of community participation in tourism development. *Tourism Management, 27*(3), 493–504. doi:10.1016/j.tourman.2004.12.004

Udaya Sekhar, N. (2003). Local people's attitudes towards conservation and wildlife tourism around Sariska Tiger Reserve. India. *Journal of Environmental Management, 69*(4), 339–347. doi:10.1016/j.jenvman.2003.09.002

Walpole, M. J., & Goodwin, H. J. (2001). Local attitudes towards conservation and tourism around Komodo National Park. *Environmental Conservation, 28*(2), 160–166. doi:10.1017/S0376892901000169

West, P., Igoe, J., & Brockington, D. (2006). Parks and Peoples: The Social Impact of Protected Areas. *Annual Review of Anthropology, 35*(1), 251–277. doi:10.1146/annurev.anthro.35.081705.123308

Whitener, E. M., Brodt, S. E., Korsgaard, M. A., & Werner, J. M. (1998). Managers as initiators of trust: An exchange relationship framework for understanding managerial trustworthy behavior. *Academy of Management Review, 23*(3), 513–530. doi:10.5465/amr.1998.926624

Wiratno. (2018). Sepuluh Cara baru Kelola Kawasan Konservasi di Indonesia: Membangun "Organisasi Pembelajar." *Direktorat Jenderal KSDAE Kementerian Lingkungan Hidup Dan Kehutanan.* Retrieved from http://ksdae.menlhk.go.id/assets/publikasi/10_Cara_BaruKelolaKK.pdf.

Yuningsih, Y., & Herdiansyah, H. (2019). The role of management Institution of the Ciremai Mountain National Park in encouraging public participation. *Journal Sosial Dan Humaniora, 9*(1), 26–38. https://doi.org/dx.doi.org/10.31940/soshum.v9i1.1171

Yusran, Y., Sahide, M. A. K., Supratman, S., Sabar, A., Krott, M., & Giessen, L. (2017). The empirical visibility of land use conflicts: From latent to manifest conflict through law enforcement in a national park in Indonesia. *Land Use Policy, 62,* 302–315. doi:10.1016/j.landusepol.2016.12.033

Zhang, H., & Xu, H. (2019). A structural model of liminal experience in tourism. *Tourism Management, 71,* 84–98. (October 2018),doi:10.1016/j.tourman.2018.09.015

Impacts of tourism on support for conservation, local livelihoods, and community resilience around Maasai Mara National Reserve, Kenya

Kathleen Krafte Holland, Lincoln R. Larson, Robert B. Powell, W. Hunter Holland, Lawrence Allen, Moriaso Nabaala, Salaton Tome, Simon Seno and James Nampushi

ABSTRACT

The tourism enterprise-based approach to conservation can diversify livelihood opportunities in and around protected areas while incentivizing conservation support and building community resilience. However, the use of tourism as a tool for integrated conservation and development (ICD) has produced mixed results, underscoring the need for adaptive management that accounts for local context. We used a mixed-methods approach to examine how individual and community-level involvement in tourism influenced support for conservation, reliance on the natural environment, and perceptions of governance and community resilience in communities around Maasai Mara National Reserve, Kenya. During January 2017, we surveyed households in three communities (n = 197) with varying levels of tourism involvement and conducted focus groups with village elders (n = 28). Individuals and communities who engaged in tourism-related livelihoods expressed stronger support for conservation and reduced reliance on the natural environment. Links between tourism and environmental, social, and economic resilience varied. Results suggest the presence of a tourism sweet spot where moderate levels of tourism facilitate alignment of conservation and community development goals. Communities that achieve equitable access to and engagement in tourism and associated benefits may be more resilient and better positioned to capitalize on linkages between tourism-based livelihoods and biodiversity conservation.

Introduction

Natural resource conservation and tourism-based community development, once viewed as opposing forces, are increasingly embraced as synergistic and compatible goals (Alpert, 1996;

Ferraro & Hanauer, 2014; Oldekop et al., 2016; Stronza et al., 2019; West et al., 2006). The tourism enterprise-based approach to conservation, inspired by the concept of integrated conservation and development (ICD, Newmark & Hough, 2000), is thought to incentivize support for conservation within local communities by providing jobs in tourism or other alternative livelihoods that replace more destructive and extractive traditional livelihood activities (Fennell & Weaver, 2005; Salafsky et al., 2001; Sene-Harper et al., 2019; Wright et al., 2016). Recent research by Naidoo et al. (2019) and others highlight the various mechanisms through which protected areas (PAs), integrated with tourism development, can positively increase household wealth, income generation, and other material benefits (Mbaiwa & Stronza, 2010; Stronza et al., 2019; Stronza & Pêgas, 2008). The ICD approach can also increase the resilience of communities (Holladay & Powell, 2013; Jamaliah & Powell, 2018; Powell et al., 2018) by reinforcing positive linkages between livelihoods and biodiversity conservation (Stone & Nyaupane, 2016, 2018).

However, some research evaluating the effectiveness of ICD through tourism has revealed mixed or even negative outcomes for local communities. Potential pitfalls of tourism-based ICD projects include unequal distribution of economic benefits (Coria & Calfucura, 2012; Stronza et al., 2019), emphasis on indirect benefits that do not directly impact the lives of local residents (Munanura et al., 2020), and poor governance that prevents local involvement in tourism planning and decision-making (Coria & Calfucura, 2012; Powell et al., 2018; Sabuhoro et al., 2021). ; Weak linkages between tourism and biodiversity can also detract from conservation goals (Kiss, 2004; Salafsky & Wollenberg, 2000; Stone & Nyaupane, 2016), often resulting in further degradation of natural resources as park-proximate development spurs human population growth (Karanth & DeFries, 2011). These problems are compounded when tourism enterprises do not align with cultural norms and traditions (Nyaupane & Thapa, 2004; West et al., 2006), or when training opportunities are absent and capacity for local engagement in the tourism industry is low (Aref & Redzuan, 2009).

In their assessment of enterprise-based conservation strategies that succeed and fail, Salafsky et al. (2001) highlighted one consistent component of success: an adaptive management process that enables communities and industry partners to systematically test assumptions and learn from results as projects unfold. Tourism-based ICD approaches can work, but positive outcomes depend on a thorough understanding of local context, perceptions, and norms (Salafsky et al., 2001). Our study sought to test the efficacy of tourism-based ICD around the Maasai Mara National Reserve in Kenya, a region where wildlife tourism is a critical economic driver. Following recommendations by Stronza et al. (2019), we used a mixed-methods approach in three different communities, each with varying levels of tourism development, to explore the impacts of tourism involvement on four key outcomes of ICD: local support for conservation, reliance on the natural environment, and perceptions of inclusive governance and community resilience (Campbell et al., 2010; Marcus, 2001). Because communities consist of groups of individuals that collectively define community trajectories and experiences, both individual and collective perceptions matter. Therefore, following suggestions by Ban et al. (2013), we examined these relationships based on an individual's direct involvement in the tourism industry (individual level) and levels of tourism engagement within the broader community (community level). Specifically, this research sought to answer the following research questions about the tourism enterprise-based conservation model in the Maasai Mara:

> Q1: How does tourism involvement at the individual level influence support for conservation, reliance on the natural environment, and perceptions of local governance and community resilience (environmental, social, and economic)?

> Q2: How does tourism involvement at the community level influence support for conservation, reliance on the natural environment, and perceptions of local governance and community resilience (environmental, social, and economic)?

Literature review

Theoretical basis of integrated conservation development through tourism

In the context of protected areas, the premise underlying ICD is based on several assumptions about tourism development (Fennell & Weaver, 2005; Sene-Harper et al., 2019). First is the assumption that tourism provides alternative livelihood opportunities for local communities living within or adjacent to the PAs. Second, that these alternative livelihood opportunities will be more financially beneficial than traditional livelihood opportunities (thus alleviating poverty). Third, that these financial benefits will incentivize local communities to forego traditional and often more extractive practices, ultimately reducing reliance on the natural environment for livelihoods. And fourth, that local communities will recognize that the protected area and its resources are the attractant (or destination) that ultimately drives tourism, and that future economic benefits from tourism depend on the conservation of these resources. These conditions and assumptions, although often not explicitly acknowledged, stem from the application of Rational Choice Theory, which suggests that individuals are "rational actors" and that decision-making (in this case, support for or opposition to tourism and protected areas) is driven by the desire to maximize utility (typically economic gain) to the individual performing an action (e.g., Herrnstein, 1990).

However, research has repeatedly demonstrated that human decision-making in a conservation context is not only driven by the desire to maximize individual utility, but also by a host of other considerations related an individual's broader social-ecological environment (Ban et al., 2013). For instance, individual actions may be influenced by external constraints and extrinsic motivators (Ajzen, 1991; Ryan & Deci, 2000), personal identity (Stets & Burke, 2000), and values (Stern et al., 1999). Additional factors emerging from higher levels of social structure, including social norms (Chen et al., 2009; Stern et al., 1999) and various forms of social (Pretty & Smith, 2004) and community capital (Stone & Nyaupane, 2016), also impact individual and collection action. The tourism-conservation link is therefore affected by both the rational choices of individuals and aggregated perceptions and behaviors of individuals that operate at the community level (Waylen et al., 2010). Together these forces interact to influence public support for tourism and protected areas and, ultimately, the efficacy of ICD projects.

Conditions that support ICD through tourism

As ICD initiatives and global tourism centered on protected areas grow in popularity, researchers are beginning to identify conditions and programmatic approaches that may simultaneously lead to improved support for conservation and poverty alleviation (Balmford et al., 2015; Naidoo et al., 2019; Stone & Nyaupane, 2016, 2018). Rural communities around protected areas are often considered resource dependent. Defined by reliance on a narrow range of natural resources, this dependence can threaten sustainability and lead to social and economic stresses within livelihood systems (Adger, 2000). When tourism in an area increases, the growth can create new economic opportunities and alternative occupations that replace activities which might produce more lasting environmental damage (Sene-Harper et al., 2019; Weaver & Lawton, 2007). If carefully planned and implemented, tourism could therefore support ICD by diversifying livelihoods and decreasing local residents' reliance on the natural environment (Mbaiwa & Stronza, 2010), ultimately fostering support for protected area management (Badola, 1998; Kideghesho et al., 2007).

In the context of global protected areas, support for conservation is frequently measured as local support for the nearby park and the associated park management policies (Black & Cobbinah, 2018; Larson et al., 2016; Scherl & Edwards, 2007). If local residents object to park management policies, conflict results and conservation goals are compromised (Faasen & Watts,

2007; Rastogi et al., 2012; Sene-Harper et al., 2019). Public support for parks is often linked to perceptions of and support for tourism and tourism-related benefits, especially when residents recognize the park as a critical tourism resource (Gadd, 2005; Kideghesho et al., 2007; Walpole & Goodwin, 2001). Support for conservation-centered policies and park management practices may therefore be strongest in communities where the tangible benefits of tourism for local residents are most evident (Naidoo et al., 2019).

Support for conservation can also be strengthened when local residents feel they have a voice in local governance and decision-making processes. Perceptions of local empowerment and involvement, key components of political capital, are frequently associated with support for conservation and tourism around protected areas (Stone & Nyaupane, 2016). While the value of inclusive and collaborative governance of natural resources has been acknowledged for some time (Berkes, 2017), the concept of inclusive tourism development has only recently gained traction (Scheyvens & Biddulph, 2018). Such approaches typically ensure that local residents, many of whom have been historically marginalized, are included in the production and consumption of tourism. Inclusive tourism development provides unique pathways to social integration and advancement (Scheyvens & Biddulph, 2018), which can in turn strengthen support for key tourism resources (e.g., protected areas) and ICD initiatives (Muganda et al., 2013).

In addition to benefits that accrue to individuals, perceptions of community-level benefits yield important insights regarding the potential sustainability of tourism efforts. As global change accelerates, tourism is increasingly viewed as a strategy for enhancing resilience, or a community's capacity to adapt and respond to change (Cheer & Lew, 2017; Lew et al., 2016). Within the context of nature-based tourism, resilience has several components (Holladay & Powell, 2013, 2016). Environmental resilience refers to a community's ability to maintain its ecosystem productivity and function (Adger, 2000; Gunderson, 2000); it accounts for factors such as naturalness, control of infrastructure development, ability to withstand drought and other stressors, and maintenance of biological diversity (Holladay & Powell, 2013; 2016; Jamaliah & Powell, 2018). Economic resilience relates to the variety, quality, and range of economic opportunities community members are able to access (Adger, 2000; Holladay & Powell, 2013; Holling, 2001). Communities highly reliant on one resource are vulnerable to severe economic losses when external forces impact that particular resource (Thomas & Twyman, 2005). Diversity of livelihood options, through market integration, can therefore provide communities with expanded economic security (Adger, 2000). Social resilience stems from factors such as social trust, networks, collective learning, and equity (Adger, 2000; Pelling & High, 2005). Social capital is a key element of resilience, enabling local residents and institutions to connect and foster links between the ecological and social components of local systems (Berkes, 2009; Holladay & Powell, 2013).

These dimensions of resilience relate to the natural, financial, and social aspects of the Community Capitals Framework, which has been used by Stone and Nyaupane (2016, 2018) to understand linkages between biodiversity conservation and community development. Most scholars agree that enhanced community resilience is key to the long-term success of communities and conservation, especially in and around protected areas (Berkes & Ross, 2013). However, there is some debate about the extent to which tourism, a market driven phenomenon, can influence different dimensions of resilience (Amir et al., 2015; Bec et al., 2016; Cheer & Lew, 2017).

Conditions that threaten ICD through tourism

Despite optimism surrounding ICD through tourism, others have argued that integrated conservation and tourism-based poverty alleviation has many potential pitfalls to implementation that negatively impact both conservation and development outcomes (Agrawal & Redford, 2006; Salafsky et al., 2001). For example, Kiss (2004) found that ICD efforts focused on providing alternative livelihoods (often linked to tourism) were somewhat effective at reducing poverty, but largely failed at reducing threats to biodiversity or enhancing conservation efforts. Too often, ICD projects that emphasize tourism quickly morph into mass tourism operations that jeopardize environmental sustainability (Stronza et al., 2019; Weaver, 2001). In cases were ICD projects with a tourism emphasis have been deemed successful, that success has often taken a long time to achieve. For example, Baral et al. (2007) described how an ICD project in Nepal over-emphasized economic development activities in early stages, only achieving a tenuous balance with conservation about a decade after initiation.

Even when tourism effectively supports conservation goals, development outcomes and social benefits may be compromised (Stronza et al., 2019). In many tourism enterprise-based approaches to conservation, local residents are merely spectators who have little direct involvement or influence in either the tourism industry or management of the protected area on which it depends (Coria & Calfucura, 2012; Karanth & DeFries, 2011; Munanura et al., 2020; Nyaupane & Thapa, 2004). These disparities often result in inequitable distribution of tourism revenue and other economic incentives (Coria & Calfucura, 2012; Eagles et al., 2002) that are exacerbated by low levels of tourism-related training and capacity building in host communities (Aref & Redzuan, 2009; Munanura et al., 2020). In some cases, local voices are excluded in planning and decision-making processes (Berkes, 2004; Stone & Nyaupane, 2016). This can lead to tourism products and management outcomes that do not adequately reflect the cultural norms and traditions of local communities (West & Carrier, 2004). For all of these reasons, many ICD projects cited as success stories have actually done little to improve the well-being of local residents or alter their conservation-related attitudes and behaviors (Kiss, 2004). Consequently, numerous questions remain surrounding the capacity of tourism to serve as a facilitator, or inhibitor, of conservation and community development (Stronza et al., 2019; Wilkie et al., 2006).

Methods

To examine the effects of tourism enterprise-based conservation around a protected area, we explored the linkages between tourism involvement, conservation support, reliance on the natural environment, and community resilience in the Maasai Mara National Reserve, Kenya. Following recommendations by Stronza et al. (2019), we used a mixed-methods approach in three different communities, each with varying levels of tourism development, to explore the impacts of tourism on conservation and community development at multlple scales.

Study site

The Maasai Mara National Reserve (MMNR) in Kenya is a world-renowned protected area that has global significance because of its concentration of flora and fauna, including charismatic predator species such as lions, leopard, cheetah, and hyenas. MMNR serves as a centerpiece for tourism in Kenya, leading to a high volume of tourists in the reserve (Kenya Tourism Federation, 2012). In 1984, a section of the Reserve was designated to provide watering points for local communities and their livestock, creating the present conservancy lands north and west of the present MMNR (Allen et al., 2019). Today this area known as the Mara Triangle is managed by the Mara Conservancy, while the remainder of the Reserve is managed by Narok County Government. In a conservancy setting, local residents can become members, and then use shared land, receive a portion of community benefits (often monetary), and participate in the management process.

MMNR is part of the Mao ecosystem, which includes the Mao forest, rivers, agricultural land and savannahs. Most wildlife populations are facing a significant decline of up to 70% (Ogutu et al., 2011), prompting concern among conservationists regarding future management (Kenya Tourism Federation, 2012). In addition to ecosystem degradation, there has also been an increase in human-wildlife conflict around the reserve (Hallo et al., 2011; Ndegwa Mundia & Murayama, 2009). The Maasai people have inhabited the region around MMNR for centuries and have traditionally adopted a nomadic pastoral lifestyle that supports the long-term viability of rangelands. Recently, the Maasai have adopted a more sedentary agro-pastoral lifestyle characterized by more permanent homesteads (Ogutu et al., 2009). This shift has resulted in increased reliance on the land around MMNR, threats to natural resources, and conflicts with wildlife (Allen et al., 2019). Fencing of land, large-scale agriculture (e.g. wheat farming), and tourism development (e.g. safari camp resorts) that draws on scarce water resources and exacerbates resource exploitation all fuel additional conservation-related conflict around the reserve (Ogutu et al., 2011; Serneels & Lambin, 2001).

We selected three communities surrounding MMNR for participation in this study based on the presence of tourism in each location (Figure 1). Talek experiences high levels of tourism involvement relative to other local communities due to its geographic proximity to the reserve and its role as a major access point (or gate) for tourists. There are several lodges within close proximity to Talek that provide tourism-related employment opportunities. Aitong is located farther from the reserve near several community conservancies (e.g. Mara North Conservancy and Olare Orok Conservancy). Many community members in Aitong are also members of these conservancies and have access to conservancy land. Community members in Aitong are generally less involved in the tourism industry than their neighbors in Talek, but tourism-related employment opportunities still exist. Loita is located farthest from MMNR and is surrounded by forested land, providing a slightly different ecological setting than its neighboring communities. Due to its more remote location and limited access to passable roads, community members in Loita have the fewest tourism-related opportunities among the three communities. All three communities are heavily reliant on agriculture and livestock and members of all three communities experience frequent interactions and conflicts with wildlife.

Data collection

Surveys of residents (quantitative)

We conducted surveys in January 2017 in the three community areas around the MMNR: Talek (high tourism presence), Aitong (moderate tourism presence), and Loita (low tourism presence). Within each of these community areas, we systematically sampled at least 50 homesteads (bomas). Research assistants began data collection at randomly selected points within each community to ensure appropriate geographic representation. From those points, when approaching each boma from the road (or access point), the first dwelling encountered was used as a starting point. From there, every third dwelling (every second if the boma was small) was approached for participation in the study. We targeted 50 bomas per village to ensure that a sufficient number of diverse community perspectives would be represented; 81 households were surveyed in the Loita area, 56 in the Aitong area, and 60 in the Talek area, resulting in a total of 197 responses from individuals in different households across the three communities. The most senior member of each household (as identified by residents) was selected to participate in the study. Research assistants from Maasai Mara University, who were knowledgeable of the local setting and conversant in Swahili and Maa (the traditional local language), administered the surveys orally due to low literacy rates in the region. Prior to data collection, they received training pertaining to appropriate data collection processes and techniques, sample selection, and ethical considerations from the project team.

Conversations with village elders (qualitative)

We also conducted qualitative focus groups with participants identified by community members as "elders," or local leaders (7 in Loita, 8 in Aitong, and 13 in Talek). Focus group meetings took

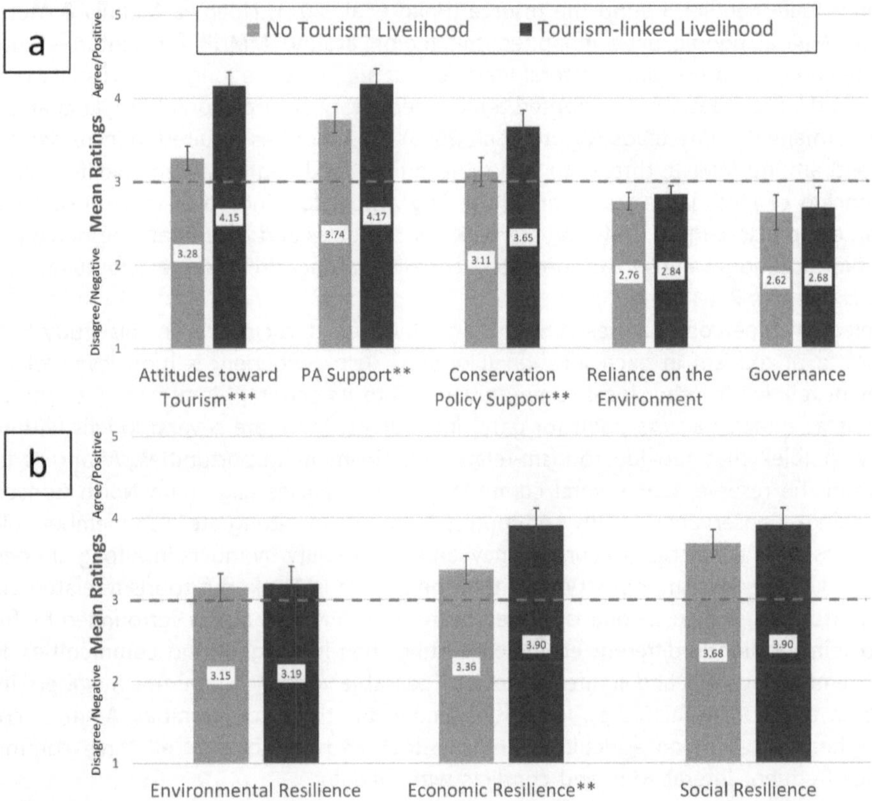

Figure 1. Map of Maasai Mara National Reserve (MMNR), Kenya, and surrounding area. Dark grey shading represents MMNR and light grey shows the surrounding conservancies. Locations of communities included in the study are indicated by black dots.

place at a neutral location within the community and lasted approximately 120-180 minutes each. Participants were informed of meeting times and locations by research assistants, who were members of the community and familiar with local social networks and procedures. The questions for the focus groups were designed to complement the household surveys and directed the conversation towards key research objectives and constructs. Each focus group was led by a Kenyan researcher who was conversant in Swahili and Maa. Each focus group session was voice recorded and observed by three additional researchers who noted non-verbal conversational points of tension (Creswell & Creswell 2007).

Questionnaire & instrument development

To understand the socio-economic characteristics of our focal communities, our questionnaire gathered demographic information such as age, gender, income, number of adults and children in the home, education, and years in community. To assess individuals' potential involvement in the tourism industry, we asked participants to describe how they primarily make a living (with tourism as one of the options) and if they derive any income from tourism. We also asked participants to respond to this statement: "I rely on tourism for my livelihood." Individuals who indicated livelihoods linked to tourism (i.e., high tourism involvement) were coded as the "tourism" group. Individuals who indicated weak or no livelihood linkages to tourism were coded as the "no tourism" group. This information also helped to validate our selection of the three different villages based on level of local involvement in tourism, indicating the proportion of residents in each community who were heavily reliant on the tourism industry. We then used subjective high (Talek), medium (Aitong), and low codes (Loita) to approximate the level of tourism involvement at the community level. To evaluate perceptions of tourism in each community, we also assessed residents' *attitudes towards tourism* through a series of seven questions pertaining to the impact of tourism in the area (Table S1). This scale was adapted from Holladay & Powell (2013) and used a Likert-type scale of 1 (strongly disagree) to 5 (strongly agree).

With respect to key dependent variables, we assessed individuals' support for conservation using questions pertaining to support for MMNR (*PA support*, 3 items, Table S1) and support for conservation policies related to wildlife management (*conservation policy support*, 3 items, Table S2). Though based on approaches utilized in previous research (Larson et al., 2016; Walpole & Goodwin, 2001), all items related to policy support were developed specifically for this study and assessed using a scale of 1 (strongly disagree) to 5 (strongly agree). Survey respondents were asked to rank their *reliance on the natural environment* for resources such as fire wood, water, building materials, fruits and vegetables, and household supplies on a scale of 1 (not at all reliant) to 5 (highly reliant, Table S3). We assessed broader beliefs about the inclusivity of local environmental *governance* using 5 items, adapted from Holladay and Powell (2013), on a five-point scale ranging from 1 (strongly disagree) to 5 (strongly agree, Table S2).

Perceptions of community resilience were measured using scales developed and tested by Holladay and Powell (2013, 2016) and Powell et al. (2018). These scales were selected because they were designed to assess perceptions of community resilience specifically in a nature-based tourism context. The scales focused on *environmental resilience* (4 items), *economic resilience* (4 items), and *social resilience* (5 items, Table S4). All community resilience scales were assessed using a five-point scale ranging from 1 (strongly disagree) to 5 (strongly agree).

Focus groups with village elders covered the same themes as the survey instrument, with the facilitators asking questions related to each construct to direct the conversation. For example, to gain insight into the role tourism plays in communities we asked, "What types of impacts do you see in the community from tourism in the area?" To gain insight into perceptions of environmental governance in communities we asked, "How much input are you able to provide to the county government about decisions that are going to impact your community?" For more details about the focus group questions, see Table S5.

Data analysis

Surveys of residents (quantitative)

We analyzed quantitative data using IBM SPSS statistical package version 25. After data cleaning, we followed recommendations by Jackson (2005) and used a Principal Component Analysis (PCA) to examine scales used to measure key constructs. Through this process we developed psychometrically-sound index scores with the most parsimonious list of survey items. Items with factor loadings less than 0.4 were eliminated, and reliability analyses confirmed Cronbach's alpha values of 0.6 or greater (Tables S1-S4). Frequency distributions and descriptive statistics were calculated for all variables. To compare dependent variables (e.g., support for conservation, reliance on the environment, and perceptions of local governance and community resilience) reported by individuals with tourism-linked livelihoods to those not involved in tourism, we conducted a series of independent-samples t-tests with family-wise error rate adjusted using the Bonferroni correction. To facilitate interpretation of mean comparisons, we calculated the *Cohen's d* measure of effect size for each statistically significant result: small (0.2), medium (0.5), or large (0.8, Lakens, 2013). We then used a factorial ANOVA to examine relationships among key dependent variables and individual tourism involvement (tourism or no tourism) as well as community tourism involvement (high, medium, or low). We examined interactions between individual and community-level tourism involvement for all dependent variables using Tukey's post hoc comparisons. Finally, we conducted separate one-way analysis of variance (ANOVA) to determine if there were significant differences in key dependent variables among communities with varying levels of tourism presence (high, medium, or low). In all ANOVA tests, we used partial eta squared to assess effect sizes for significant findings.

Conversations with village elders (qualitative)

To analyze the focus group data, conversations were translated and transcribed by a Kenyan researcher. Following transcription, three researchers analyzed qualitative data by conducting an independent analysis of the transcripts using a priori themes to determine how comments supported or refuted our quantitative findings (Wisdom & Creswell, 2013). Coded themes corresponded to the quantitative variables measured in the survey (e.g., attitudes toward tourism, support for conservation, perceptions of local governance and community resilience). Next, each researcher compared notes to identify sources of variation until consensus was reached in our interpretation of the data. Any discrepancies were addressed by referring to original transcripts and contexts. Participant quotes as well as researchers' interpretation of the quotes were reviewed and agreed upon by the three researchers to ensure that overall synthesis of focus group conversations accurately represented the collective sentiments expressed during the sessions.

Results

Surveys of residents: Individual and community level analysis

Across the entire survey sample, the mean age of respondents was approximately 39 years old (slightly higher in Aitong) and respondents had lived in their community an average of 29 years (slightly lower in Aitong, Table 1). The sample was majority male (59%). On average, respondents had attended 5.8 years of primary school. Households contained an average of four adults and four children, with the most children residing in households in Aitong (Table 1).

When examining results at the individual level, engagement in tourism was generally associated with higher scores on key dependent variables (Figure 2). Pooling data across all communities and comparing scores via *t*-tests, we found that individuals whose livelihood was linked to tourism in some way reported higher scores for *attitudes towards tourism* ($t = 7.85$, $p < 0.001$), *PA support* ($t = 3.52$, $p = 0.001$), *conservation policy support* ($t = 4.17$, $p < 0.001$), and perceptions of *economic resilience* ($t = 3.81$, $p < 0.001$) than individuals who were not engaged in tourism

Table 1. Summary of demographic attributes of local residents surveyed in low (Loita), moderate (Aitong), and high (Talek) tourism communities surrounding Maasai Mara National Reserve, Kenya (n = 197).

	Low (Loita) (n = 81)		Mod (Aitong) (n = 56)		High (Talek) (n = 60)		Total (n = 197)	
Demographic Variables	*M*	*SD*	*M*	*SD*	*M*	*SD*	*M*	*SD*
Age	37.4	12.86	44.0	14.79	37.3	10.30	38.6	12.67
Years in community	35.2	14.20	12.2	8.26	29.3	17.15	28.9	16.96
Gender (% Male)	63.9	–	62.5	–	51.7	–	59.3	–
Education (years attended school)	6.9	7.52	4.3	6.11	4.9	6.15	5.8	6.92
How many adults (over 18) live in household	3.9	2.42	3.9	1.74	3.6	2.00	3.8	2.15
How many children (under 18) live in household	3.4	2.20	5.0	2.96	4.2	3.54	4.0	2.89
Percentage of community involved in tourism	28.4	–	41.1	–	56.7	–	40.6	–

(Table S6). We observed few differences based on individual engagement in tourism for *reliance on the environment*, and perceptions of *governance*, *social resilience*, and *environmental resilience* (Figure 2, Table S6).

Factorial ANOVA results revealed that community-level effects of tourism were also significant and more variable than relationships at the individual level (Table 2). For every dependent variable except *environmental resilience*, the community effect was significant. In many cases interactions between the individual and community-level tourism variables were also significant (Table 2), underscoring the need to consider potentially different associations between tourism and conservation at multiple scales. To better understand the influence of community context on the dependent variables, we explored specific differences in perceptions among the three different communities based on their level of involvement tourism.

Only 28 percent of respondents in Loita indicated any involvement in tourism. The percentage of respondents involved in tourism in Aitong (41%) and Talek (57%) were higher, supporting the

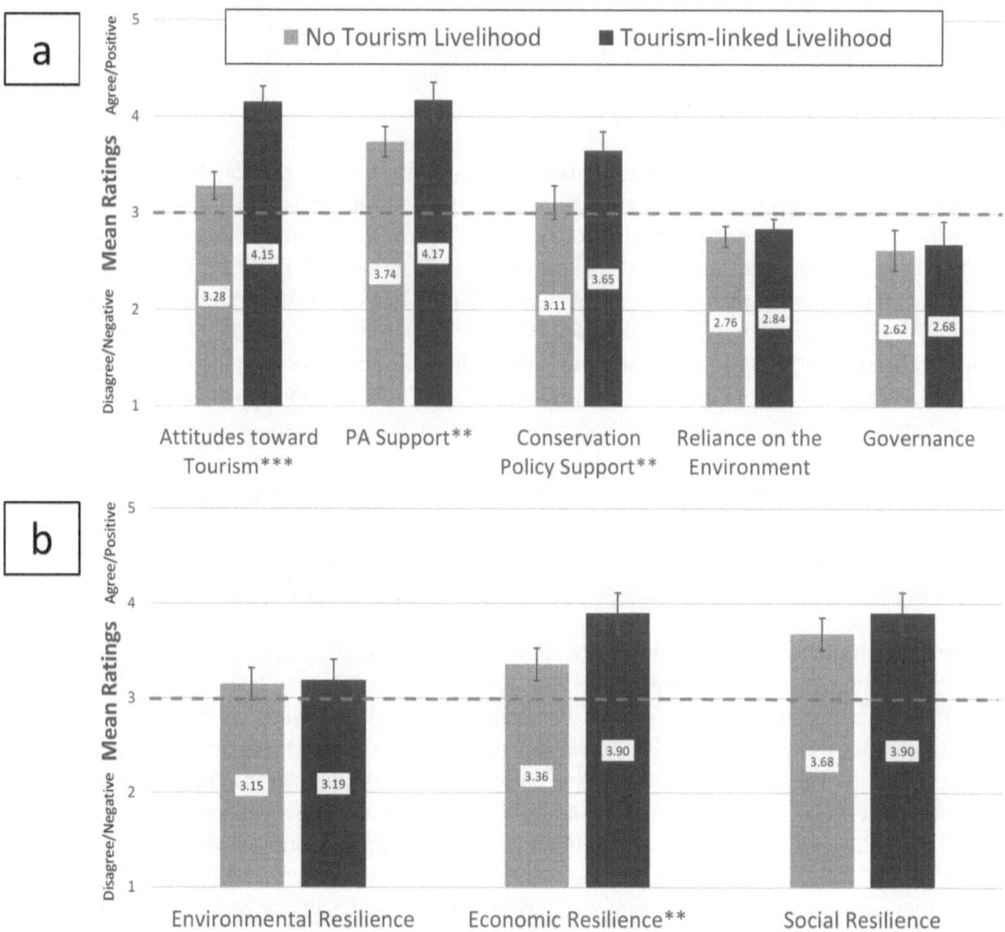

Figure 2. Mean ratings (with 95% CI) for key dependent variables for two groups of survey respondents around Maasai Mara National Reserve, Kenya: individuals with livelihoods not linked to tourism (n = 102) vs. individuals whose livelihood is directly or indirectly linked to tourism (n = 86). Dependent variables in (a) include attitudes towards tourism, support for the protected area (MMNR), support for general wildlife conservation policies, reliance on the natural environment, and perceptions of inclusive local governance. Dependent variables in (b) include perceptions of environmental, economic, and social resilience. Mean ratings on all scales ranged from 1 = Strongly Disagree (or negative) to 5 = Strongly Agree (or positive). *,**, and *** denote statistical significance of t-test comparing mean scores at $p < 0.05$, 0.01, 0.001, respectively; *,**, and *** also correspond with effect sizes described by Cohen's d as small (0.2), medium (0.5), or large (0.8).

Table 2. Results of factorial ANOVA showing positive associations among individual and community-level involvement in tourism and key dependent variables in communities around the Maasai Mara National Reserve, Kenya (n = 197).

		Individual-level Tourism Involvement		Community-level Tourism Involvement		Interaction	
Dependent Variable	*Adj.* R^2	*F*	*Part.* η^2	*F*	*Part.* η^2	*F*	*Part.* η^2
Attitudes towards Tourism	0.400	62.50***	0.28	6.60**	0.08	15.24*	0.16
PA Support	0.144	7.63**	0.04	7.16**	0.08	4.08*	0.04
Conservation Policy Support	0.237	9.86**	0.06	12.74***	0.13	6.60***	0.07
Reliance on Environment	0.444	20.03***	0.10	69.58***	0.44	0.64	0.01
Governance	0.143	0.33	0.00	13.65*	0.16	1.10	0.02
Environmental Resilience	0.359	7.86**	0.04	50.70***	0.36	0.53	0.01
Economic Resilience	0.093	11.72**	0.06	0.95	0.01	3.20*	0.04
Social Resilience	0.088	2.16	0.01	9.55***	0.10	1.37	0.02

*,**,*** denote statistical significance of ANOVA *F*-test for main effects and interactions at $p < .05$, .01, and .001, respectively. Significant results indicate tourism involvement at either the individual level [no tourism livelihood = 102 respondents, tourism-linked livelihood = 86], the community level [Loita(low) = 81, Aitong(Moderate) = 56, Talek(High) = 60], or some combination of the two (interaction term) are significantly associated with the dependent variable. Effect size reported as partial eta-square (η^2) with cutoffs of 0.01 = small effect, 0.06 = medium effect, and 0.14 = large effect. Blue cells indicate significantly higher scores reported for individuals/communities with greater involvement in tourism; orange cells indicate significantly lower scores for individuals/communities with greater involvement in tourism; yellow cells indicates mixed results.

low, medium, and high tourism designations that we established *a priori* for each community. *Attitudes towards tourism* also differed across the three communities, with the most positive attitudes in Aitong, the moderate tourism community. Attitudes towards tourism were slightly less positive in Talek and significantly less positive in Loita ($F(2,169)=6.52$, $p=0.002$, Figure 3a, Table S7).

Support for protected area management was similar in Talek and Aitong, but significantly lower in Loita, the low tourism community ($F(2,187)=8.92$, $p<0.001$, Table S7). Similarly, *support for general conservation policies* was significantly higher in Talek and Aitong than Loita ($F(2,178)=17.94$, $p<0.001$, Table S7). These patterns suggested that, at the community level, higher levels of involvement in tourism resulted in stronger support for wildlife conservation policies and stronger support for the Maasai Mara National Reserve, specifically (Figure 3a). *Reliance on the natural environment* was highest in Loita and lowest in Talek; residents in Aitong were also less likely to rely on the natural environment than residents of Loita ($F(2,190)=58.43$, $p<0.001$, Table S7). Overall, individuals living in communities with higher levels of tourism involvement reported lower reliance on the natural environment to support livelihoods (Figure 3a). Positive perceptions of local environmental *governance* were strongest in Aitong, the moderate tourism community, and weakest in Talek, the high tourism community ($F(2,154)=13.10$, $p<0.001$, Table S7). Even Loita, the low tourism community, reported higher average governance scores that Talek (Figure 3a).

Environmental resilience was highest in Loita and lowest in Talek, with a large effect size ($F(2,188)=40.40$, $p<0.001$, Table S7). *Economic resilience* was comparable across all communities, but increased slightly with growing tourism presence, resulting in a significant difference between Loita and Talek ($F(2,189)=2/56$, $p=0.080$, Table S7). *Social resilience* ratings revealed a pattern that closely aligned with *governance* ratings, with the highest values in Aitong ($F(2,189)=8.07$, $p<0.001$, Table S7). Overall, individuals living in communities with higher levels of tourism involvement were generally more likely to perceive economic resilience but less likely to perceive environmental resilience (Figure 3b). Perceptions of social resilience were highest in the moderate tourism community (Figure 3b).

Conversations with village elders in Loita (low tourism community)

Discussions with village elders in Loita confirmed that reliance on the environment was high in this region and that natural resource conservation, though important, also created potential

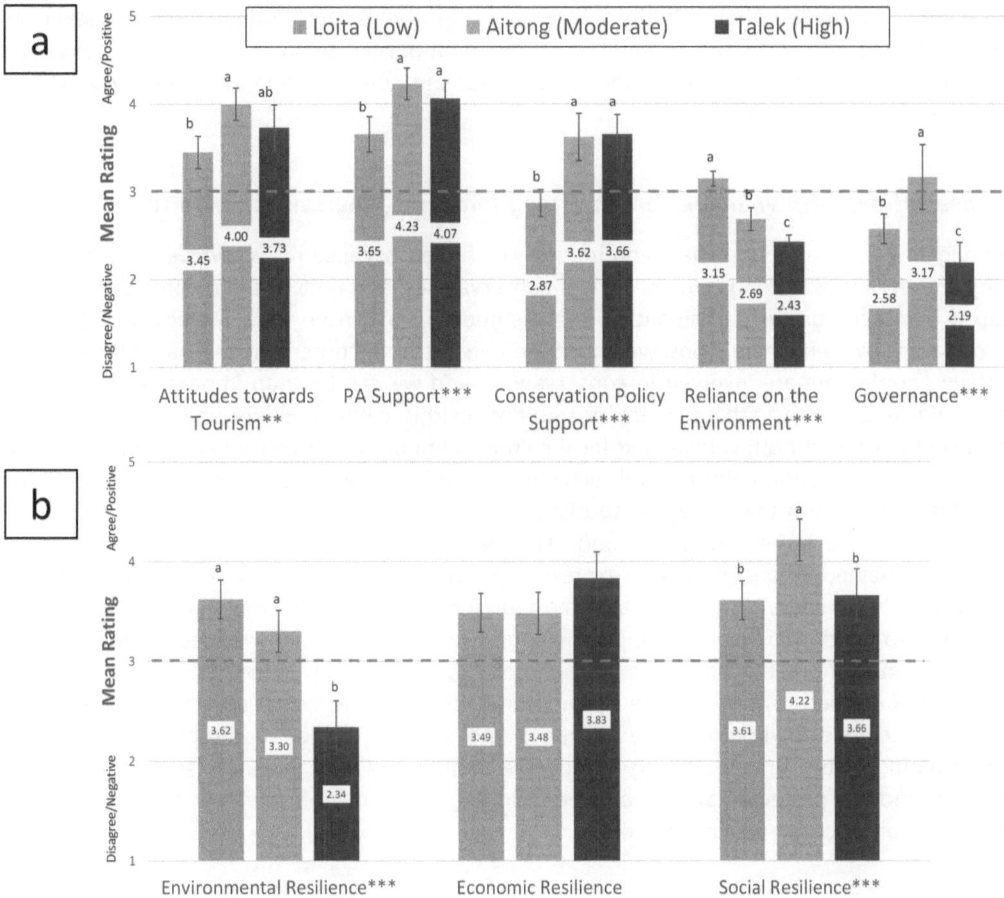

Figure 3. Mean ratings (with 95% CI) for key dependent variables among three communities around Maasai Mara National Reserve, Kenya, with different levels of tourism involvement: Loita (low tourism, n = 81), Aitong (moderate tourism, n = 56), and Talek (high tourism, n = 60). Dependent variables in (a) include attitudes towards tourism, support for the protected area (MMNR), support for general wildlife conservation policies, reliance on the natural environment, and perceptions of inclusive local governance. Dependent variables in (b) include perceptions of environmental, economic, and social resilience. Mean ratings on all scales ranged from 1 = Strongly Disagree (or negative) to 5 = Strongly Agree (or positive). *,**, and *** denote statistical significance of ANOVA comparing mean scores at $p < 0.5$, 0.01, 0.001, respectively. Different superscripts represent means that are significantly different based on Tukey's post hoc mean comparison tests (similar superscripts suggest no mean difference).

problems. One elder remarked, "the lands here are community lands and we are able to get resources we need from them as a group rather than as individuals. The forest should be held intact." Because of its location near forested land, Loita is home diverse natural resources and wildlife, resulting in an environmentally resilient community but one with high rates of human-wildlife conflict. The elders explained, "we have a greater number of problems with wildlife here because the forest is here and is healthy."

With respect to links between tourism and conservation, an elder summarized: "it is understood that from a broad view, wildlife is beneficial. But specifically here, no." Elders explained that they have relatives in other areas who are working in the Mara, and they can see the benefits for them. However, "the money and benefits that the Mara can provide to some people do not have an impact here." Yet, there is some optimism that these benefits will eventually reach the area: "We hope that at some point we can be able to benefit the way other people are. That is somewhat keeping people from destroying the wildlife. They want things to stay the same so maybe they can benefit from it in the future." Many elders believed the externally-imposed and tourism-centered conservation policies popular elsewhere in Kenya were not effective in Loita: "We do not receive benefits and compensation like other areas do. The policies do not work for us here."

Conversations with village elders in Aitong (moderate tourism community)

Aitong was unique among the communities we studied because of its involvement in community conservancies. Aitong borders several conservancies, and 86% of the community (including all of our focus group participants) were members of a conservancy. Members of the focus group were overwhelmingly positive about the benefits from conservancies. One elder explained, "Many people here are involved in conservancies and we can have an impact on the land and the wildlife that live with us which is tradition in this culture." Elders unanimously believed tourism focused on both wildlife and local culture is important for conservation. One participant explained that "Maasai culture is still playing a big role in conservation, everyone is fascinated by Maasai dress and that brings in tourists."

However, with respect to conservation support, the management of the MMNR and conservancies remained a concern. Elders expressed positive views of conservancies but noted some problems: "revenues are too little and most of the time there are meetings and we are not included so we don't know what is happening." Overall, elders acknowledged some social benefits stemming from tourism across the conservancies (e.g., new schools, improved access to water) and claimed to have a voice in some of these decisions, which reflects stronger perceptions of inclusive environmental governance and social resilience in Aitong. Although some resentment existed among Aitong elders, the community was generally supportive of both tourism and the protected area. They were also eager to remain involved in conservancies and conservation decision-making processes.

Conversations with village elders in Talek (high tourism community)

Elders acknowledged that tourism was a key industry in Talek and an important economic driver. As in Aitong, community members credit their rich culture with attracting tourists to the area: "Our cultural values help protect wildlife. It [Maasai culture] is an attraction to tourists and they are bringing in income." Elders expressed concern that people who are not Maasai are pretending to be in order to benefit from tourism, and they articulated a need for a "protection mechanism for culture" in the community. Many elders also acknowledged that, despite the potential benefits of tourism, local people are rarely employed in MMNR at lodges or in management positions.

Community members in Talek believed that they could benefit more from tourism if the community was more involved in the management of the conservancies and MMNR itself. One elder suggested, "there needs to be a committee that represents the reserve, conservancies, and communities who come together in a discussion like this to make decisions." Such comments demonstrate the absence of a positive and productive relationship between the community and the county government charged with managing MMNR. Greater involvement in tourism and close proximity to the reserve often exacerbated the negative relationship between members of the community and management authorities. One participant explained, "There is a lot of harassment on the part of the county. There is a fine for cattle in the reserve but rather than giving the fine, the officers are expecting a bribe from us." Elders in the focus group strongly believed the county government and the Kenya Wildlife Service were more concerned with wildlife than with people. "If someone is killed by an animal they don't respond – if an animal is killed there are helicopters all over the place." Levels of frustration with MMNR management and decision making were noticeably higher in Talek than any of the other communities. One comment captured this sentiment particularly well: "the Mara is dead. The community is not involved. There are too many camps and lodges and population of people and livestock has increased."

Discussion

Our study of communities around Maasai Mara National Reserve (MMNR), Kenya, explored answers to ongoing questions about the efficacy of tourism enterprise-based approaches to conservation, illuminating the complex relationships between tourism, conservation support, and community resilience at multiple scales (Coria & Calfucura, 2012, Naidoo et al., 2019). Benefits of tourism reported by individuals did not always translate across communities, showing that consideration of larger cultural context is critical when assessing conservation outcomes (Waylen et al., 2010). While our mixed methods comparison of individuals and communities suggested that tourism involvement can generate positive outcomes for conservation, the long-term viability of this linkage may depend on how tourism affects perceptions of inclusive governance and social resilience. Our results revealed several patterns that could help define and inform the role of tourism as an ICD strategy around protected areas.

Tourism as a facilitator of support for conservation

Many studies have established links between local livelihoods and support for conservation (Badola, 1998; Kideghesho et al., 2007; Larson et al., 2016; Mbaiwa & Stronza, 2010). Individuals who are highly reliant on the natural environment may be less likely to support environmental protections and restrictions imposed by authorities because of potential threats to their food and income security (Munanura et al., 2014, 2020). On the other hand, when people see tangible economic benefits from tourism that stem from the protection of wildlife and natural resources, they may be more inclined to support both parks and the policies that govern them (Balmford et al., 2015; Naidoo et al., 2019; Sekhar, 1998). Our study adds to a growing body of literature suggesting that tourism involvement can help to influence local perceptions, fostering a strong environmental ethic and generating more support for protected areas and conservation-oriented policies (Gadd, 2005; Kideghesho et al., 2007; Stronza et al., 2019).

We found that individuals whose livelihoods were directly linked to tourism were significantly more likely to support conservation policies than those who were not dependent on tourism. We also observed stronger support for conservation-related policies (i.e. policies related to wildlife management or wildlife damage compensation) in communities experiencing higher levels of tourism. Support for MMNR itself was similar and generally positive across all communities. These collective results illustrate the positive conservation impacts that tourism can have at the individual and community level, especially when the tourism-conservation linkage supports development goals (Oldekop et al., 2016; Stone & Nyaupane, 2016). Results also align with other research showing that strong public support for parks and protected areas may persist even when positive impacts on local livelihoods are not evident (Martin et al., 2018; Walpole & Goodwin, 2001).

Tourism as a sustainable livelihood strategy

We found that in communities where the presence of tourism was stronger, individuals tended to be less reliant on the natural environment. Our findings support previous assertions that tourism, when used as an alternative livelihood option, may represent a potential strategy for decreasing reliance on the natural environment by increasing access to other economic opportunities (Nyaupane & Poudel, 2011; Salafsky et al., 2001; Stronza et al., 2019; Stronza & Gordillo, 2008). By reducing pressure on natural resources within a protected area, tourism-focused livelihoods have the potential to build support for biodiversity conservation - a vital theme in livelihood-linkage frameworks (Nyaupane & Poudel, 2011; Salafsky et al., 2001).

Positive attitudes towards tourism in both Loita and Aitong, where reliance on the natural environment remains high, indicates individuals in these communities believe tourism has the potential to improve livelihoods. Communities where positive tourism attitudes are present may be good candidates for expanded tourism development. However, as other researchers have noted (Lele et al., 2010; Sene-Harper et al., 2019), tourism cannot be assumed to work equally across a protected area. Disparities in factors such as access, resource quality, and infrastructure affect the competitiveness and potential of a tourism destination. Therefore, tourism should not be viewed as the only enterprise-based alternative livelihood strategy (Salafsky et al., 2001). A large proportion of residents in all our study communities around MMNR (including Talek, the high tourism community) continue to rely on traditional occupations (e.g., agriculture, livestock) for sustenance. Acknowledging local needs and priorities and effectively leveraging those rural livelihoods alongside tourism can help to enhance resilience and reduce community vulnerability (Mbaiwa, 2011; Powell et al., 2018; Sene-Harper et al., 2019; Wright et al., 2016).

Although tourism may decrease direct reliance on the environment for natural resource extraction and use, tourism development relies heavily on effective management of wildlife-rich protected areas that support high-quality nature-based tourism opportunities (Balmford et al., 2015; Joshi et al., 2017). Protected areas with high levels of biodiversity that are managed for biodiversity protection tend to attract more visitors and generate more revenue (Chung et al., 2018). However, extreme visitation levels also lead to higher levels of resource degradation (Cheer et al., 2019; Larson & Poudyal, 2012). If tourism in a region is not managed in a sustainable way, the linkages between tourism and biodiversity conservation may backfire, leading to negative outcomes (Skibins et al., 2016). As a result, the value of tourism as a form of ICD should not be taken for granted.

Tourism as a mechanism for enhancing community resilience

The concept of resilience, which is typically viewed as a positive indicator of community function and future conservation success, is increasingly recognized as a critical outcome of sustainable tourism (Berkes & Ross, 2013; Cheer & Lew, 2017; Jamaliah & Powell, 2018; Powell et al., 2018). Components of community resilience identified in the tourism literature (Cheer & Lew, 2017) generally align with previously identified themes of tourism-conservation linkages such as community capitals (Stone & Nyaupane, 2016, 2018) and local empowerment and capacity building (Nyaupane & Poudel, 2011). Examining relationships between tourism involvement and resilience at the individual and community level, we found that tourism was associated with resident perceptions of community resilience in multiple ways. At the individual level, individuals with direct links to tourism-based livelihoods reported higher mean scores on all components of community resilience, and significantly higher social resilience scores, than individuals not involved in tourism.

However, when considering aggregated perceptions of resilience at the community level, different patterns emerged. Mean scores pertaining to economic resilience showed no major differences across communities. This supports earlier findings that, although tourism is an important source of income and revenue generation for certain individuals in park-proximate communities, it is rarely the only livelihood strategy fueling local economies. In many cases, a more diverse suite of job opportunities outside of the tourism sector may foster greater economic diversity and resilience (Adger, 2000).

Environmental resilience was highest in the low tourism communities, and many elders in Loita acknowledged the strengths of their unique natural environment and their traditional connection with the landscape. In the other communities, participants described the damage that heavy, tourism-based resource exploitation produced in the Mara's unique ecosystems. Such "overtourism" threatens the social and ecological resilience of park-proximate communities around the world (Cheer et al., 2019), and illustrates how linkages between tourism-centered livelihoods and conservation can break down and generate negative outcomes if tourism is not effectively managed (Larson & Poudyal, 2012).

Social resilience and perceived inclusivity in local environmental governance, which reflect collaborative capacity and local engagement with and control over decision-making processes, are typically key predictors of conservation success (Armitage et al., 2009; Bennett et al., 2019; Berkes, 2004). When present, these factors also help to fuel local empowerment through tourism (Boley et al., 2014; Stronza et al., 2019). We found that individual involvement in tourism yielded few differences in perceptions about social resilience and local environmental governance. However, at the community level, both of these factors were more prominently recognized in Aitong, the moderate tourism community. According to elders in Aitong, community conservancies played an important role in this process, providing an opportunity for community input and increasing support of management decisions. Other studies underscore the value of local empowerment and engagement in decision-making with respect to conservation and tourism development (Boley et al., 2014; Scanlon & Kull, 2009; Snyman, 2012). In fact, community-driven interventions often create the most positive direct linkages between tourism and conservation (Nyaupane & Poudel, 2011; Stronza et al., 2019).

Despite sharing some positive sentiments regarding tourism and conservation efforts, elders in all communities still craved more opportunities for input and engagement in decisions related to MMNR and tourism development. These concerns were most pronounced in Talek (the high tourism community), mirroring governance index ratings and suggesting the influx and inequitable distribution of wealth linked to tourism was exacerbating power disparities and fracturing traditional community processes. In Loita, the low tourism community, elders knew that other communities around MMNR were receiving benefits from tourism and they were hopeful that similar opportunities would open for them in the future. Similar situations where tourism thrives but only a small portion of local residents derive benefits and/or have access to tourism-related opportunities have fueled conflicts in other contexts across many continents (Bruyere et al., 2009; Karanth & DeFries, 2011; Larson et al., 2016; Lee & Jamal, 2008). Around MMNR, as in other locations (Sebele, 2010; Wali et al., 2017), these conflicts might be resolved when tourism benefits and opportunities are more equitably distributed and local residents are empowered to participate in decision-making processes.

Limitations and future research

Our analyses, which revealed a strong influence of both individual and aggregated community level perceptions on the tourism-conservation connection, highlights the need for research at multiple scales, including multi-levels models with larger samples to identity correlates of success (Stronza et al., 2019). While we focused on communities around a government-managed protected area, other types of protected areas, including those that are privately-owned and those built around traditional ecological knowledge and governance structures, could yield unique opportunities and benefits (Roe & Elliott, 2004; Serenari et al., 2017). The geographical scope of tourism-related benefits should also be considered. In Loita, the community farthest from MMNR, tourism involvement and benefits were lower than in other communities. Other research indicates that distribution of tourism-related benefits to communities well outside of protected areas is often negligible (den Braber et al., 2018; Larson et al., 2016), creating potential sources of regional conflict.

We measured outcomes based on local residents' perceptions, a key source of evidence in conservation decision-making (Bennett, 2016). However, future studies could focus on more objective indicators of conservation and/or development goals (e.g., Naidoo et al., 2019), including overt measurement of human behavior (Nilsson et al., 2020). More comprehensive, and perhaps qualitative measures of factors such as conservation support, inclusive governance, and community resilience could be integrated into future work (Stone & Nyaupane, 2018). Because community involvement in local tourism and conservation governance is a key predictor of success, research should also focus on the engagement process and ways it might be strengthened to accommodate more diverse stakeholders (Mbaiwa & Stronza, 2010; Stronza et al., 2019). This might include longitudinal work assessing the long-term impact of particular interventions, including those involving community conservancies (Stronza et al., 2019). Potential for conflict among individuals and within communities due to inequitable tourism benefit distribution could be measured and taken into account. To address these conflicts, researchers and practitioners should continue to explore how ICD through tourism influences the well-being of human and ecological communities that reside close to parks and protected areas (Wilkie et al., 2006).

Conclusion

Previous studies exploring connections between tourism-based livelihoods and conservation have often found the strongest linkages in more developed tourism sites (Chung et al., 2018; Nyaupane & Poudel, 2011). Our results from MMNR in Kenya support this assertion to some degree, but we also discovered one major caveat: high levels of tourism might undermine support for conservation and community resilience, particularly when governance structures are not inclusive and distribution of tourism-related benefits and opportunities are viewed as inequitable. At either end of the tourism involvement spectrum, inequities and disparities (real and perceived) fueled potential conflict, threatened capacity building around tourism-focused livelihoods, and jeopardized community resilience – a key to sustaining positive links between tourism and conservation. While tourism might enhance development of financial capital, such inequities and disparities impede the formation of other forms of community capital (e.g., human, social, political, natural) that reinforce positive linkages and sustainable tourism growth (Stone & Nyaupane, 2016, 2018).

Attitudes towards tourism and support for conservation were most positive in Aitong, the moderate tourism community. Although challenges related to tourism-benefit-distribution were still present, these challenges were not eroding perceptions of inclusive governance and community resilience. Communities like Aitong may represent sweet spots where tangible benefits linked to tourism are recognized by individuals, and social and political structures support opportunities for expanded local involvement and equitable benefit distribution at the community level (Mbaiwa & Stronza, 2010; Rasoolimanesh et al., 2017; Stronza et al., 2019). In these situations, well-managed tourism development can achieve key goals of ICD: fostering support for protected area conservation and enhancing community resilience. Other studies in Africa have examined the "spiraling up" of community capitals linked to participation in tourism, highlighting potential mechanisms to help explain these synergistic gains for individuals and communities (Stone & Nyaupane, 2018). In places like MMNR, we urge park managers, tour operators, and local communities to strive to find a tourism sweet spot where local residents are empowered, economic inequality is reduced, and natural resource conservation is embraced. In such situations, links between tourism, protected areas, and local livelihoods become more direct and resilient, aiding in the joint achievement of conservation and development goals.

Disclosure statement

No potential conflict of interest was reported by the authors.

Funding

We would like to thank the Narok County Government, Kenya for its financial support and technical assistance in making this project possible. We also want to thank Clemson University Institute for Parks for its financial support and the International Conservation Caucus Foundation for its assistance in organizing this project. Finally, we also want to acknowledge that a Kenyan Government research permit was obtained from the National Commission for Science, Technology, and Innovation (NACOSTI). We thank NACOSTI for its assistance and support.

ORCID

Lincoln R. Larson http://orcid.org/0000-0001-9591-1269
Robert B. Powell http://orcid.org/0000-0003-2775-2571
W. Hunter Holland http://orcid.org/0000-0002-8557-1640

References

Adger, W. N. (2000). Social and ecological resilience: Are they related?*Progress in Human Geography*, *24*(3), 347–364. https://doi.org/10.1191/030913200701540465

Agrawal, A., & Redford, K. H. (2006). *Poverty, development, and biodiversity conservation: Shooting in the dark?* (Working Paper No. 26). Wildlife Conservation Society.

Ajzen, I. (1991). The theory of planned behavior. *Organizational Behavior and Human Decision Processes*, *50*(2), 179–211. https://doi.org/10.1016/0749-5978(91)90020-T

Allen, L., Holland, K. K., Holland, H., Tome', S., Nabaala, M., Seno, S., & Nampushi, J. (2019). Expanding staff voice in protected area management effectiveness assessments within Kenya's Maasai Mara National Reserve. *Environmental Management*, *63*(1), 46–59. https://doi.org/10.1007/s00267-018-1122-6

Alpert, P. (1996). Integrated conservation and development projects. *BioScience*, *46*(11), 845–855. https://doi.org/10.2307/1312970

Amir, A. F., Ghapar, A. A., Jamal, S. A., & Ahmad, K. N. (2015). Sustainable tourism development: A study on community resilience for rural tourism in Malaysia. Procedia - Social and Behavioral Sciences, *168*, 116–122. https://doi.org/10.1016/j.sbspro.2014.10.217

Aref, F., & Redzuan, M. R. B. (2009). Community capacity building for tourism development. *Journal of Human Ecology*, *27*(1), 21–25. https://doi.org/10.1080/09709274.2009.11906187

Armitage, D. R., Plummer, R., Berkes, F., Arthur, R. I., Charles, A. T., Davidson-Hunt, I. J., Diduck, A. P., Doubleday, N. C., Johnson, D. S., Marschke, M., McConney, P., Pinkerton, E. W., & Wollenberg, E. K. (2009). Adaptive co-management for social-ecological complexity. *Frontiers in Ecology and the Environment*, *7*(2), 95–102. https://doi.org/10.1890/070089

Badola, R. (1998). Attitudes of local people towards conservation and alternatives to forest resources: A case study from the lower Himalayas. *Biodiversity and Conservation*, *7*(10), 1245–1259. https://doi.org/10.1023/A:1008845510498

Balmford, A., Green, J. M., Anderson, M., Beresford, J., Huang, C., Naidoo, R., Walpole, M., & Manica, A. (2015). Walk on the wild side: Estimating the global magnitude of visits to protected areas. *PLoS Biology*, *13*(2), e1002074. https://doi.org/10.1371/journal.pbio.1002074

Ban, N. C., Mills, M., Tam, J., Hicks, C. C., Klain, S., Stoeckl, N., Bottrill, M. C., Levine, J., Pressey, R. L., Satterfield, T., & Chan, K. M. (2013). A social-ecological approach to conservation planning: Embedding social considerations. *Frontiers in Ecology and the Environment*, *11*(4), 194–202. https://doi.org/10.1890/110205

Baral, N., Stern, M. J., & Heinen, J. T. (2007). Integrated conservation and development project life cycles in the Annapurna Conservation Area, Nepal: Is development overpowering conservation?*Biodiversity and Conservation*, *16*(10), 2903–2917. https://doi.org/10.1007/s10531-006-9143-5

Bec, A., McLennan, C. L., & Moyle, B. D. (2016). Community resilience to long-term tourism decline and rejuvenation: A literature review and conceptual model. *Current Issues in Tourism*, *19*(5), 431–457. https://doi.org/10.1080/13683500.2015.1083538

Bennett, N. J. (2016). Using perceptions as evidence to improve conservation and environmental management. *Conservation Biology : The Journal of the Society for Conservation Biology*, *30*(3), 582–592. https://doi.org/10.1111/cobi.12681

Bennett, N. J., Di Franco, A., Calò, A., Nethery, E., Niccolini, F., Milazzo, M., & Guidetti, P. (2019). Local support for conservation is associated with perceptions of good governance, social impacts, and ecological effectiveness. *Conservation Letters, 12*(4), e12640. https://doi.org/10.1111/conl.12640

Berkes, F. (2004). Rethinking community-based conservation. *Conservation Biology, 18*(3), 621–630. https://doi.org/10.1111/j.1523-1739.2004.00077.x

Berkes, F. (2009). Evolution of co-management: Role of knowledge generation, bridging organizations and social learning. *Journal of Environmental Management, 90*(5), 1692–1702. https://doi.org/10.1016/j.jenvman.2008.12.001

Berkes, F. (2017). Environmental governance for the anthropocene? Social-ecological systems, resilience, and collaborative learning. *Sustainability, 9*(7), 1232. https://doi.org/10.3390/su9071232

Berkes, F., & Ross, H. (2013). Community resilience: Toward an integrated approach. *Society & Natural Resources, 26*(1), 5–20. https://doi.org/10.1080/08941920.2012.736605

Black, R., & Cobbinah, P. B. (2018). Local attitudes towards tourism and conservation in rural Botswana and Rwanda. *Journal of Ecotourism, 17*(1), 79–105. https://doi.org/10.1080/14724049.2016.1258074

Boley, B. B., McGehee, N. G., Perdue, R. R., & Long, P. (2014). Empowerment and resident attitudes toward tourism: Strengthening the theoretical foundation through a Weberian lens. *Annals of Tourism Research, 49*, 33–50. https://doi.org/10.1016/j.annals.2014.08.005

Bruyere, B. L., Beh, A. W., & Lelengula, G. (2009). Differences in perceptions of communication, tourism benefits, and management issues in a protected area of rural Kenya. *Environmental Management, 43*(1), 49–59. https://doi.org/10.1007/s00267-008-9190-7

Campbell, B. M., Sayer, J. A., & Walker, B. (2010). Navigating trade-offs: Working for conservation and development outcomes. *Ecology and Society, 15*(2), 16. https://doi.org/10.5751/ES-03380-150216

Cheer, J. M., & Lew, A. A. (Eds.). (2017). *Tourism, resilience and sustainability: Adapting to social, political and economic change*. Routledge.

Cheer, J. M., Milano, C., & Novelli, M. (2019). Tourism and community resilience in the Anthropocene: Accentuating temporal overtourism. *Journal of Sustainable Tourism, 27*(4), 554–572. https://doi.org/10.1080/09669582.2019.1578363

Chen, X., Lupi, F., He, G., & Liu, J. (2009). Linking social norms to efficient conservation investment in payments for ecosystem services. *Proceedings of the National Academy of Sciences, 106*(28), 11812–11817. https://doi.org/10.1073/pnas.0809980106

Chung, M. G., Dietz, T., & Liu, J. (2018). Global relationships between biodiversity and nature-based tourism in protected areas. *Ecosystem Services, 34*, 11–23. https://doi.org/10.1016/j.ecoser.2018.09.004

Coria, J., & Calfucura, E. (2012). Ecotourism and the development of indigenous communities: The good, the bad, and the ugly. *Ecological Economics, 73*, 47–55. https://doi.org/10.1016/j.ecolecon.2011.10.024

Creswell, J. W., & Creswell, J. D. (2017). Research design: Qualitative, quantitative, and mixed methods approaches. Sage publications.

den Braber, B., Evans, K. L., & Oldekop, J. A. (2018). Impact of protected areas on poverty, extreme poverty, and inequality in Nepal. *Conservation Letters, 11*(6), e12576. https://doi.org/10.1111/conl.12576

Eagles, P. F., McCool, S. F., & Haynes, C. D. (2002). *Sustainable tourism in protected areas: Guidelines for planning and management (No. 8)*. IUCN.

Faasen, H., & Watts, S. (2007). Local community reaction to the 'no-take' policy on fishing in the Tsitsikamma National Park. Ecological Economics, *64*(1), 36–46. https://doi.org/10.1016/j.ecolecon.2007.06.026

Fennell, D., & Weaver, D. (2005). The ecotourism concept and tourism-conservation symbiosis. *Journal of Sustainable Tourism, 13*(4), 373–390. https://doi.org/10.1080/09669580508668563

Ferraro, P. J., & Hanauer, M. M. (2014). Quantifying causal mechanisms to determine how protected areas affect poverty through changes in ecosystem services and infrastructure. *Proceedings of the National Academy of Sciences, 111*(11), 4332–4337. https://doi.org/10.1073/pnas.1307712111

Gadd, M. E. (2005). Conservation outside of parks: Attitudes of local people in Laikipia. *Environmental Conservation, 32*(1), 50–63. https://doi.org/10.1017/S0376892905001918

Gunderson, L. H. (2000). Ecological resilience-in theory and application. *Annual Review of Ecology and Systematics, 31*(1), 425–439. https://doi.org/10.1146/annurev.ecolsys.31.1.425

Hallo, B., Oba, G., & Guyo, H. (2011). *Livestock based perceptions of range quality change at landscape level among borana community of Northern Kenya*. Animal Production Society of Kenya, 78.

Herrnstein, R. J. (1990). Rational choice theory: Necessary but not sufficient. *American Psychologist, 45*(3), 356–367. https://doi.org/10.1037/0003-066X.45.3.356

Holladay, P. J., & Powell, R. B. (2013). Resident perceptions of social-ecological resilience and the sustainability of community-based tourism development in the Commonwealth of Dominica. *Journal of Sustainable Tourism, 21*(8), 1188–1211. https://doi.org/10.1080/09669582.2013.776059

Holladay, P. J., & Powell, R. B. (2016). Social-ecological resilience and stakeholders: A qualitative inquiry into community-based tourism in the Commonwealth of Dominica. *Caribbean Studies, 44*(1–2), 3–28. https://doi.org/10.1353/crb.2016.0000

Holling, C. S. (2001). Understanding the complexity of economic, ecological, and social systems. *Ecosystems*, *4*(5), 390–405. https://doi.org/10.1007/s10021-001-0101-5

Jackson, J. E. (2005). *A user's guide to principal components* (Vol. 587). John Wiley & Sons.

Jamaliah, M. M., & Powell, R. B. (2018). Ecotourism resilience to climate change in the Dana Biosphere Reserve. Journal of Sustainable Tourism, *26*(4), 519–536. https://doi.org/10.1080/09669582.2017.1360893

Joshi, O., Poudyal, N. C., & Larson, L. R. (2017). The influence of sociopolitical, natural, and cultural factors on international tourism growth: A cross-country panel analysis. *Environment, Development and Sustainability*, *19*(3), 825–838. https://doi.org/10.1007/s10668-016-9767-x

Karanth, K. K., & DeFries, R. (2011). Nature-based tourism in Indian protected areas: New challenges for park management. *Conservation Letters*, *4*(2), 137–149. https://doi.org/10.1111/j.1755-263X.2010.00154.x

Kenya Tourism Federation. (2012). *Maasai Mara National Reserve Management Plan 2012-2022*.

Kideghesho, J. R., Røskaft, E., & Kaltenborn, B. P. (2007). Factors influencing conservation attitudes of local people in Western Serengeti. *Biodiversity and Conservation*, *16*(7), 2213–2230. https://doi.org/10.1007/s10531-006-9132-8

Kiss, A. (2004). Is community-based ecotourism a good use of biodiversity conservation funds?*Trends in Ecology & Evolution*, *19*(5), 232–237. https://doi.org/10.1016/j.tree.2004.03.010

Lakens, D. (2013). Calculating and reporting effect sizes to facilitate cumulative science: A practical primer for t-tests and ANOVAs. *Frontiers in Psychology*, *4*, 863. https://doi.org/10.3389/fpsyg.2013.00863

Larson, L. R., & Poudyal, N. C. (2012). Developing sustainable tourism through adaptive resource management: A case study of Machu Picchu. Journal of Sustainable Tourism, *20*(7), 917–938. https://doi.org/10.1080/09669582.2012.667217

Larson, L. R., Conway, A. L., Krafte, K. E., Hernandez, S. M., & Carroll, J. P. (2016). Community-based conservation as a potential source of conflict around a protected area in Sierra Leone. *Environmental Conservation*, *43*(3), 242–252. https://doi.org/10.1017/S0376892916000096

Lee, S., & Jamal, T. (2008). Environmental justice and environmental equity in tourism: Missing links to sustainability. *Journal of Ecotourism*, *7*(1), 44–67. https://doi.org/10.2167/joe191.0

Lele, S., Wilshusen, P., Brockington, D., Seidler, R., & Bawa, K. (2010). Beyond exclusion: Alternative approaches to biodiversity conservation in the developing tropics. *Current Opinion in Environmental Sustainability*, *2*(1–2), 94–100. https://doi.org/10.1016/j.cosust.2010.03.006

Lew, A. A., Ng, P. T., Ni, C. C., & Wu, T. C. (2016). Community sustainability and resilience: Similarities, differences and indicators. *Tourism Geographies*, *18*(1), 18–27. https://doi.org/10.1080/14616688.2015.1122664

Marcus, R. R. (2001). Seeing the forest for the trees: Integrated conservation and development projects and local perceptions of conservation in Madagascar. *Human Ecology*, *29*(4), 381–397. https://doi.org/10.1023/A:1013189720278

Martin, A., Myers, R., & Dawson, N. M. (2018). The park is ruining our livelihoods. We support the park! Unravelling the paradox of attitudes to protected areas. *Human Ecology*, *46*(1), 93–105. https://doi.org/10.1007/s10745-017-9941-2

Mbaiwa, J. E. (2011). Changes on traditional livelihood activities and lifestyles caused by tourism development in the Okavango Delta. Tourism Management, *32*(5), 1050–1060. https://doi.org/10.1016/j.tourman.2010.09.002

Mbaiwa, J. E., & Stronza, A. L. (2010). The effects of tourism development on rural livelihoods in the Okavango Delta, Botswana. *Journal of Sustainable Tourism*, *18*(5), 635–656. https://doi.org/10.1080/09669581003653500

Muganda, M., Sirima, A., & Ezra, P. M. (2013). The role of local communities in tourism development: Grassroots perspectives from Tanzania. *Journal of Human Ecology*, *41*(1), 53–66. https://doi.org/10.1080/09709274.2013.11906553

Munanura, I. E., Backman, K. F., Sabuhoro, E., & Bernhard, K. P. (2020). The potential of tourism benefits to reduce forest dependence behavior of impoverished residents adjacent to Volcanoes National Park in Rwanda. *Tourism Planning & Development*, *17*(5), 475–496. https://doi.org/10.1080/21568316.2019.1640282

Munanura, I. E., Backman, K., Moore, D., Hallo, J., & Powell, R. B. (2014). Household poverty dimensions influencing forest dependence at Volcanoes National Park, Rwanda: An application of the sustainable livelihoods framework. *Natural Resources*, *05*(16), 1031–1047. https://doi.org/10.4236/nr.2014.516087

Naidoo, R., Gerkey, D., Hole, D., Pfaff, A., Ellis, A. M., Golden, C. D., Herrera, D., Johnson, K., Mulligan, M., Ricketts, T. H., & Fisher, B. (2019). Evaluating the impacts of protected areas on human well-being across the developing world. *Science Advances*, *5*(4), eaav3006. https://doi.org/10.1126/sciadv.aav3006

Ndegwa Mundia, C., & Murayama, Y. (2009). Analysis of land use/cover changes and animal population dynamics in a wildlife sanctuary in East Africa. *Remote Sensing*, *1*(4), 952–970. https://doi.org/10.3390/rs1040952

Newmark, W. D., & Hough, J. L. (2000). Conserving Wildlife in Africa: Integrated Conservation and Development Projects and Beyond: Because multiple factors hinder integrated conservation and development projects in Africa from achieving their objectives, alternative and complementary approaches for promoting wildlife conservation must be actively explored. *BioScience*, *50*(7), 585–592.

Nilsson, D., Fielding, K., & Dean, A. J. (2020). Achieving conservation impact by shifting focus from human attitudes to behaviors. *Conservation Biology : The Journal of the Society for Conservation Biology*, *34*(1), 93–102. https://doi.org/10.1111/cobi.13363

Nyaupane, G. P., & Poudel, S. (2011). Linkages among biodiversity, livelihood, and tourism. *Annals of Tourism Research*, *38*(4), 1344–1366. https://doi.org/10.1016/j.annals.2011.03.006

Nyaupane, G. P., & Thapa, B. (2004). Evaluation of ecotourism: A comparative assessment in the Annapurna Conservation Area Project. Journal of Ecotourism, *3*(1), 20–45. https://doi.org/10.1080/14724040408668148

Ogutu, J. O., Owen-Smith, N., Piepho, H. P., & Said, M. Y. (2011). Continuing wildlife population declines and range contraction in the Mara region of Kenya during 1977–2009. *Journal of Zoology*, *285*(2), 99–109. https://doi.org/10.1111/j.1469-7998.2011.00818.x

Ogutu, J. O., Piepho, H. P., Dublin, H. T., Bhola, N., & Reid, R. S. (2009). Dynamics of Mara–Serengeti ungulates in relation to land use changes. *Journal of Zoology*, *278*(1), 1–14. https://doi.org/10.1111/j.1469-7998.2008.00536.x

Oldekop, J. A., Holmes, G., Harris, W. E., & Evans, K. L. (2016). A global assessment of the social and conservation outcomes of protected areas. *Conservation Biology : The Journal of the Society for Conservation Biology*, *30*(1), 133–141. https://doi.org/10.1111/cobi.12568

Pelling, M., & High, C. (2005). Understanding adaptation: What can social capital offer assessments of adaptive capacity?*Global Environmental Change*, *15*(4), 308–319. https://doi.org/10.1016/j.gloenvcha.2005.02.001

Powell, R. B., Green, T. F., Holladay, P. J., Krafte, K. E., Duda, M., Nguyen, M. T., Spencer, J. H., & Das, P. (2018). Examining community resilience to assist in sustainable tourism development planning in Dong Van Karst Plateau Geopark. Tourism Planning & Development, *15*(4), 436–457. https://doi.org/10.1080/21568316.2017.1338202

Pretty, J., & Smith, D. (2004). Social capital in biodiversity conservation and management. *Conservation Biology*, *18*(3), 631–638. https://doi.org/10.1111/j.1523-1739.2004.00126.x

Rasoolimanesh, S. M., Jaafar, M., Ahmad, A. G., & Barghi, R. (2017). Community participation in World Heritage Site conservation and tourism development. *Tourism Management*, *58*, 142–153. https://doi.org/10.1016/j.tourman.2016.10.016

Rastogi, A., Hickey, G. M., Badola, R., & Hussain, S. A. (2012). Saving the superstar: A review of the social factors affecting tiger conservation in India. *Journal of Environmental Management*, *113*, 328–340. https://doi.org/10.1016/j.jenvman.2012.10.003

Roe, D., & Elliott, J. (2004). Poverty reduction and biodiversity conservation: Rebuilding the bridges. *Oryx*, *38*(2), 137–139. https://doi.org/10.1017/S0030605304000249

Ryan, R. M., & Deci, E. L. (2000). Self-determination theory and the facilitation of intrinsic motivation, social development, and well-being. *American Psychologist*, *55*(1), 68–78. https://doi.org/10.1037/0003-066X.55.1.68

Sabuhoro, E., Wright, B., Munanura, I. E., Nyakabwa, I. N., & Nibigira, C. (2021). The potential of ecotourism opportunities to generate support for mountain gorilla conservation among local communities neighboring Volcanoes National Park in Rwanda. *Journal of Ecotourism*, *20*(1), 1–17. https://doi.org/10.1080/14724049.2017.1280043

Salafsky, N., & Wollenberg, E. (2000). Linking livelihoods and conservation: A conceptual framework and scale for assessing the integration of human needs and biodiversity. *World Development*, *28*(8), 1421–1438. https://doi.org/10.1016/S0305-750X(00)00031-0

Salafsky, N., Cauley, H., Balachander, G., Cordes, B., Parks, J., Margoluis, C., Bhatt, S., Encarnacion, C., Russell, D., & Margoluis, R. (2001). A systematic test of an enterprise strategy for community-based biodiversity conservation. *Conservation Biology*, *15*(6), 1585–1595. https://doi.org/10.1046/j.1523-1739.2001.00220.x

Scanlon, L. J., & Kull, C. A. (2009). Untangling the links between wildlife benefits and community-based conservation at Torra Conservancy, Namibia. *Development Southern Africa*, *26*(1), 75–93. https://doi.org/10.1080/03768350802640107

Scherl, L. M., & Edwards, S. (2007). Tourism, indigenous and local communities and protected areas in developing nations. In R. Bushell, & P. F. J. Eagles (Eds.), *Tourism and protected areas: Benefits beyond boundaries* (pp. 71–88). CAB International.

Scheyvens, R., & Biddulph, R. (2018). Inclusive tourism development. *Tourism Geographies*, *20*(4), 589–609. https://doi.org/10.1080/14616688.2017.1381985

Sebele, L. S. (2010). Community-based tourism ventures, benefits and challenges: Khama rhino sanctuary trust, central district, Botswana. *Tourism Management*, *31*(1), 136–146. https://doi.org/10.1016/j.tourman.2009.01.005

Sekhar, N. U. (1998). Crop and livestock depredation caused by wild animals in protected areas: The case of Sariska Tiger Reserve, Rajasthan, India. *Environmental Conservation*, *25*(2), 160–171. https://doi.org/10.1017/S0376892998000204

Sene-Harper, A., Matarrita-Cascante, D., & Larson, L. R. (2019). Leveraging local livelihood strategies to support conservation and development in West Africa. *Environmental Development*, *29*, 16–28. https://doi.org/10.1016/j.envdev.2018.11.002

Serenari, C., Peterson, M. N., Wallace, T., & Stowhas, P. (2017). Private protected areas, ecotourism development and impacts on local people's well-being: A review from case studies in Southern Chile. *Journal of Sustainable Tourism*, *25*(12), 1792–1810. https://doi.org/10.1080/09669582.2016.1178755

Serneels, S., & Lambin, E. F. (2001). Proximate causes of land-use change in Narok District, Kenya: A spatial statistical model. *Agriculture, Ecosystems & Environment*, *85*(1–3), 65–81. https://doi.org/10.1016/S0167-8809(01)00188-8

Skibins, J. C., Powell, R. B., & Hallo, J. C. (2016). Lucky 13: Conservation implications of broadening 'Big 5' flagship species recognition in East Africa. *Journal of Sustainable Tourism*, *24*(7), 1024–1040. https://doi.org/10.1080/09669582.2015.1101126

Snyman, S. (2012). Ecotourism joint ventures between the private sector and communities: An updated analysis of the Torra Conservancy and Damaraland Camp partnership, Namibia. *Tourism Management Perspectives*, *4*, 127–135. https://doi.org/10.1016/j.tmp.2012.07.004

Stern, P. C., Dietz, T., Abel, T., Guagnano, G. A., & Kalof, L. (1999). A value-belief-norm theory of support for social movements: The case of environmentalism. *Human Ecology Review*, *6*(2), 81–97.

Stets, J. E., & Burke, P. J. (2000). Identity theory and social identity theory. *Social Psychology Quarterly*, *63*(3), 224–237. https://doi.org/10.2307/2695870

Stone, M. T., & Nyaupane, G. P. (2016). Protected areas, tourism and community livelihoods linkages: A comprehensive analysis approach. *Journal of Sustainable Tourism*, *24*(5), 673–693. https://doi.org/10.1080/09669582.2015.1072207

Stone, M. T., & Nyaupane, G. P. (2018). Protected areas, wildlife-based community tourism and community livelihoods dynamics: Spiraling up and down of community capitals. *Journal of Sustainable Tourism*, *26*(2), 307–324. https://doi.org/10.1080/09669582.2017.1349774

Stronza, A. L., Hunt, C. A., & Fitzgerald, L. A. (2019). Ecotourism for conservation?*Annual Review of Environment and Resources*, *44*(1), 229–253. https://doi.org/10.1146/annurev-environ-101718-033046

Stronza, A., & Gordillo, J. (2008). Community views of ecotourism. *Annals of Tourism Research*, *35*(2), 448–468. https://doi.org/10.1016/j.annals.2008.01.002

Stronza, A., & Pêgas, F. (2008). Ecotourism and conservation: Two cases from Brazil and Peru. *Human Dimensions of Wildlife*, *13*(4), 263–279. https://doi.org/10.1080/10871200802187097

Thomas, D. S., & Twyman, C. (2005). Equity and justice in climate change adaptation amongst natural-resource-dependent societies. *Global Environmental Change*, *15*(2), 115–124. https://doi.org/10.1016/j.gloenvcha.2004.10.001

Wali, A., Alvira, D., Tallman, P., Ravikumar, A., & Macedo, M. (2017). A new approach to conservation: Using community empowerment for sustainable well-being. *Ecology and Society*, *22*(4), 6. https://doi.org/10.5751/ES-09598-220406

Walpole, M. J., & Goodwin, H. J. (2001). Local attitudes towards conservation and tourism around Komodo National Park. *Environmental Conservation*, *28*(2), 160–166. https://doi.org/10.1017/S0376892901000169

Waylen, K. A., Fischer, A., Mcgowan, P. J., Thirgood, S. J., & Milner-Gulland, E. J. (2010). Effect of local cultural context on the success of community-based conservation interventions . *Conservation Biology : The Journal of the Society for Conservation Biology*, *24*(4), 1119–1129. https://doi.org/10.1111/j.1523-1739.2010.01446.x

Weaver, D. B. (2001). Ecotourism as mass tourism: Contradiction or reality?*The Cornell Hotel and Restaurant Administration Quarterly*, *42*(2), 104–112. https://doi.org/10.1016/S0010-8804(01)80022-7

Weaver, D. B., & Lawton, L. J. (2007). Twenty years on: The state of contemporary ecotourism research. *Tourism Management*, *28*(5), 1168–1179. https://doi.org/10.1016/j.tourman.2007.03.004

West, P., & Carrier, J. G. (2004). Ecotourism and authenticity: Getting away from it all?*Current Anthropology*, *45*(4), 483–498. https://doi.org/10.1086/422082

West, P., Igoe, J., & Brockington, D. (2006). Parks and peoples: The social impact of protected areas. *Annual Review of Anthropology*, *35*(1), 251–277. https://doi.org/10.1146/annurev.anthro.35.081705.123308

Wilkie, D. S., Morelli, G. A., Demmer, J., Starkey, M., Telfer, P., & Steil, M. (2006). Parks and people: Assessing the human welfare effects of establishing protected areas for biodiversity conservation. *Conservation Biology : The Journal of the Society for Conservation Biology*, *20*(1), 247–249. https://doi.org/10.1111/j.1523-1739.2005.00291.x

Wisdom, J., & Creswell, J. W. (2013). *Mixed methods: Integrating quantitative and qualitative data collection and analysis while studying patient-centered medical home models*. Agency for Healthcare Research and Quality, U.S. Dept. of Health & Human Services.

Wright, J. H., Hill, N. A., Roe, D., Rowcliffe, J. M., Kümpel, N. F., Day, M., Booker, F., & Milner-Gulland, E. J. (2016). Reframing the concept of alternative livelihoods. *Conservation Biology : The Journal of the Society for Conservation Biology*, *30*(1), 7–13. https://doi.org/10.1111/cobi.12607

Community-based ecotourism and bushmeat consumption dynamics: implications for conservation and community development

Moren Tibabo Stone and Lesego Senyana Stone

ABSTRACT

This paper uses qualitative research methods guided by the social exchange theory and the Community-Based Natural Resource Management concept to explore the contribution of community-based ecotourism to bushmeat consumption/production using the Chobe Enclave Conservation Trust, located adjacent to the Chobe National Park, Botswana. Data were collected between May and August 2018. Using semi-structured in-depth interviews, results indicate that, bushmeat consumption became prevalent through trophy hunting tourism. With the hunting ban in 2014, game hunting licenses was stopped, cutting institutionalized bushmeat supply and widening trade-offs between community livelihoods improvement and conservation. Socially, hunting tourism has contributed to affinity for bushmeat taste. Bushmeat contributes to food security for a significant number of community members. Institutionally, bushmeat's contribution to community livelihoods is perceived varyingly by different stakeholders - some see it as a threat to wildlife resources, a key driver to poaching and extinction of species resulting in unmanageable stakeholders' interests while others see it as a means to improve their livelihoods and dietary needs. Overall, the nutritional role and financial contribution of bushmeat are prevalent in rural people's livelihoods. The paper contributes through the development of a model that aims to improve bushmeat production and community participation in natural resources conservation

Introduction

Ecotourism development has been appraised for its potential to change communities' perceptions on the use of natural and cultural resources and its ability to act as a catalyst for positive change on natural/cultural resources management (Villiers, 2011; Stone, 2015a). Despite this, there are also contestations over its value and the impacts it has on people and places affected by its delivery (Sharpley, 2008; Duffy, 2013). Currently, calls have been made to assess the performance and delivery of ecotourism to conservation and community livelihoods enhancements (e.g. Stone & Stone, 2020). For instance, the *Journal of Ecotourism* published a special issue that seeks to advance the field by examining social, economic and ecological aspects of ecotourism on several scales, including regions from around the world. The International Ecotourism Society (TIES) also held the Ecotourism and Sustainable Tourism Conference in April 2015 under the theme: *"The Future We Want: Effecting Positive Change through Ecotourism"* in Quito, Ecuador. Although the Southern African Development Community (SADC) region has experienced increased ecotourism projects through the community-based natural resources management (CBNRM) regime, it has also experienced increased bushmeat consumption, and wildlife poaching (Mfunda & Røskaft, 2016). Research indicates that conservation policies that deny local communities' access to wildlife areas for hunting and for practising wildlife-based traditional rituals often lead to increased poaching incidences (Campbell et al. 2001; Gaodirelwe, Masunga & Motsholapheko, 2018). In Botswana, the 2014 trophy hunting ban, which promoted photographic tourism over hunting and the curtailing of the supply of bushmeat to communities may be considered a case in point. Through joint ventures with private tourism enterprises, communities had agreements with their partners to supply communities with bushmeat from hunting quotas sold to them (the private enterprise). Endorsed to promote rural development and the conservation of natural resources (Muposhi, Gandiwa, Bartels & Makuza, 2016), trophy hunting is "a specific form of wildlife use that involves payment for a hunting experience and the acquisition of a trophy by the hunter" (WWF, 2005; Harris, Cooney & Leader-Williams, 2013). The trophy hunting ban on wildlife can perhaps explain the increase in poaching (illegal killing of animals) following the ban (Chevallier & Harvey, 2016; Liberg, Chapron, Wabakken, Pedersen, Hobbs & Sand, 2012).

At their meeting held in Maputo in October 2013, Southern African Development Community (SADC) ministers responsible for the environment and natural resources resolved to develop and adopt a comprehensive anti-poaching strategy and establish a coordination centre within SADC to combat poaching and illegal wildlife trade (SADC, 2014). This decision was facilitated by an increase in poaching incidences. For instance, 3668 rhinos have been poached in South Africa since 2009 and large-scale seizures of ivory, mostly originating in Africa, had exceeded sustainable limits by 2013 (SADC, 2015). Nevertheless, southern Africa still supports the largest remaining populations of elephants and rhinos in the world, challenging SADC member states to collectively engage with both the region and Asian consumer states to combat this unsustainable and illegal harvesting of wildlife resources (SADC, 2015). Numerous initiatives continue to be undertaken to combat the illegal trade and killing of wildlife; this has become a global issue that goes beyond a country or region. One such initiative is the adoption of community-based ecotourism. Since the early 1990s, rural communities in southern Africa have adopted and embarked on ecotourism enterprises through Community-Based Natural Resource Management (CBNRM) (Stone & Nyaupane, 2018). This development has brought both opportunities and challenges to communities that participate in sustainable tourism development. Critical questions of interrogation are, what emerges when communities in host destinations become active in ecotourism? Do local communities become better stewards of the wildlife and its habitat? If so, what is the process by which that happens? What are the challenges along the way as communities are differently organized?

Nevertheless, ecotourism has been touted as a "win-win solution", a "magic bullet" for getting people to value wildlife resources (Stronza, 2005). It has become a popular tool for biodiversity conservation and is based on the principle that biodiversity must pay for itself by generating economic benefits, particularly for local people (Kiss, 2004). The concept and practice of ecotourism has received so much attention that the United Nations declared 2002 the International Year of Ecotourism and marked it as a time to take collective stock of the lessons learned (TIES, 2006). Given the amplified cognizance about the prominence of host communities and environmental responsibility in tourism, ecotourism also promotes social appropriation and safeguarding the environment (TIES, 2006). This paper aims to assess how the development of community-based ecotourism contributes to ecological conservation by evaluating its contribution to bushmeat production, consumption and/or trade dynamics. Bushmeat consumption and its relationship with community-based ecotourism development have received comparatively little attention in Africa, especially southern Africa. Literature on bushmeat trade in Africa is limited to a review of the bushmeat trade in East Africa (Barnett, 2000), and sporadic studies in Serengeti National Park (Hofer et al., 2008; Loibooki et al., 2002; Marealle et al., 2010) and other sites in Tanzania (Nielsen, 2006; Wilfred & MacColl, 2010).While these studies demonstrate significant negative ecological impacts in East Africa, the drivers, impacts and interventions needed to address the bushmeat trade are not well articulated. Lack of research and actions to address the bushmeat consumption/trade to date may reflect an under-appreciation of the problem's significance among contemporary governments, wildlife agencies and non-governmental organizations and donors in the region.

This paper therefore seeks to explore the contribution of community-based ecotourism to bushmeat consumption/production and its outcomes for southern African countries with specific reference to Botswana as one country that has developed its tourism policies around the principles and values of ecotourism. The paper aims to contribute to the limited existing knowledge on the contribution of community-based ecotourism to bushmeat consumption/production, especially in Botswana where the data is scanty, to assist in the development of institutional frameworks and policies that consider community needs and conservation goals. Community and other stakeholders' perceptions of bushmeat consumption/production in community-based ecotourism at Chobe Enclave Conservation Trust (CECT) are analysed to get a better understanding. By studying the impacts of the halting of the distribution of bushmeat due to the hunting ban, the study aims to gain a better understanding of the costs and benefits associated with this move and the possible impacts on the conservation and sustainable development of wildlife resources.

Community-based ecotourism development

In community-based ecotourism, tourism sustainability encompasses "a mutually reinforcing relationship between environmental conservation, local economic livelihoods, and cultural preservation – a kind of mutualist symbiotic relationship which benefits all three" (Reimer & Walter, 2013, p. 123). Community-based ecotourism therefore thrives on the use of resources of an area to create a balance between potential and existing activities (Bhuiyan, Siwar & Ismail, 2016). It is important to view ecotourism as shifting from a definition of small-scale nature tourism to a set of principles applicable to any nature-related tourism in rural communities (Kim & Park, 2017). Nevertheless, the principal concern for ecotourism development is how to minimize negative environmental impacts and protect natural resources from degradation.

The term ecotourism began to appear in the tourism literature in the late 1980s (Weaver & Lawton, 2007). Since its first reference in the academic literature, ecotourism has been recognized by NGOs, human development agencies and governments as the most environmentally sound and locally beneficial form of tourism development (Backman & Munanura, 2017). Ecotourism is enshrined in the objectives of ecological and economic impacts for all stakeholders, however, some authors question whether these goals are being achieved (Fletcher, 2009; Herath, 1997; Lindberg & McKercher, 1997). Despite its challenged goals, over the last thirty years, sub-Saharan Africa has adopted ecotourism as the primary focus for tourism development (Backman & Munanura, 2017). One of the main catalysts for the achievement of the Millennium Development Goals (and Sustainable Development Goals) and rural development was seen to be ecotourism (Das & Chatterjee, 2015).

In terms of ecotourism development, Africa is particularly exceptional because of its wildlife diversity and the abundance of charismatic wildlife species (Lindsey et al., 2007). Over the last 30 years, Africa has experienced significant success in ecotourism development (Backman & Munanura, 2017). Evidence of such success is revealed through substantial revenue generation, foreign investment, community-based ecotourism establishments and other socio-economic benefits which have led to significant demand for ecotourism (Boyd & Butler, 1996). Multiple community-based conservation initiatives have emerged from ecotourism with both ecological and socio-economic benefits to many African rural communities (Backman & Munanura, 2017) demonstrating its viability. As the need for recreation among potential visitors has increased alongside rising income, community-based ecotourism has been gaining popularity among rural communities (Kim & Lee, 2017). Despite the tremendous reported success of community-based ecotourism development in Africa, little is known on the relationship of ecotourism and bushmeat consumption. Contextualizing the relationship between community-based ecotourism and bushmeat consumption is critical for understanding interaction dynamics, which may have some environmental and livelihoods consequence outcomes.

To contextualize and conceptualize this paper, in addition to ecotourism and CBNRM principles, this article adopts the social exchange theory (SET) to understand the relationship between community-based ecotourism development and bushmeat consumption. According to SET a community is likely to support tourism development when positive (economic, environmental, and sociocultural) outcomes are realised (Paraskevaidis & Andriotis, 2017) and shun those where negative results are experienced (Bagherian et al., 2009). In line with SET, the CBNRM programme assumes that local communities are keen to participate in natural resource conservation when the use of such resources improve their livelihoods (Blackie, 2019: Government of Botswana, 2007). SET involves a cost-benefit analysis made by people to participate or not in an activity compared to other possibilities (Nunkoo, 2016). As a result, SET argues that the relationship between community-based ecotourism development and bushmeat consumption (and CBNRM) should improve local people's livelihoods and lessen costs incurred by locals to be supportive. This is especially important for countries such as Botswana where programmes like CBNRM and ecotourism have been adopted to promote economic benefits from the conservation of wildlife in a 'win-win' situation and attain sustainability (Mogomotsi, et al., 2020). The assumption is that people living close to wildlife resources like those at CECT experience a number of benefits and costs for conservation, hence one of the core tenets of CBNRM is that benefits should outweigh costs for people to support conservation activities and sustainable development (Zuze, 2009).

Bushmeat

Information on the significance of bushmeat hunting is scanty in southern Africa, the common perception is that it is done for subsistence purposes and occurs on a limited scale (Barnett 2000; Hayward, 2009).While it has been evident that bushmeat is not formally controlled, it touted as positively contributing significantly to food security, often representing the most important source of protein in rural areas (Nasi et al., 2008) and cash, often used to purchase other foods and essentials during lean harvests (Blackie, 2019; Mbaiwa, 2018). For many communities across Africa, "many sedentary rural communities eat wildlife as a primary or supplementary source of animal protein" (Wilke et. el., 2016, p.403). In most African states, wildlife is owned by the government and local communities living with this wildlife normally have no rights to it and thus most hunt illegally for food or to sell (Gibson, 1999) since they believe they have rights according to customary laws (Wilkie, Redford & McShane, 2010). In Zimbabwe, Lindsey, et al., (2011) discovered that unemployed, poor, young men with limited access to food carried out most of the illegal hunting in the Savé Valley Conservancy area; the bushmeat, which they sold to wealthy people, enabled them to acquire food and money. For some southern African countries, the bushmeat trade poses a threat to conservation efforts, with the development of a commercial trade relying on meat from protected areas (Barnett, 1998; Lindsey et al., 2011). In Botswana the government banned trophy hunting in 2014 due to declines in wildlife populations claims (based on wildlife surveys), some experts attribute this decline to illegal bushmeat (Rogan et al., 2015, 2017; Chase, 2011; DWNP, 2013).

Generally, the rate at which animals are being hunted raises concern at local, national and international levels (Festus & Omoboye, 2014). Unregulated bushmeat hunting is often regarded as being unsustainable, causing widespread wildlife population declines (Fa et al., 2000; Rogan et al., 2017). Several studies have also confirmed that bushmeat hunting is a major cause of species endangerment (see Raven et al, 2010; Jack, 1999; Adetunberu, 2001). Uncontrolled hunting, over-hunting and illegal hunting are not sustainable for wild animals (Festus & Omoboye, 2014). Therefore, uncontrolled hunting is perilous to ecotourism (Tall & Gueye, 2004) as it uses the same resources being hunted as ecotourists' attractions. As such, ecotourism also becomes a player in the killing of wild animals in the form of hunting tourism. While some researchers support ecotourism through hunting tourism as long as it contributes to community livelihoods improvement (e.g. Brown & Williams, 2003; Mbaiwa, 2018), others are against the practice as it is feared that it endangers wildlife species (Baker, 1997; Finch, 2004). Therefore, in view of these debates, hunting tourism is polarized, with animal rights groups and protectionists on one side and hunters and pragmatic conservationists on the other (Hutton & Leader-Williams, 2003; Loveridge, Reynolds, & Milner-Gulland, 2006).

Conceptualizing and contextualizing the problem

The idea to initiate this study came from critical observations and available literature (e.g. Mfunda & Røskaft, 2016; Mbaiwa, 2011, Stone & Stone, 2011) which indicates that in most Sub-Saharan-African countries today, communities are engaged in many natural resources based development projects that are intended to sustain conservation and improve livelihoods. Nevertheless, communities appear to be losing the battle against the loss of wildlife and its habitat, and natural resources seem to be dwindling both in numbers and diversity (Stone & Nyaupane, 2018), thereby affecting communities' livelihoods. There has been an endemic outbreak of famine in the region that is linked to the declining state of the environment, as reflected by the systematic deterioration of food security. For example, more than 10 million people are now affected in drought-stricken areas of Djibouti, Ethiopia, Kenya, Somalia, Malawi, Zimbabwe and Uganda and the situation is worsening (United Nations, 2018). There is no country in the entire African continent that is truly industrialized and each of the 54 independent countries derive much of their economic resources (food, cash crops, livestock, minerals, fuel energy, and hydro-electricity) directly or indirectly from the natural environment (Lee, 2003). Thus, this strong dependence on natural resources has serious implications on the economic performance of many developing countries, if environmental performance declines.

Regardless of the environmental degradation in the region, each country has its development, environmental and conservation policies that are geared towards ameliorating the situation. Natural resources, more especially wildlife in Botswana, like in most sub-Saharan countries, are dwindling in both absolute numbers and species diversity (Chase, 2011). This problem occurs despite the government's efforts of setting aside land for wildlife management. Botswana's 39% of her land surface is used for conservation and tourism development purposes (Botswana, 2010). Chase's (2011) study found that eleven species have declined by 61% since a 1996 survey in the Okavango Delta. Ostrich (*Struthio camelus)* numbers dropped by 95%, from 11,893 to 497; 90% of wildebeest (*Connochaetes taurinus*) have also been wiped-out, along with 84% of the population of the antelope tsessebe (*Damaliscus lunatus)* ; 81% of warthogs (*Phacochoerus aethiopicus*) and kudus (*Tragelaphus Strepsiceros*), and nearly two-thirds of giraffes (*Giraffa camelopardalis*) (Chase, 2011). Consequently, these figures show that the Okavango Delta ecosystem has suffered "catastrophic" species loss over the past 15 years. In the Great Elephant Census carried across 18 African countries, Chase et.al. (2016) found that elephant populations are declining by 8% per annum and dropped by 144,000 between 2007 and 2014, due mainly to poaching for ivory (Poulsen et.al., 2017). In northern Botswana, Chase, Schlossberg, Sutcliffe & Seonyatseng

(2018) discovered that despite the country having the largest elephant population of any country (about 30% of all savannah elephants and 58.9% of the African elephant population), since 2010, aerial surveys indicate that overall the population has stabilised and may be gradually declining. However, although populations may be increasing along the Zimbabwe border as a whole elephant populations are not (Chase, Schlossberg, Sutcliffe & Seonyatseng, 2018). It can also be argued that the same diminishing species have made Botswana one of the popular African wildlife-based tourism destinations.

Nonetheless, the Okavango Delta ecosystem is not the only tourist destination in Africa facing the loss of natural bounty. It is reported that in the Maasai Mara, Kenya, the numbers of impala (*Aepyceros melampus*), warthog (*Phacochoerusaethiopicus*), giraffe (*Giraffa camelopardalis*) and hartebeest (*Alcelaphus bucelaphus*) have declined by more than 70% over a period of three decades (Pinnock, 2011). The causes for the observed declines in sub Saharan Africa are not well understood but collectively postulated to be related to a number of factors including changing perennial rivers flood regimes, habitat fragmentation, drought and erratic rainfall, habitat over-use through anthropogenic pressures, unregulated hunting and poaching, human population and demographic trends (Allison & Moore, 2010; Chase, 2011; Mbaiwa, 2011; Moswete, Thapa & Child, 2012; Salafsky & Wollenberg, 2000; Stricklans-Murnro, 2010; Mbaiwa & Stronza, 2010). In view of this scenario, it is likely that present wildlife development programs in Botswana may fail unless a change in approach is adopted by the local populace.

The decline in wildlife resources occurs even when intervention measures have been sought and put in place. Such include community wildlife-based ecotourism as a management tool in wildlife management areas. The emergence of community-based ecotourism and its incorporation in conservation is seen as an alternative form of sustainable development. Thus, others argue that community-based ecotourism has been increasingly fronted as a key instrument for maintaining protected areas globally (Tumusiime & Vedeld, 2012). The adoption of community-based ecotourism as a conservation and development measure with existing local historical, socio-economic, and institutional landscapes is associated with the promotion of local community participation in resource management. This integrated approach is an appealing concept and is often cited as the rationale for promoting conservation and rural development as it is difficult to ignore (Tumusiime & Vedeld, 2012).

Conservationists and development planners believe there are lessons to be learnt from the decline of wildlife populations in the sub Saharan Africa region (e.g. Chase, 2011; Pinnock, 2011) but there has been no methodical assessment of the relationship between community-based ecotourism and bushmeat consumption, thus this research is geared to bridge this research gap. This is especially true in Botswana where bushmeat was a benefit for most community-based ecotourism projects, through joint venture partnerships with private partners who buy community hunting quotas.

Methods

Study area

This study design is informed by a case study approach, in which the focus is the Chobe Enclave Conservation Trust (CECT), a community that is engaged in community-based ecotourism. The case study approach was chosen "to generate an in-depth, multi-faceted understanding of a complex issue in its real-life context" (Crowe, Cresswell, Robertson, Huby, Avery & Sheikh (2011, p.1), in this case bushmeat consumption and production at CECT. The CECT which started operating in 1993 is in the Chobe District, northern Botswana. Bordering Zambia on the north, Namibia on the north and west and Zimbabwe on the east, the Chobe District's administrative centre is Kasane (Chobe District Council [CDC], 2003). Settlements in the district are divided into two cluster communities: east (Kazungula, Lesoma, Pandamatenga) and west (Mabele, Kavimba,

Kachikau, Satau, Parakarungu); the western cluster is known as the Chobe Enclave and forms the CECT (Garekae, Lepetu & Thakadu, 2020). CECT was selected for this study because it is the first CBNRM and regarded as one of the successful projects in Botswana, which can provide an ideal setting as it practices both community-based ecotourism and bushmeat transactions. CECT is a community-based organization (CBO), shared by five villages, namely Mabele, Kavimba, Kachikau, Satau and Parakarungu. The villages (see Figure 1) are located on a belt that runs along the Chobe Basin forming an enclave. The Enclave villages are located within two controlled hunting areas; CH1 and CH2 where photographic and trophy hunting tourism are practiced. However, during the time data were collected, only photographic tourism was practiced since trophy hunting was banned in 2014 (it was reinstated in May 2019). In addition to CH1 (leased to a private company - Linyanti Bush Camp) (see Figure 1), CECT owns a brick moulding workshop which sells bricks to locals, six tractors used to plough for locals, two administrative offices leased to private business operators and a lodge (Ngoma Safari lodge) run in partnership with a private investor – the investor employs locals and will run the lodge for a period of 20 years, (Mbaiwa & Tshamekang, 2012; Stone, 2013; Stone & Nyaupane, 2018). CECT has natural resource rights on CH1, therefore it is responsible for the monitoring of natural resources in the area. The Chobe enclave is home to diverse fauna and flora (accommodated at Chobe National Park [CNP]), dense miombo woodlands (Chobe Forest Reserve [CFR]) and transboundary water bodies (Chobe River) (Garekae, Thakadu & Lepetu, 2017). All these resources are vital to communities' livelihoods: the (CFR) provides adjacent communities with non-timber forest products (NTFPs) used for medicinal purposes, building materials, household energy and for agriculture, the CNP provides sources of bushmeat and tourism revenue which assists in the enhancement of community livelihoods while the Chobe River is used for flood recession farming, edible plants and fish (Garekae, Thakadu & Lepetu, 2017, 2020). The Chobe district has a large population of wildlife, especially elephants and buffalos. The wildlife, particularly the elephants migrate freely to and from the Hwange National Park in Zimbabwe and the Caprivi Strip in Namibia (CDC, 2010).

The estimated population of the enclave community is 4,108 (Kachikau: 1, 356, Kavimba: 549, Mabele: 773, Parakarungu: 845 and Satau: 605) respectively (Botswana, 2011). The Chobe Enclave is inhabited by people from diverse ethnic backgrounds, with the BaSubiya constituting most of the inhabitants followed by the BaTawana, BaMbukushu and BaYeyi. The BaSubiya are found in almost all of the Chobe Enclave settlements more especially Parakarungu and Satau (CDC, 2003). CECT is run by a board of trustees elected from each of the five villages. With 15 board members (two elected by the general membership from each village and the five village chiefs), the board is elected every 3 years (Stone, 2015b). The board works closely with all the village development committees (VDCs) with income generated by CECT being allocated to VDCs to decide on development projects. The CECT community has a mixed economy based on three main domains: subsistence livestock rearing, crop production, and wage employment from tourism and other activities. The cattle population owned by the CECT community is estimated at around 9000 (Stone, 2013).

Data collection

Due to the exploratory nature of the research, it was guided by a qualitative research approach. Qualitative research facilitates the in-depth understanding of motivations, attitudes and reasons behind human behaviours (Kaae & Traulsen, 2015), hence this was considered a suitable approach to adopt in trying to get a better in-depth understanding of community-based ecotourism and bushmeat dynamics at CECT. Primary and secondary data sources were used. Primary data sources include individual semi-structured in-depth interviews with key government officers, tour operators, tour guides, lodge owners, villages' chiefs, village development committee (VDC) chairpersons, and household heads. Semi-structured in-depth interviews were chosen

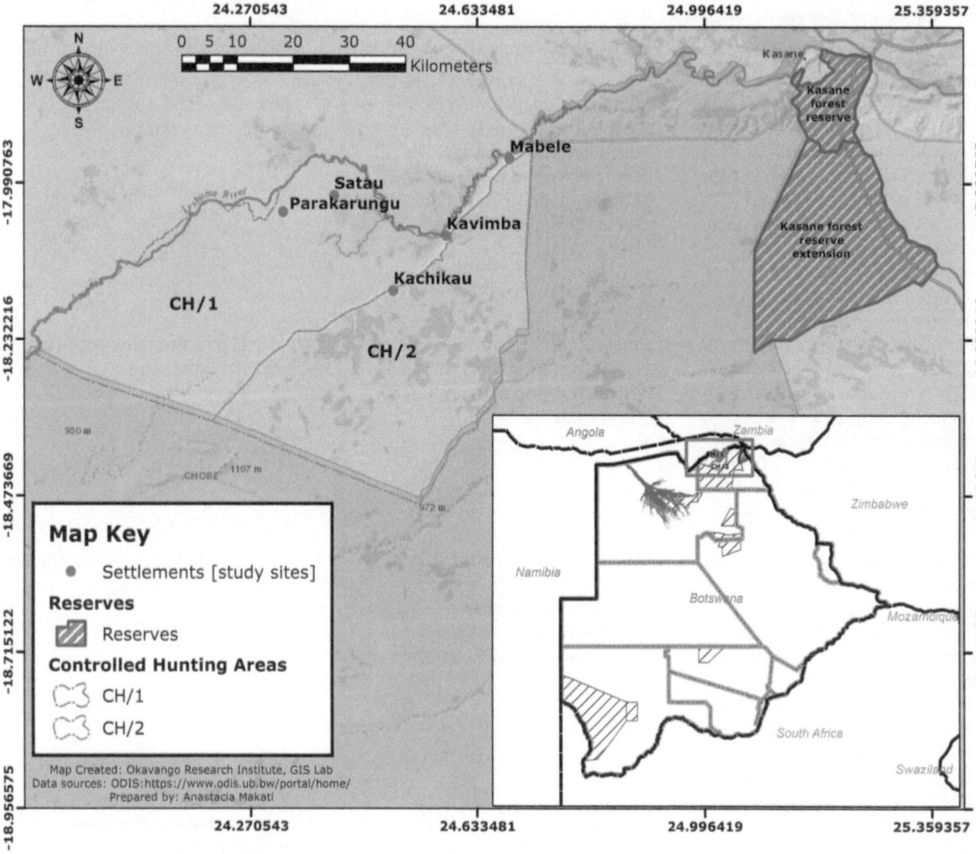

Figure 1. Map of Botswana showing study areas.

to allow for a deeper understanding of community and other stakeholders views on ecotourism development and bushmeat consumption. Interviews took place between the months of May-August 2018. A household list from CECT was used to randomly draw household interviewees; thirty-two household interviews were conducted. Key informants interviewed included government representatives from the department of Wildlife and Tourism (DWNP) (1), the Department of Tourism (1) and the Botswana Police Service (BPS) (1), Botswana Tourism Organization (1), Crop and Animal Production (1), the Land Board. Government officials were selected to represent government agents actively involved in tourism, conservation and livelihoods planning. In addition, tour operators (5), tour guides (3), lodge owners (2), villages' chiefs (5), village development committee (VDC) chairpersons (5) and a butchery representative (1) were also interviewed. Key informants were purposively selected from various stakeholders based on their knowledge about ecotourism and bushmeat dynamics. In total 27 key stakeholders and 32 household heads were included in the study. Due to the qualitative nature of the study, the sample size was guided by the attainment of saturation (Patton, 1990), with interviews being stopped when no new information was acquired from respondents. Interviews were audio recorded, where interviewees gave consent, where none was given, shorthand entries were performed. Before data were collected, a study permit was sought and granted by the Ministry of Ministry of Environment, Natural Resources Conservation and Tourism.

Secondary data sources included published and unpublished articles. Unpublished reports dealing with CBNRM at CECT, annual reports and projects carried out by CECT were obtained from the CECT office while legal and illegal hunting statistics and the trophy hunting quota

Table 1. Codes and Themes Development.

Code development (open code)	Emergent sub-themes (winnow codes)	Final research themes (final themes)
Community relocation from the park, Commencement of Chobe National Park, CECT commencement, change from predominantly mobile to sedentary lifestyle, alienation of the CECT community from natural resources, historically conflict-ridden practices reinforced a culture of unending conflicts, CBNRM introduced a bottom up approach in resources management,	Chronology of historic events led to CECT participation in ecotourism and bushmeat production, unending conflicts affected CECT community and park relations, CBNRM made CECT to participate in ecotourism and bushmeat transactions	The historical commencement of Ecotourism and bushmeat consumption at CECT
CBNRM facilitated CECT formation as a legitimate community institution, CECT's constitution strengthened community participation in ecotourism and bush meat production, existence of CECT made wildlife meat consumption acceptable, community wildlife quota allocations.	Formal recognition of CECT as a legitimate community institution, production and consumption of bush meat	Bushmeat legitimization, production, consumption issues
Bushmeat management strategies are developed outside the CECT communities, gaps between strategy developers and the CECT community, CECT act as a recipient and expected to implement management strategies, CECT not involved on deciding composition of wildlife quota, wildlife quota decided by the Department of Wildlife & National Parks.	CECT passive participation in wildlife decision making, one-sided decision-making process adopted by the DWNP, government unilaterally controls and owns wildlife resources	Unilateral and exclusionary bushmeat intervention strategies
Difficulties to control bushmeat production, CECT transactions with bush meat created conducive environment for poaching, CECT transactions with bush meat favoured poachers to front poached meat as legally obtained meat, intercepted bush meat claimed to be livestock meat, difficulties to monitor and control bushmeat transactions within CECT, creation of external demand for bushmeat consumption	Uncontrolled and unintended bushmeat production within the CECT community, ecotourism created a market for bushmeat consumption beyond the borders of the CECT villages, demand – supply of bushmeat created a problem of controlling bushmeat consumption	Uncontrolled and unintended bushmeat affair
The hunting ban affected community-based ecotourism project adversely, outcomes include a negative impact on the financial inflow, costs of living with wildlife continued and escalated, hunting ban cut the supply for bushmeat, hunting ban affected community food security, hunting ban affected protein supply to CECT community.	The hunting ban prevented sustained growth in the community-based ecotourism industry, blanket ban affected bushmeat production	Effect of "blanket hunting ban" on bushmeat production
Contested data that led to the blanket hunting ban, lack of data creates knowledge gap, reliance on poor wildlife inventory technique's results, lack of transparency and presentation of reliable information, ill-informed methodology, flawed snapshot research, lack of knowledge on long-term wildlife trends or time series data on wildlife populations	Lack of reliable baseline data, lack of long-term wildlife trends or time series data on wildlife populations	Challenged wildlife baseline data

system were obtained from the BPS and the DWNP respectively. Published reports on CBNRM and CBNRM in Botswana and at CECT were obtained through google scholar.

Data analysis

An analytical procedure that is based on techniques proposed by extant interpretive studies (Miles & Huberman, 1994) was adopted to guide data analysis. First, data from household and key informant interviews were transcribed and translated from *Setswana* to the English language verbatim. The interview and translation process were easy to handle as the researchers are fluent in both languages. The holistic content perspective was the analytical procedure adopted that predominantly entailed the researcher exploring the central meaning of participants' narratives (Lieblich, Tuval-Mashiach, & Zilber, 1998) from both the household and key informant interviews. All transcripts were systematically explored with the objective of understanding participants' overall perceptions', narratives and contextual meanings. This procedure facilitated and enhanced appreciation of the community's construction of its views on ecotourism development and bushmeat consumption. Analytical notes and memos for each transcript were developed and emergent themes were documented using key words and phrases as recommended by Polkinghorne (1988). Thereafter, subsequent steps involved cautiously re-reading of the transcripts to code different segments of each transcript with a certain theme. This coding process was guided by Miles and Huberman (1994) and involved the identification of words, phrases, sentences or paragraphs that convey a specific message relating to community's perceptions. This exercise necessitated the process of code development (see Table 1). All the transcripts as well as field notes were coded thematically and produced themes as discussed in the results section. The code development continued until all transcripts were read. The data analysis process was guided by the open-code technique (i.e. codes emerging from data also known as in vivo coding). Analytical notes and memos were written alongside transcripts with the aim of understanding each transcript's overall insight and contextual meanings of factors that contribute to ecotourism and bush meat engagements. This process led to the emergence of subthemes that have explanatory value (Kirby & McKenna, 1989). The last step involved clustering coded fragments of sub- themes into progeny groupings according to their respective descendant themes in line with Pritchard et al's (2011) recommendations. After open coding was performed, focused coding was then applied to winnow down the codes as suggested by Charmaz (2006). Subsequently, group-like codes were treated as common themes. Emergent sub- themes were developed during the code development process. Finally, themes were derived from the merged subthemes as summarized in Table 1. The exercise aided a better understanding of the relationship between ecotourism and bushmeat production dynamics.

As reflected on Table 1, six main themes were derived from household and stakeholder interviews. These include the historical commencement of Ecotourism and bushmeat consumption at CECT, Bushmeat legitimization, production and consumption issues, unilateral and exclusionary bushmeat intervention strategies, uncontrolled and unintended bushmeat affair, effect of "blanket hunting ban" on bushmeat production and lack of wildlife baseline data. These are discussed below.

Results

The historical commencement of ecotourism and bushmeat production at CECT

Results from village chiefs, household heads and a representative from the Department of Wildlife and National Parks indicate that before the commencement of community-based ecotourism enterprise the CECT community had a predominantly agro-based economy, based on livestock and crop productions. Household heads and village chiefs also highlighted that following the ban of subsistence wildlife hunting and the coerced removal of the CECT community from the Chobe National Park in the late 1950s by the colonial government, the CECT community was alienated from its rich natural resources. The restraints on subsistence wildlife hunting coerced the community to convert to predominately agro-pastoralism resulting in sedentary lifestyles. Subsequently, the over dependence on existential agriculture resulted in increased cattle numbers, leading to competition with wildlife for space and forage resources. According to the Department of Wildlife and National Parks respondent, poaching was never a pronounced problem until communities were prohibited from hunting wild animals. Poaching became prevalent; this was triggered by wildlife wandering beyond its borders and damaging local communities' crops and livestock now located outside the national park. The demarcation of land for national parks and for agriculture brought separations in resources management. Those who were caught hunting wildlife for subsistence were labelled as poachers and were prosecuted, however, poaching never stopped. Consequently, these historically conflict-ridden practices reinforced a culture of unending conflicts between the CECT community and Chobe National Park management. The state-of-affairs continued beyond independence in 1966, until the realization that separating communities from their natural resources does not work in the early 1990s. The government changed the approach and adopted community-based natural resource management (CBNRM) – regarded as a bottom up approach in which communities are given user rights to manage natural resources in their areas of jurisdiction. This new regime in natural resources management introduced community-based ecotourism. As per the requirement of CBNRM, the community had to form a Community Based Organization (CBO), devise a constitution to spell out natural resources management and formulate land use or zoning management.

Accordingly, VDC chairpersons indicate that in 1992 CECT was allocated controlled hunting areas 1 and 2 (CHA 1 & 2) as shown in Figure 1. As a result, the DWNP representative indicated that the first wildlife quota was also awarded, and the community had a choice to hunt or to sell its hunting quota to professional hunters. CECT opted to sell the quota to professional hunters. Subsequently, according to the representative from DWNP, village chiefs and VDC members, quotas have been sold to professional hunters with the proceeds going to the community. Figure 2 shows the hunting quota allocated to the CECT community between 2005 and 2013. While professional hunters' interests are not entirely on carcasses, the community drafted an agreement that binds professional hunters to bring carcasses of hunted animals to the community. This arrangement helped cut costs of bringing the meat from expanse hunting grounds to the community in villages. However, according to the DWNP representative's view, ecotourism has complicated bushmeat dealings in the CECT village as legal and illegal obtained meat cannot be differentiated. As stated by a DWNP official,

> "While many stakeholders celebrate the participation of CECT in hunting tourism, managing illegal bushmeat has proven to be a mammoth task the DWNP has to control as it is fronted as having been legally attained from the community quota system"

While the CECT community benefitted from bushmeat, the community also gained financially from professional hunters buying community wildlife quotas as shown in Table 2 which shows financial gains attained by CECT through all its activities and wildlife quotas from 2010 to 2018.

According to household respondents from CECT, an arrangement was made with private tourism enterprises that bought the community's allocated hunting quota to ensure that bushmeat from hunts be equitably distributed among households in all participating village. This development formalized the consumption of bushmeat amongst the CECT community reversing fortress conservation approaches that regarded wildlife resources as untouchable by communities living adjacent to protected areas. While the community benefitted from the proceeds of selling the wildlife quota, it also benefitted from the availability of bushmeat. Each household had the liberty to do whatever it wanted with its bushmeat allocation. Some consumed it while fresh, others dried it to preserve it for future use, while those with refrigerators would freeze it. The

Table 2. CECT Revenue Generation from 2010 - 2018.

Year	Total Revenue generated in BWP (Botswana Pula)	Portion of revenue generated from hunting in BWP (Botswana Pula)	Proportion of revenue generated from hunting (in %)
2010	6 379330.80	4 842 554,52	76%
2011	5 770 992.06	3 435 950,82	60%
2012	5 647402.14	3 915 385,80	69%
2013	7 097 536.47	4 147 508,52	58%
2014	3 152 725.71	0	0
2015	3 385 782.88	0	0
2016	4 062 524.57	0	0
2017	4 140 387.75	0	0
2018	4 571 897.92	0	0

bulk of the meat came from elephant carcasses, leading to large quantities of the meat being consumed fresh and some being dried for future use. Some sold their dried meat to other community members and to others outside the CECT community. This development availed money to those who otherwise would not have access to it. This was one of the most tangible benefits realized by the community. This is supported by one of the participating villages' chief who observed that:

> "In reality, apart from bushmeat benefits that directly and positively impact on household food security, local communities realize very little direct benefit from community-based ecotourism projects, but rather bear the costs of living near large wildlife populations. Apart from benefiting from the meat, very little has been realized to empower local communities in tourism development".

The chief's statement shows some appreciation of the bushmeat's direct trickling down effects on households' food security. In line with the chief's statement, a respondent from the Department of Wildlife and National Parks also highlighted the importance of bushmeat availability to the community by suggesting that:

> "Often too many projects funded by international donors focus on conservation activities, with insufficient attention directed towards the livelihoods of local communities, thereby ignoring the key pillars and best practice in sustainable conservation. The accessibility of bushmeat to CECT communities made a very pronounced benefit that is visible to everyone"

The respondent from the Department of Wildlife and National Parks highlighted that apart from conservation objectives, the primary concern in any activity where people are involved is communities' livelihoods and well-being. Therefore, making bushmeat available was needed as a way of easing conflicts and improving community livelihoods. Nevertheless, the unprecedented circumstance of interrupting bushmeat production came to being when a hunting moratorium was instigated by the government. Therefore, with the 2014 hunting ban, revenue generation declined due to the halting of the hunting quota (see Table 2). This decline affected community livelihoods.

Bushmeat legitimization, production and consumption issues

The formation of CECT as a community-based ecotourism project facilitated the recognition of CECT as a formal and legitimate institution that deals with the production, distribution and consumption of bushmeat within the participating villages. This development influenced the bushmeat supply – demand dynamics.

One community respondent summarized this development as follows:

> "Before the commencement of CECT there were no dealings with bushmeat production and consumption in the CECT villages. It was unthinkable to deal with wildlife meat in the open. We would illegally hunt, yes, … and if lucky you would secretly bring something home for the family … Even in situations whereby wildlife got hit by vehicles within the vicinity of CECT villages, we all knew that the carcass belonged to the government. The DWNP would collect such carcass, extract valuable body parts and/or burn it rather than

giving it to communities, but with the existence of CECT wildlife meat was surprisingly made available to us".

Most community respondents shared the same sentiments; that they never thought at one point their community-based ecotourism project will grow to the extent that it can make bushmeat available to the community. While some respondents appreciated that their project has contributed positively to improvements in their livelihoods, others raised concerns as they see this development as being inconsistent with conservation objectives. To illustrate this point, one respondent made this observation:

> "The legitimization of killing wildlife for its meat and trophy contributes to the abuse of our wildlife resources because we now regard it as a commodity that belongs to us. When we see it [wildlife] as a community to which we are a part of, we may begin to use it with love and respect. When we start to kill it [wildlife] for meat, then we should consider ourselves as sophisticated predators due to our technological advancements because we easily locate and select the best species we prefer and kill wildlife that is mostly healthy, productive and young, while other predators target weak, old, sick and easy-to-kill animals with less energy. As a result, wildlife resources are expected to stay the same, and at the same time keep supporting communities whose values change rapidly".

Results from household heads show that there are community members who question the rationality of making bushmeat available to communities through ecotourism. Consequently, there are questions about the conservation outcomes of community-based ecotourism policies as they are often applied with little consideration of the long-term human-wildlife interactions that may be affected by changing community needs over time.

While some household representatives appreciate positive changes brought by their ecotourism project, others think that the ecotourism project further promotes exploitation by those who are externally linked to the project in the form of professional hunters, who are able to buy the community's wildlife quota at ease because of the means they have. They claim that, if it wasn't for the needs of professional hunters, bushmeat could not have been made available to local communities, thus it is availed as a ploy to meet the needs and wants of external communities. One respondent alluded to this sentiment when she said:

> "While we are made to believe that bushmeat has contributed to improvements in our livelihoods, we are never told how much professional hunters are making from the wildlife trophies they are taking away, they only leave us meat. Why can't we access the same market where wildlife trophies are sold, then we can benefit twice directly, from the trophy sales and meat?"

The CECT community has differing perceptions on the legitimization, production and consumption of bushmeat as facilitated by their project, reflecting different sentiments on how ecotourism impacts community livelihoods.

Unilateral and exclusionary bushmeat intervention strategies

According to VDC chairpersons, although the CECT community is regarded as a legitimate stakeholder in wildlife resources management, bushmeat management strategies are developed outside the communities in which they are to be implemented. This development has created a gap between strategy developers and the CECT community. The community is the recipient and is expected to implement the management strategies, however, the wildlife quota system is decided by the DWNP without involving the CECT community. The community did not know why and how the composition of the quota is decided. This process does not instil community ownership and involvement in decision making. There seems to be a mismatch between community perceptions regarding the species they preferred to be included in the quota and what they were given. The community preferred those species that heighten human-wildlife conflicts to be included in the quota, however, in some cases the quota is inconsistent with their expectations, thereby making them wonder how the quota is decided. As one community member stated'

> "We have no say in the hunting quota… we would prefer for the hunting quota to include those species that affect our livelihood; however, the government has control of everything".

According to the DWNP representative, the quota is determined based on the availability of species and not on the magnitude of human-wildlife conflicts caused. Therefore, the DWNP stressed that wildlife quotas are used as a conservation management tool to control wildlife species numbers and are not determined based on community needs and wants. This disparity between DWNP and community expectations aggravates the perception that the government unilaterally controls and owns wildlife resources, in the process excluding the participation of communities. A more community-based participatory and inclusive approach in deciding community wildlife quota would reconcile community perceptions and the one-sided decision-making process the DWNP has adopted. The CECT community feels excluded in decision-making that affects their life, yet they are expected to implement policy devices designed without their input.

Uncontrolled and unintended bushmeat affair

The adoption of trophy hunting tourism created uncontrolled and unintended bushmeat production within the CECT community. The chiefs, DWNP and police representative indicate that before the tourism hunting ban in 2014 it was very difficult to control bushmeat production, especially in the form of biltong. The bushmeat transactions within CECT created a conducive environment for poachers to bring dried meat to the CECT community and claim that it is part of the meat obtained through community-based ecotourism. This development favoured poachers as they would front poached meat as legally obtained meat. Communities could dry bushmeat for future use; the dried bushmeat could last for almost a year. This preservation measure made it easy for the community to have access to bushmeat even outside the hunting season. According to a representative from the Botswana Police Service, poachers illegally hunt wild animals, dry the meat at the hunting grounds and bring it to their respective villages, thereby making it difficult for customers to separate the legally and illegally obtained meat. The station commander of the Kachikau Police Station alluded to this difficulty and indicated that sometimes they confiscated the meat for forensic testing in cases where suspects are caught in possession of illegal bushmeat from species not legally distributed from the ecotourism project. Results also indicate that other culprits claim that the meat in their possessions comes from their livestock (interview with Kachikau Police Station, 2018). This development makes it difficult to monitor and control bushmeat transactions within the CECT community. Table 2 summarizes offences committed and reported to the Kachikau Police station from 2000 to 2016. These wildlife offenses are linked to the community-based ecotourism project formal transactions with bushmeat within the CECT villages. A total of 142 offences were committed ranging from hunting game without a license, unlawful possession of government trophy, hunting game at night and hunting game during closed seasons. The most committed crime as reflected in Table 3, is hunting game without a license, followed by the unlawful possession of government trophy, respectively.

According to household heads, the availability of bushmeat created a market as far as Kasane Township (i.e. a nearby town. See Figure 1); this is corroborated by the police as well. Kasane dwellers travel to the CECT community to buy bushmeat. While this development promotes modest revenue generation within the CECT community, community-based ecotourism has created a market for bushmeat consumption beyond the CECT villages. The external demand for bushmeat consumption has created an outside market, exerting more pressure on suppliers to meet market demands. One community member asserts that,

"Bushmeat obtained through CECT has allowed us to have access to meat and to make money by selling meat to other people … some in Kasane. I use the money to buy food and other things I need."

According to the Kachikau police, this demand – supply of bushmeat has created a problem of controlling bushmeat consumption; this can easily fuel poaching to satisfy the growing market. The police representative indicates that

"The hunting ban has somehow made our job easier because it is now easier to determine cases of poaching as opposed to when bushmeat was available through ecotourism".

Police case registries indicate that species that are mostly poached in the CECT area are buffalos, impalas, kudus, zebras, eland and warthogs for meat purposes. These local preferences for certain species for bushmeat are consistent with overall national preferences as summarized in Table 4. According to Gaodirelwa, Masunga & Motsholapheko (2018), the lack of game meat in many rural areas lead to poaching incidences by some community members due to reliance on bush meat to meet commercial and subsistence needs.

Generally, at the national level, the expectation would be that poaching incidents should be decreasing, but as reflected in Table 4, this is not the case. This happens even in communities where CBNRM has been adopted, despite expectations that it instils conservation attitudes to the general community. In 2014 when trophy hunting tourism was banned reported incidents of elephant poaching stood at 535 cases. This worrisome situation may not necessarily be linked to bushmeat production but to the ivory trade, hence different approaches are needed to provide intervention measures. However, extant data from Statistics Botswana (2014) indicates that between 2009 and 2013, more than 75% of the wildlife species poached in Botswana were killed for bushmeat.

"Blanket hunting ban" on bushmeat production

In 2014, the government of Botswana issued a directive that banned trophy hunting tourism country wide; hence hunting licences were no longer issued. Hunting was banned due to declining wildlife populations in Botswana owing to an array of factors, including hunting, poaching, human encroachment, habitat fragmentation, drought, and veldt fires (Chase, 2011). The 'blanket ban' on hunting tourism affected community-based ecotourism projects adversely. According to the CECT representative, such outcomes include the negative impact on the trust's finances inflow for conservation and community livelihoods while the costs of living with wildlife continued in the form of wildlife foraging and preying on crops and livestock. The representative stated:

"The hunting ban was implemented without any consultations with us. It has affected us negatively and as a result CECT has seen a reduction in revenue. This is because we no longer get a hunting quota … we also have no access to bushmeat". These two actions have affected our community negatively.

The DWNP representative on the other hand indicated that while the consumptive uses of wildlife are necessary to generate enough funding to support meaningful conservation and

Table 3. Offences committed in the CECT area related to Bushmeat Consumption.

Offence Committed	2010	2011	2012	2013	2014	2015	2016	Total
Hunting Game Animal without a License	1	–	5	9	4	5	5	**29**
Hunting a game animal during closed season	2	–	3	–	–	–	–	**5**
Unlawful Possession of Government Trophy	1	1	–	1	2	4	4	**13**
Hunting Game Animals at Night	–	1	–	2	1	3	–	**7**
Hunting Protected Game Animal without License	–	–	–	1	–	3	2	**6**
Unlawful Possession of Elephant Tusks	2	1	3	–	–	–	1	**7**
Total	**6**	**3**	**11**	**13**	**7**	**15**	**12**	**67**

Source: Kachikau Police Station Case Registries from 2000 – 2016.

Table 4. Poaching statistics at National Level from 2010 to 2018

Species	2010	2011	2012	2013	2014	2015	2016	2017	2018	Total
Buffalo	14	18	18	22	17	5	2	1	3	100
Eland	7	3	16	11	8	10	16	8	7	86
Elephant	26	121	190	87	111	3	*	*	156	694
Gemsbok	12	6	35	25	22	3	16	3	15	137
Impala	19	54	36	36	9	23	21	13	24	235
Kudu	22	58	64	68	18	30	21	11	28	320
Zebra	13	10	12	22	18	3	0	2	2	82
Total	113	270	371	271	203	77	76	38	235	1654

*No elephant statistics were provided for the years 2015 – 2017.
Source: (DWNP, 2020; Chase, et al., 2018).

community livelihoods, the hunting ban cut the supply of bushmeat, threatening community food security, especially protein. However, community members indicated that while the community no longer gets bushmeat through CECT, butcheries in Kasane are selling bushmeat. This development has brought mixed reactions, within the CECT community, with some wondering where butcheries source bushmeat and whether they are not affected by the same directive that stopped bushmeat production. Others associate these butcheries with elitism that takes advantage of the bushmeat market created by community-based ecotourism project. Informal interviews with butcheries that sell bushmeat indicate that they have licenses to trade with bushmeat on selected wildlife species. The butcheries source bushmeat from certain local and regional game ranches.

Nevertheless, some community members appreciate the butcheries endeavour to keep supplying bushmeat to compensate for its non-availability from the community-based ecotourism project. The CECT community strongly feels that the hunting ban should have targeted only those species that were dwindling in numbers and that the quota system could have continued with those species with high numbers, especially elephants and buffaloes that keep raiding their crops. As highlighted by one community member,

> "We are now stuck with a situation where some animals like elephants are in such great numbers that they are now destroying our livelihoods, and we gain nothing from them".

The hunting ban has prevented sustained growth in the community-based ecotourism industry. The blanket ban of bushmeat is seen by some as a demonstration that community-based ecotourism failed to attain conservation objectives.

Challenged wildlife baseline data

In reviewing secondary data sources, indications are that while certain studies (e.g. Kalikawe, 2001; Chase 2011, 2016; Stone & Nyaupane, 2018) suggest that Botswana's biodiversity is threatened by the growing human population, the overexploitation of wild animals, fuel wood extraction, expanding agriculture, drought, and uncontrollable fires, there is no concrete baseline data to support these claims. The blanket ban on hunting tourism is also contested as it was mainly informed by Chase's (2011) study, which was also reliant on poor wildlife inventory techniques, as strongly suggested by Mbaiwa' s (2018) study. It is interesting to note that even the Government of Botswana (2020), responsible for the hunting ban in 2014, now indicates that its decision to embark on a hunting moratorium was not based on any scientific evidence. Community members also feel the ban was uncalled for, as one respondent questioned the ban's rationale as follows:

> "It was not guided by considerations of which species were in decline and which ones were increasing and a nuisance to us, it is not clear what guided it".

Introducing community-based trophy hunting ecotourism without having wildlife baseline data can work against conservation efforts. The lack of a comprehensive wildlife baseline data system that can inform wildlife stakeholders on the status quo of wildlife resources creates a knowledge gap that affects both wildlife managers and community-based ecotourism projects that need such information to inform decision-making. Mbaiwa (2018) further argues that other stakeholders with interest in bushmeat, such as academics, did not buy into the government's decision to totally halt bushmeat production as the whole process was ill-informed by methodologically flawed snapshot research thereby calling for knowledge on long-term wildlife trends or time series data on wildlife populations in Botswana. Empirical data is required on wildlife population, abundance and distribution before any management decisions can be made.

Discussions

Based on this paper's results, indications are that the government's decision to decide on wildlife quotas allocations to community projects, the blanket ban on hunting tourism and the failure to thwart unintended consequences of bushmeat production contribute to the view that local community institutions like CECT appear to be weak in participating in natural resources management. By not involving communities who are the immediate custodians of resources, communities become passive participants thereby undermining their role as key stakeholders. Despite the assumption that CBNRM devolves power over natural resources to communities, the case of Botswana indicates that government through its agencies such as DWNP retain decision-making powers on issues such as hunting quotas determination and allocation (Mogomotsi, Stone, Mogomotsi, & Dube, 2020: Stone, 2015). CBNRM advanced bottom-up approaches to development, its proponents believe it addresses the failures of top down, exclusionary state-controlled policies typically linked to protected areas, by ensuring conservation offers benefits to the local people to solicit their support (Khumalo & Yung, 2015). However, in Botswana the experience on the ground denies communities access to their property rights, defeating the principles of ecotourism development. As attested by Minin et al., (2016), allowing local stakeholders, such as private landowners and communities to retain property rights over these species is a necessary precursor for them to justify offsetting direct and opportunity costs of conservation. Efforts should be geared towards ecotourism activities that promote environmental, socio-cultural and economic sustainability. In line with ecotourism principles, a kind of mutualistic and/or symbiotic relationship reinforcing relationships between environmental conservation, local economic livelihoods, and cultural preservation should be fostered.

In view of the demand - supply dynamics of bushmeat production and consumption at CECT, on the surface the scale, vigour and penetration of the bushmeat trade might be viewed in a highly positive light, as one of the great success stories of autonomous food production and a testimony to the resilience and self-sufficiency of a community-based ecotourism project. However, on the other hand, unintentionally, it created problems such as poaching to meet the demand for bushmeat consumption linked to the community project. Therefore, while community-based ecotourism at CECT can be hailed for making bushmeat available to poor households to complement food deficiency, it can also be blamed for creating a conducive environment that makes it difficult to control illegal bushmeat prevalence. Although CECT was formed to curb illegal hunting, improve community livelihoods and promote rural development (Stone & Nyaupane, 2018), one can deduce from results that through bushmeat transactions the CECT has created a market for bushmeat and consequently the poaching of wildlife to satisfy demand. This is supported by evidence from the police where "poaching" incidences have continued despite the hunting ban. Furthermore, bushmeat transactions within CECT and the preservation methods used to keep the meat for a longer time period have made it difficult for the police to distinguish between legally and illegally sourced bushmeat.

The consumption of bushmeat by low density populations like CECT may be deemed sustainable; however, the demand caused by increasing numbers of urban consumers like those in Kasane town might have created a substantial market, possibly contributing to unsustainable levels of illegal hunting to satisfy the demand. In line with Mangel et al. (1996) and Wilkie & Carpenter (1999) studies that argue that large bodied animals with low reproductive rates are most susceptible to over-exploitation compared with small species that apparently can tolerate relatively intensive hunting, this study's results show that the illegal hunting of large species like elephant and buffaloes are not easy to monitor as there is no baseline data. The management of natural resources is dependent on the availability of resources baseline data. As long as urban populations continue to grow and economies grow, and as long as there is lack of research that can quantify and establish the link between urban dwellers and bushmeat, informed and appropriate action is unlikely to be undertaken to change the bushmeat demand-supply landscape. Nevertheless, it is important to note that, even when human population densities are low as is the case at CECT, illegal hunters can eliminate large, slow reproducing species (Fitzgibbon et al. 1995; Minin, Leader-Williams & Bradshaw, 2016). Consequently, sustainable communities such as CECT will be denied participation in trophy hunting tourism, a form of tourism that has the potential to change their livelihoods and mitigate them from perennial human-wildlife conflicts.

A blanket ban on hunting tourism consequently means that communities such as CECT will be left with nothing else but costs of living with wildlife resources as wild animals continue to forage on their crops and prey on their livestock. This outcome therefore has the potential to polarise community livelihoods from conservation leading to crisis management interventions that may not necessarily resonate well with the community's welfare. Research indicates that despite the declines in wildlife species that resulted in the blanket hunting ban, some wildlife species are increasing in numbers at an alarming rate (Chase, 2011; Department of Wildlife and National Parks, 2013). However, this was not considered in the application of the blanket ban on hunting tourism. In view of this paper's results we propose different possible scenarios that are likely to occur between biodiversity conservation and community livelihood development mediated by hunting tourism and bushmeat consumption as depicted by Figure 2. The conservation and development possible outcomes that are likely to occur are positive-positive, positive-negative, negative-positive and negative-negative, with the positive-positive outcome being the ideal while the negative-negative is the worst possible outcome.

The model conceptualizes community-based ecotourism (hunting tourism) and livelihoods sustainability, that it can perhaps best be understood by considering different outcome probabilities in view of reconciling the demands of conservation and livelihoods. Therefore, the acceptance of tourism by conservation spaces requires adaptation to new situations and challenges that come with this conservation and development evolution nexus. Hence, the proposed model is an extension on the debate about how to conceptualize hunting tourism and community livelihood relationships. A number of such conceptualisations of understanding under the banner of 'conservation and development nexus' (e.g. Budowski, 1976; Carter, 1994; Salafsky & Wollenberg, 2000; Nyaupane & Poudel, 2011) attempt to provide guidance about 'managing' physical and social relationships in response to growing environmental and development complexities (Stone & Nyaupane, 2017), but the debate is inconclusive as finding a remedy is still elusive. In this model, one ideal way of achieving conservation and community development sustainability (+ + scenario) is through game farming which can be achieved through cluster farms (by clustering neighbouring community owned farms to make one big game farm). This can lead to the continuation of the provision of bushmeat and reduce the reliance on government.

While community-based ecotourism through trophy hunting has the potential to assist communities and/or countries to achieve biodiversity conservation goals, Minin et al., (2016) cautions that the contribution of trophy hunting tourism to conservation is often contentious for various reasons, such as the uncertainty over the sustainability of offtake rates. These concerns arise because quotas and offtakes are not often based on scientific assessments, as is the case at

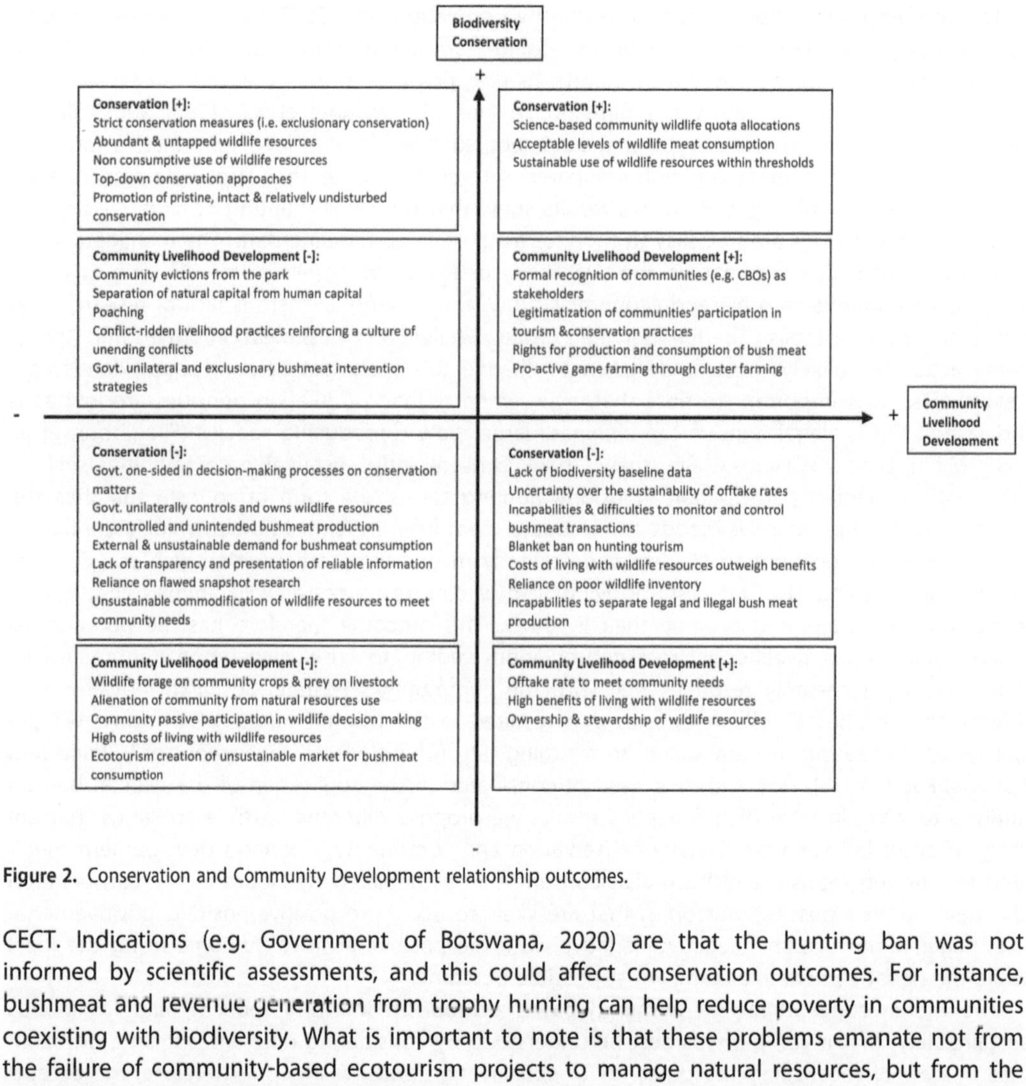

Figure 2. Conservation and Community Development relationship outcomes.

CECT. Indications (e.g. Government of Botswana, 2020) are that the hunting ban was not informed by scientific assessments, and this could affect conservation outcomes. For instance, bushmeat and revenue generation from trophy hunting can help reduce poverty in communities coexisting with biodiversity. What is important to note is that these problems emanate not from the failure of community-based ecotourism projects to manage natural resources, but from the structural implementation being deficient of scientific information as well as side lining certain stakeholders in decision making that in turn affects the whole ecotourism and bushmeat landscape dynamics. All these deficiencies led to difficulties in putting up measures to control and manage legal bushmeat obtained from community-based ecotourism against bushmeat obtained through poaching. It is however important to note that bushmeat production is a complex phenomenon that links food security, development and conservation; therefore, it demands multifaceted interventions (Rogan et al. 2017). Therefore, conservation and tourism development must adhere to the ideals of sustainability to achieve biodiversity conservation, improved rural livelihoods and sustainable economic development in developing countries.

Research indicates that problems experienced by many countries in southern Africa result from the major role played by governments in decision-making regarding the utilisation of natural resources (Ngwira, Kolawole & Mbaiwa, 2013) as is the case with Botswana where communities have limited control on their resources and how they are governed. The hunting ban and the subsequent trophy hunting prohibition without consulting the community as affected stakeholders is a case in point, indicating the limited rights given to communities through management agreements and administrative decrees and not through legislation and rights-based approaches (Blackie, 2019). Similar arrangements exist in countries such as Namibia where ownership of wildlife belongs to the state (Sullivan, 2002). However, despite the challenges, CBNRM in Namibia has facilitated the distribution of game meat to rural communities, increases in wildlife numbers leading to more game available for trophy hunting and community harvesting and the promotion of other income generating opportunities for rural communities such as the access to veld products, production of essential oils from indigenous plants and the sale of crafts (Nuulimba & Taylor, 2015). Like Blackie (2019), results from this study indicate that by halting the provision of bushmeat, local communities became more vulnerable because they were not provided with alternative activities to address the impacts of the government's action on jobs and revenue generation made by CECT. In line with the SET, this study determined that by ensuring local communities derive benefits from the utilisation of wildlife resources such as bushmeat, employment and revenue, they have positive reaction, however, once costs are realised (such as the halting of the distribution of bushmeat, loss in food security, reduced revenue generation and lost employment opportunities), support may be be lost hence the communities' involvement in illegal activities to gain access to resources they are being denied.

Conclusion

Results from this study indicate that bushmeat production through community-based ecotourism projects would prosper more if practiced in a way that facilitates harmony between community livelihoods and conservation, through adaptive management practices. The development of policy devices should also endeavour to further harmonize community development and conservation, by promoting all stakeholders' participation to instil conservation community ownership of community-based ecotourism projects and trust. The ban on trophy hunting and its curtailing of bushmeat distribution to communities was premised on evidence of declines in wildlife species (Chase, 2011; Department of Wildlife and National Parks, 2012), however, wildlife species such as elephants, buffalos, and zebras were increasing exponentially at the time and beyond the hunting (Department of Wildlife and National Parks, 2012). The decision therefore had an impact on community livelihoods and caused resentments amongst the community members. Policy development requires the protection of not only the biodiversity that provides sustenance for community development, but also the management and use of this biodiversity to meet the basic needs of local people. To achieve this, and through the application of science, there is a need for constant appraisal of the impacts of existing systems and the designing of new systems with stakeholders, experiments from which we can more effectively quantify performance and learn ways to manage such complex systems. Debates on conservation and development have been around for some time and there is no silver bullet to solve or reconcile the disparities (i.e. trade-offs) that come as a result of the interactions of the two. Continuous realignment through adaptive conservation and development management styles is fundamental. For instance, the Department of Wildlife and National Parks Act of 1992 (Cap38:01) section 45 gives the minister powers to ban trophy hunting for periods not more 12 months at a time; the ban can be extended as evidenced by the four extensions in 2015, 2016, 2017 and 2018. This is often done without consultations with host communities who bear the brunt of such decisions. For success to occur all stakeholders need to participate in the design of measures that promote win-win outcomes.

The adoption of community-based ecotourism as a form of adaptive management is consistent with trying to find a balance between the use of natural resources and community development. As this case study has revealed, it is not that ecotourism is failing to bring better results and solve problems at hand, the fundamental problems are lack of baseline data and knowledge on long-term wildlife trends or time series data on wildlife populations, methodologically flawed snapshot research and the failure by community-based ecotourism projects to keep data that can be quantified to measure the impacts of bushmeat production on community nutrition and/

or food security. Without these data bases and/or knowledge, conservation and community development will continue to grapple for sustenance (Costanza et al., 2017), that one key limiting factor in sustaining natural capital is lack of knowledge on how ecosystems function and how they support human wellbeing.

As results from this study indicate, the hunting ban was implemented even though the data that informed it was contested by some stakeholders. There was no effort taken to dispel contested stakeholders' perceptions, but rather a unilateral decision persisted over and above other stakeholders' discontent. Furthermore, the blanket ban impacted the community negatively since it included even wildlife species that had increased and continue to, leading to increased human-wildlife conflicts. A truly transparent, neutrally engaging, informed, representative, cordial, and impartial decision on hunting ban must serve all stakeholders in the conservation and development nexus discourse and must take into the lived experiences of all affected stakeholders. Such ill-informed decisions can translate to stakeholder conflicts exacerbated by scarcity and restrictions on bushmeat consumption, heightening the costs of living with wildlife resources and leading to increased poaching incidences and the non-achievement of conservation goals. This paper therefore concludes by advocating for science that can inform all stakeholders. Science makes trade-offs explicit, facilitates management and planning discourse and enables stakeholders to make sound value judgments (Costanza et al. 2017). Ill-informed decisions reliant on faulty information may lead to a conservation and development crisis management. The paper contributes to practice through its call for the development of adaptive management approaches that will improve bushmeat production by allowing locals to develop game farms that can supply them with bushmeat. The development of a framework to be used is an initiative that can influence policy and take situations on the ground into account. Furthermore, all stakeholders must participate in decision-making processes about natural resources conservation to ensure all gain in the process. As advanced by SET, it is important to ensure that communities get economic, social and ecological benefits for them to fully support wildlife conservation.

Funding

University of Botswana's Office of Research and Development;

References

Bagherian, R., Samah, B., Samah, A., & Ahmad, S. (2009). Factors influencing local people'participation in watershed management programs in Iran. *American-Eurasian Journal of Agricultural & Environmental Science*, *6*(5), 532–538.

Baker, J. (1997). Development of a model system for touristic hunting revenue collection and allocation. *Tourism Management*, *18*(5), 273–286. https://doi.org/10.1016/S0261-5177(97)00016-2

Barnett, R. (2000). *Food for Thought: The Utilisation of Wild Meat in Eastern and Southern Africa. TRAFFIC East/ Southern Africa.*,

Blackie, I. (2019). The impact of wildlife hunting prohibition on the rural livelihoods of local communities in Ngamiland and Chobe District Areas, Botswana. *Cogent Social Sciences*, *5*(1), 1558716. https://doi.org/10.1080/23311886.2018.1558716

Botswana (2010). *Community-based natural resources management in Botswana practitioners manual*. World Wide Fund.

Botswana (2011). *Botswana population & housing census. Government Printers.*,

Bhuiyan, M., Siwar, C., & Ismail, S. (2016). Sustainability Measurement for Ecotourism Destination in Malaysia: A Study on Lake Kenyir. *Social Indicators Research*, *128*(3), 1029–1045. https://doi.org/10.1007/s11205-015-1068-5

Brown, D., & Williams, A. (2003). The case for bushmeat as a component of development policy: issues and challenges. *The. International Forestry Review*, *5*(2), 148–155. https://doi.org/10.1505/IFOR.5.2.148.17414

Budowski, G. (1976). Tourism and environmental conservation: Conflict, coexistence, or symbiosis? *Environmental Conservation*, *3*(1), 27–31. https://doi.org/10.1017/S0376892900017707

Chobe Enclave Conservation Trust (CECT) (2014). *Income Generated from Wildlife Hunting: 2010 – 2014*. Unpublished Annual Report

Costanza, R., Groot, R., Braat, L., Kubiszewski, I., Fioramonti, L., Sutton, P., Farber, S., & Grasso, M. (2017). Twenty years of ecosystem services: How far have we come and how far do we still need to go? *Ecosystem Services*, (*28*), 1–16. https://doi.org/10.1016/j.ecoser.2017.09.008

Crowe, S., Cresswell, K., Robertson, A., Huby, G., Avery, A., & Sheikh, A. (2011). The case study approach. *BMC Medical Research Methodology*, *11*(1), 100 https://doi.org/10.1186/1471-2288-11-100

Das, M., & Chatterjee, B. (2015). Ecotourism: A panacea or a predicament? *Tourism Management Perspectives*, *14*, 3–16. https://doi.org/10.1016/j.tmp.2015.01.002

Duffy, R. (2013). *A trip too far: "ecotourism, politics and exploitation*. Routledge.

Department of Wildlife and National Parks. (DWNP (2020).). *CECT revenue generation 2010-2018, Department of Wildlife and National Parks.*,

Department of Wildlife and National Parks (2014). *Part III: Game Animals. Chobe Enclave Conservation Trust Wildlife Hunting Quota: 2005 – 2013 Report, Unpublished Report.* Botswana.

Campbell, K., Nelson, V., & Loibooki, M. (2001). Sustainable use of wildland resources: Ecological, economic and social interactions. An analysis of illegal hunting of wildlife in the Serengeti National Park, Tanzania

Carter, E. (1994). Introduction. In E. Cater & G. Lowman (Eds.), *Ecotourism: A sustainable option?*. (pp. 3–16). Wiley.

Chase, M. (2011). Dry season fixed-wing aerial survey of elephants and wildlife in northern Botswana. *Elephants Without Borders report.*

Chase, M. J., Schlossberg, S., Griffin, C. R., Bouché, P. J. C., Djene, S. W., Elkan, P. W., Ferreira, S., Grossman, F., Kohi, E. M., Landen, K., Omondi, P., Peltier, A., Selier, S. A. J., & Sutcliffe, R. (2016). Continent-wide survey reveals massive decline in African savannah elephants. PeerJ., *4*, e2354 https://doi.org/10.7717/peerj.2354

Chase, M., Schlossberg, R. Sutcliffe., & Seonyatseng, E. (2018). *Dry season aerial survey of elephants and wildlife in northern Botswana.*, Elephants Without Borders and the Department of Wildlife and National Parks. ().

Chevallier, R., & Harvey, R. (2016). Is community-based natural resource management in Botswana viable? *South African Institute of International Afairs (SAIIA) Policy Insights*, *31*, 1–12.

Chobe District Council. (2003). *Chobe District Development Plan 6: 2003-2009*. Government Printer.

Chobe District Council (2010). *Chobe District Development Plan 7: 2010-2016*. Government Printer.

Department of Wildlife and National Parks (2013)., *Aerial census of animals in northern Botswana: dry season 2013*. Department of Wildlife and National Parks.

Department of Wildlife and National Parks (2012). *Aerial census of animals in Botswana: 2012 dry season*. Government Printer.

Fa, J., Yuste, J., & Castelo, R. (2000). Bushmeat markets on Bioko Island as a measure of hunting pressure; Mercados de Carne Silvestre en la Isla Bioko como Medida de la Presión por Caza. Conservation *Biology.*, *14*(6), 1602–1613. https://doi.org/10.1111/j.1523-1739.2000.99067.x

Festus, I., & Omoboye, I. (2014). Bushmeat hunting: a challenge to sustainable ecotourism. *Research on Humanities and Social Sciences*, *4*(4), 13–20.

Finch, N. (2004). The hunting and conservation debate. In G. Dryden & S. Craig-Smith (Eds.), *Safari hunting of Australian exotic wild game (pp. 19–38)*. Rural Industries Research and Development Corporation Report.

Fitzgibbon, C., Mogaka, H., & Fanshawe, J. (1995). Subsistence hunting in Arabuko-Sokoke forest, Kenya and its effects on mammal populations. *Conservation Biology*, *9*(5), 1116–1126. https://doi.org/10.1046/j.1523-1739.1995.9051116.x

Fletcher, R. (2009). Ecotourism discourse: Challenging the stakeholders theory. *Journal of Ecotourism*, *8*(3), 269–285. https://doi.org/10.1080/14724040902767245

Garekae, H., Lepetu, J., & Thakadu, O. (2020). Forest resource utilisation and rural livelihoods: insights from Chobe enclave. South *African Geographical Journal*, *102*(1), 22–40. https://doi.org/10.1080/03736245.2019.1606730

Garekae, H., Thakadu, O., & Lepetu, J. (2017). Socio-economic factors influencing household forest dependency in Chobe enclave. *Botswana. Ecological Processes*, *6*(1), 1–10.

Gibson, C. (1999). *Politicians and Poachers: the Political Economy of Wildlife Policy in Africa*. Cambridge University Press.

Government of Botswana (2020). *Botswana Controlled Hunting Programme: Science Based Conservation Strategy to Benefit and Empower Communities*. Ministry of Environment, Natural Resources Conservation and Tourism.

Government of Botswana (2007). *Community Based Natural Resources Management Policy, Gaborone*. Government Printer.

Harris, R., Cooney, R., Leader, -., & Williams, N. (2013). Application of the anthropogenic Allee effect model to trophy hunting as a conservation tool. *Conservation Biology*, *27*(5), 945–951. https://doi.org/10.1111/cobi.12115

Hayward, M. (2009). Bushmeat hunting in Dwesa and Cwebe Nature Reserves, Eastern Cape, South Africa. South *African Journal of Wildlife Research*, *39*(1), 70–84. https://doi.org/10.3957/056.039.0108

Herath, G. (1997). Ecotourism development in Australia. *Annals of Tourism Research*, *24*(2), 442–446. https://doi.org/10.1016/S0160-7383(97)80012-6

Hofer, H., Campbell, K., East, M., & Huish, S. (2008). Modelling the spatial distribution of the economic costs and benefits of illegal game meat hunting in the Serengeti. *Natural Resource Modeling*, *13*(1), 151–177. https://doi.org/10.1111/j.1939-7445.2000.tb00031.x

Hutton, J., & Leader-Williams, N. (2003). Sustainable use and incentive-driven conservation:Realigning human and conservation interests. Oryx, *37*(2), 215–226. https://doi.org/10.1017/S0030605303000395

Jack, D. (1999). Extinction versus sustainable development challenges. *Nest Forum*, *7* (1), 43–44.

Kaae, S., & Traulsen, J. (2015). Qualitative Methods in Pharmacy Practice Research. in Babar ZUD. (ed.) *Pharmacy Practice Research Methods*. Springer.

Kachikau Police Station Registries (2000). *A compilation of cases registries from 2000 – 2016*. Kachikau Police.

Khumalo, K., & Yung, L. (2015). Women, Human-Wildlife Conflict, and CBNRM: Hidden Impacts and Vulnerabilities in Kwandu Conservancy, Namibia. *Conservation and Society*, *13* (3), 232–243. https://doi.org/10.4103/0972-4923.170395

Kiss, A. (2004). Is community-based ecotourism a good use of biodiversity conservation funds? *Trends in Ecology & Evolution*, *19*(5), 232–237.

Kim, K., & Park, D. (2017). Relationships among perceived value, satisfaction, and loyalty: Community-based ecotourism in. *Korea. Journal of Travel & Tourism Marketing*, *34*(2), 171–191.

Lee, T. J. & Kim, J. (2017). The impact of Korea's nature-based recreation settings on tourists' emotions, satisfaction and subjective happiness in Chen, J.S & Prebensen, N.K. (eds). Nature Tourism, (pp. 46–57), Routledge.

Liberg, O., Chapron, G., Wabakken, P., Pedersen, H., Hobbs, N., & Sand, H. (2012). Shoot, shovel and shut up: cryptic poaching slows restoration of a large carnivore in Europe. *Proceedings of the Royal Society B: Biological Sciences*, *279*(1730), 910–915. https://doi.org/10.1098/rspb.2011.1275

Lieblich, A., Tuval-Mashiach, R., & Zilber, T. (1998). *Narrative research: Reading,analysis and interpretation*. Sage.

Lindberg, K., & McKercher, B. (1997). Ecotourism: A critical overview. *Pacific Tourism Review*, *1*, 65–79.

Lindsey, P., Romanach, S., Matema, S., Matema, C., Mupamhadzi, I., & Muvengwi, J. (2011). Dynamics and underlying causes of illegal bushmeat trade in Zimbabwe. Oryx, *45*(1), 84–95. https://doi.org/10.1017/S0030605310001274

Loibooki, M., Hofer, H., Campbell, K., & East, M. (2002). Bushmeat hunting by communities adjacent to the Serengeti National Park, Tanzania: the importance of livestock ownership and alternative sources of protein and income. *Environmental Conservation*, *29*(3), 391–398. https://doi.org/10.1017/S0376892902000279

Loveridge, A., Reynolds, J., & Milner-Gulland, E. (2006). Does sport hunting benefit conservation. In D. Macdonald & K. Service (Eds.), *Key topics in conservation biology*. Blackwell Publishing.

Marealle, W., Fossøy, F., Holmern, T., Stokke, B., & Røskaft, E. (2010). Does illegal hunting skew Serengeti wildlife sex ratios? Wildlife *Biology*, *16*(4), 419–429. https://doi.org/10.2981/10-035

Mbaiwa, J. (2018). Effects of the safari hunting tourism ban on rural livelihoods and wildlife conservation in Northern Botswana. *South African Geographical Journal*, *100*(1), 41–61. https://doi.org/10.1080/03736245.2017.1299639

Mbaiwa, J., & Tshamekang, T. (2012). Developing a viable community-based tourism project in Botswana: The case of the Chobe Enclave Conservation Trust. *World Sustainable Development Outlook*, *2012*, 519–536.

Mbaiwa, J. (2011). The effects of tourism development on the sustainable utilisation of natural resources in the Okavango Delta, Botswana. *Current Issues in Tourism*, *14* (3), 251–273. https://doi.org/10.1080/13683500.2011.555525

Mfunda, I., & Røskaft, E. (2016). Bushmeat hunting in Serengeti, Tanzania: An important economic activity to local people. *International Journal of Biodiversity and Conservation*, *2*(9), 263–272.

Miles, M., & Huberman, A. (1994). *Qualitative data analysis*. Sage Publications.

Minin, E., Leader-Williams, N., & Bradshaw, C. (2016). Banning Trophy Hunting Will Exacerbate Biodiversity Loss. *Trends in Ecology & Evolution*, *31* (2), 99–102.

Mogomotsi, P., Stone, L., Mogomotsi, G., & Dube, N. (2020). Factors influencing community participation in wildlife conservation. *Human Dimensions of Wildlife*, *25*(4), 372–386. https://doi.org/10.1080/10871209.2020.1748769

Muposhi, V., Gandiwa, E., Bartels, P., & Makuza, S. (2016). Trophy hunting, conservation, and rural development in Zimbabwe: issues, options, and implications. *International Journal of Biodiversity*, *2016*, 1–16. https://doi.org/10.1155/2016/8763980

Nasi, R., Brown, D., Wilkie, D., Bennett, E., Tutin, C., van Tol, G., & Christophersen, T. (2008). Conservation and Use of Wildlife-based Resources: The Bushmeat Crisis. Technical Series No.33. *Secretariat of the Convention on Biological Diversity, Montreal, and Center for International Forestry Research (CIFOR).*, Indonesia.

Ngwira, P., Kolawole, O., & Mbaiwa, J. (2013). Community based natural resource management. Tourism and Poverty Alleviation in Southern Africa: What Works and What Doesn't. *Work. Chinese Business Review*, *12*(12), 789–806.

Nielsen, M. (2006). Importance, cause and effect of bushmeat hunting in the Udzungwa Mountains, Tanzania: implications for community-based wildlife management. Biological *Conservation.*, *128*(4), 509–516. https://doi.org/10.1016/j.biocon.2005.10.017

Nunkoo, R. (2016). Towards a more comprehensive use of social exchange theory to study residents attitudes towards tourism. *Procedia Economics and Finance*, *39*, 588–596. https://doi.org/10.1016/S2212-5671(16)30303-3

Nuulimba, K., & Taylor, J. (2015). 25 years of CBNRM in Namibia: A retrospective on accomplishments, contestation and contemporary challenges. *Journal of Namibian Studies: History Politics Culture*, *18*, 89–110.

Nyaupane, G. P., & Poudel, S. (2011). Linkages among biodiversity, livelihood, and tourism. *Annals of Tourism Research, 38*(4), 1344–1366. https://doi.org/10.1016/j.annals.2011.03.006

Paraskevaidis, P., & Andriotis, K. (2017). Altruism in tourism: Social exchange theory vs altruistic surplus phenomenon in host volunteering. *Annals of Tourism Research, 62*, 26–37. https://doi.org/10.1016/j.annals.2016.11.002

Patton, M. (1990). *Qualitative Evaluation and Research Methods.* Sage.

Pinnock, D. (2011). Okavango wildlife threatened. *Mail and Guardian.* Accessed 26th August: http://www.elephantswithoutborders.org/downloadspapers/Okavango_wildlife_threatened.pdf

Polkinghorne, D. (1988). *Narrative knowing and the human sciences.* NY State.

Poulsen, J., Koerner, S., Moore, S., Medjibe, V., Blake, S., Clark, C., Akou, M., Fay, M., Meier, A., Okouyi, J., Rosin, C., & White, L. (2017). Poaching empties critical Central African wilderness of forest elephants. Curr Biol, *27*(4), R134–R135. https://doi.org/10.1016/j.cub.2017.01.023

Raven, P., & Berg, H. D. (2010). *Environment.*, Wiley.

Rogan, M., Lindsey, P., Tambling, C., Golabek, K., Chase, M., Collins, K., & McNutt, J. (2017). Illegal bushmeat hunters compete with predators and threaten wild herbivore populations in a global tourism hotspot. *Biological Conservation, 210*, 233–242. https://doi.org/10.1016/j.biocon.2017.04.020

Rogan, M., Lindsey, P., & McNutt, J. (2015). Illegal bushmeat hunting in the Okavango Delta. *Botswana: drivers, impacts and potential solutions.* Food and Agriculture Organization of the United Nations.

SADC (2015). *Law enforcement and anti-poaching strategy 2016 – 2021.* Botswana.

SADC (2014). *Development of a SADC Regional Anti-Poaching Strategy. Call for expression of interest assignment.*, Botswana.

Salafsky, N., & Wollenberg, E. (2000). Linking livelihoods and conservation: A conceptual framework and scale for assessing the integral of human needs and biodiversity. *World Development, 28*(8), 1421–1438. https://doi.org/10.1016/S0305-750X(00)00031-0

Sharpley, R. (2008). Ecotourism: a consumption perspective. *Journal of Ecotourism, 5*(1-2), 7–22. https://doi.org/10.1080/14724040608668444

Stone, M., & Nyaupane, G. (2017). Ecotourism influence on community needs and the functions of protected areas: a systems thinking approach. *Journal of Ecotourism, 16*(3), 222–246. https://doi.org/10.1080/14724049.2016.1221959

Stone, M., & Nyaupane, G. (2018). Protected areas, wildlife-based community tourism and community livelihoods dynamics: spiraling up and down of community capitals. *Journal of Sustainable Tourism, 26*(2), 307–324. https://doi.org/10.1080/09669582.2017.1349774

Stone, M. (2015a). Community-based ecotourism: A collaborative partnerships perspective. *Journal of Ecotourism, 14*(2-3), 166–184. https://doi.org/10.1080/14724049.2015.1023309

Stone, M., (2015b). Community empowerment through community-based tourism: the case of Chobe Enclave Conservation Trust in Botswana. In. Van der Duim, R., Lamers, M., & Van Wijk, J. *Institutional arrangements for conservation, development and tourism in Eastern and Southern Africa.* (pp. 81–100). Springer.

Stone, M. (2013). *Protected Areas, Tourism and Rural Community Livelihoods in Botswana* [Unpublished PhD. Dissertation]. Arizona State University.

Stone, L., & Stone, M. (2011). Community-based tourism enterprises: Challenges and prospects for community participation; Khama Rhino Sanctuary Trust. Journal *of Sustainable Tourism, 19*(1), 97–114. https://doi.org/10.1080/09669582.2010.508527

Stronza, A. (2008). Hosts and hosts: The anthropology of community-based ecotourism in the Peruvian Amazon. *NAPA Bulletin, 23*(1), 170–190. https://doi.org/10.1525/napa.2005.23.1.170

Sullivan, S. (2002). How Sustainable is the Communalizing Discourse of 'New Conservation'? The Masking of Difference, Inequality and Aspiration in the Fledgling 'Conservancies' of Namibia. in: Dawn Chatty and Marcus Colchester, (eds.), *Conservation and Mobile Indigenous Peoples: Displacement, Forced Settlement and Sustainable Development.*, and Oxford, Berghahn. 158–187.

Tall, S., & Gueye, M. (2004). *Local agreement in the Sahel: An Instrument for the co-governance of natural resources.*, International Institute for Environment and Development.

The International Ecotourism Society (TIES) (2006). Community Based Ecotourism: Best practice stories and resources", Digital traveler and Asia Pasific eNewsletter.

United Nations World Tourism Organization (2018). *Indicators of sustainable development for tourism destinations: A guidebook.* UNWTO.

UNWTO (2015). *UNWTO tourism highlights 2015 edition.* United Nations World Tourism Organization.

Villiers, J. (2011). Strategic alliances between communities with special reference to the twinning of South African provinces, cities and towns with international partners. Unpulished Dissertation, Business Management and Administration, University of Stellenbosch.

Weaver, D., & Lawton, L. (2007). Progress in tourism management twenty years on: The state of contemporary ecotourism research. *Tourism Management, 28*(5), 1168–1179. https://doi.org/10.1016/j.tourman.2007.03.004

Wilfred, P., & MacColl, A. (2010). Income sources and their relation to wildlife poaching in Ugalla ecosystem. *Western Tanzania. Afric. J. Environ. Sci. Technol, 4*, 886–896.

Wilkie, D. S., Wieland, M., Boulet, H., Le Bel, S., van Vliet, N., Cornelis, D., BriacWarnon, V., Nasi, R., & Fa, J. E. (2016). Eating and conserving bushmeat in Africa. *African Journal of Ecology*, *54*(4), 402–414. https://doi.org/10.1111/aje.12392

Wilkie, D., Redford, K., & McShane, T. (2010). Taking of rights for natural resource conservation: a discussion about compensation. Journal *of Sustainable Forestry.*, *29*, 1–17.

Wilkie, D., & Carpenter, J. (1999). Bushmeat hunting in the Congo Basin: an assessment of impacts and options for mitigation. *Biodiversity and Conservation*, *8*(7), 927–955. https://doi.org/10.1023/A:1008877309871

Zuze, C. (2009). *Community-based natural resources management in Botswana: Practitioners'manual*. Department of Wildlife and Natural Parks.

Tourism development discourse dynamics in a context of conflicts between mining and nature conservation in the Brazilian Cerrado Hotspot

Nayara Marques, Mozart Fazito and André Cunha

ABSTRACT

The Cerrado is a world hotspot for nature conservation and an important tourist destination in Brazil. This region has been subjected to intensive land conversion for the production of commodities, which has led to the eruption of socio-environmental conflicts between tourism and extractive industries. In this article, we explore the dynamics of tourism development discourses that underpin the conflicts generated by mining projects in the town of Cavalcante and seek to inform on more adequate policymaking and promote human development. We employed multi-methods data collection techniques and a discourse analysis approach to assess the tourism development discourses that compete in local policy arenas and reflect upon policy outcomes that result from this process. We develop a three-stage chronology to show the dynamics of tourism discourses and reveal that sustainable tourism discourse served to generate a neoliberal modernization discourse. However, growing dissatisfaction with both tourism modernization and mining has generated resistance against mining among the affected communities. We conclude that leisure tourism and tourism modernization should be separated in public planning processes in contexts where socio-environmental conflicts are present, and that governmental and institutional mediation require a certain level of trust to ensure more balanced human development in such situations.

1. Introduction

The Anthropocene is marked by a biodiversity crisis (Hoffman, 2010) that strongly affected the Cerrado biome, in Brazil – a world hotspot for biodiversity conservation (Myers et al., 2000). Since the 1970s, this biome has been subjected to intensive land conversion, from natural vegetation to clear fields for commodities production. Over 60% of Brazilian soybean production is grown in the Cerrado, an area that is also home to large mineral deposits, such as niobium, iron, manganese and gold, which are globally valued Brazilian products (Klink & Machado, 2005; Reis, 2017).

The Cerrado is also a nature-based tourism destination due to its important protected natural sites. The area is home to a large diversity of local or traditional populations, with more than 300 ethnic groups (Porto-Gonçalvez, 2019) and endemic species on the brink of extinction (Strassburg et al., 2017). In this scenario, the production of global commodities, such as mining, agriculture, and cattle raising compete with nature-based tourism and nature conservation. This impacts biodiversity and local communities, thus leading to conflicts both in and surrounding the protected areas. In Brazil, more than 40% of these conflicts occur in the Cerrado biome (Porto-Gonçalvez, 2019).

Recent literature shows that mining is a prominent cause of conflicts (see Leonard & Lebogang, 2018; Similä & Jokinen, 2018), which are not only related to mining *vs.* tourism, but also to disputes within the tourism industry itself (Dwyer et al., 2016). An in-depth inspection of these conflicts can provide useful information to produce more adequate contextualized development policies.

To respond to this agenda, the objective of this article is to explore the tourism development discourse dynamics that underpin the socio-environmental conflicts that have arisen from the recent mining projects implemented in the surroundings of the Chapada dos Veadeiros National Park (CVNP). This park is located in one of the most important remnants of the Cerrado hotspot, found in the central region of Brazil.

We employed a qualitative approach based on discourse analysis, as it has previously been effective at unveiling details of policy-making processes in tourism development contexts (Bramwell, 2006; Markwick, 2000; Qian et al., 2017). Data was collected through semi-structured in-depth interviews with key actors, observation and document analysis, from June 2017 to August 2018. As a result, we present the dynamics of four storylines that interact in the local policy arenas: alternative tourism, sustainable tourism, tourism modernization and community approach. Attention is paid particularly to manipulation of the sustainable tourism discourse to legitimize private economic interests and empower the neoliberal modernization discourse while avoiding transformations in the *status quo*. The next sections explore the concept of socio-environmental conflicts adopted for this research and the development theories that depart from neoliberal modernization.

2. Socio-environmental conflicts in mining and tourism contexts

Research on conflicts arising from tourism contexts has attracted the attention of academics in recent years. Previous research has examined the importance of hearing indigenous communities to establish successful tourism products (Fletcher et al., 2016), how to achieve sustainable goals in mining heritage tourism (D. Cole, 2004), and how tourism enterprises can benefit from a greater concern about human rights in terms of reputation and increased income (S. Cole, 2014). Previous authors have also reflected upon the need to reinvigorate the concepts of public interest (Dredge, 2010) and sustainability (Devine & Ojeda, 2017) to incorporate indigenous and native sovereignty so that tourism does not intensify the conflicts, inequalities, violence and dispossession that derive from neoliberalism.

Moreover, previous authors have identified the ability of tourism to promote collaboration among divided communities and bring peace (Causevic & Lynch, 2013). In the context of Latin America, tourism has been classified as an industry analogous to extractivism, as it relies on state orchestrated expropriation of natural resources and creates enclosure and dispossession, which result in widespread environmental degradation and ecological insecurity for indigenous communities (Loperena, 2017). Drawing on a case study in Costa Rica, Hunt et al. (2015) argue that ecotourism has become part of the "neoliberal conservation toolkit" and does little to address the poverty and disparities it purports to address. Similarly, conflicts over heritage tourism are quite volatile and caused by the dominance of economic interests over other interests (Porter &

Salazar, 2005), which leads to missed opportunities to learn from natural and cultural sites to build more conscious, sustainable and free societies (Fazito et al., 2017).

Limited research has been conducted to understand the mining impacts on specific protected environments and tourism sites, principally in the Global South (Leonard, 2016, Leonard & Lebogang, 2018). Several mining impacts on tourism were identified in a case study in South Africa, such as water, noise and visual pollution, disruption of community values and habits. This has also been characterized as a corrupt process marked by weak community participation and lack of transparency in local development decisions (Leonard & Lebogang, 2018). Another case study in South Africa compares the quality of job creation between mining and tourism-related activities in relation to sustainability. It shows that mining offers short-term jobs, and causes strong environmental degradation, while tourism jobs are more sustainable and dependant on a protected environment. However, tourism businesses still need to deal with low-paid seasonal jobs (Leonard, 2016). The fragile institutional governance in Africa creates a lack of trust, justice and inclusiveness that prevent the development of more sustainable tourism which could address poverty and preserve the environment at the same time (Siakwah et al., 2019). However, studies in Latin America have demonstrated that the sustainable tourism discourse has a "chameleon flexibility" in local policy arenas, causing it to adopt a more economic-driven rationality when challenged by mining (Fazito et al., 2016). As a result, it moves the tourism assets from the enjoyment of nature preservation to mining-related business. Thus, in the economic dynamics, leisure tourism loses out while business tourism benefits from a boom in mining, thereby creating conflicts within the tourism industry (Dwyer et al., 2016). Authors have also demonstrated that sustainable tourism organisations have a key role in resisting "unsustainable development" impacts due to their social relations within the local communities (Hales & Larkin, 2018).

Conflicts surrounding mining projects in tourism development contexts tend to be extremely polarizing. However, depending on factors like location and type of mining and tourism, they can vary from rival, nonrival and even to complementary discourses (Similä & Jokinen, 2018). Therefore, in this article the concept of socio-environmental conflicts is understood to mean a context in which a given social group explores a particular natural resource and benefits directly from exploring it, yet is not directly affected by the negative impacts that result from this exploration. Whereas, on the other hand, the groups that are directly affected by the exploration do not benefit (Little, 2001).

Almeida et al. (2017) developed a framework for sources of conflict in tourism development contexts and identified the importance of trust in the institutional mediation of such conflicts. The authors identified a lack of tourism studies and production of empirical evidence related to conflict analysis in tourism planning. Tourism development contexts are also in need of more critical political economy analyses in order to thoroughly understand conflicts and propose planning responses to neoliberal discourses (Bianchi, 2009). New development research frontiers require analyses of the political economy and ecology, among others, to address "a desperate need to close the policy–action gap" (Saarinen et al., 2017, p. 8). The next section explores the discourses-praxis of tourism development.

3. Tourism development discourses: departing from neoliberal modernization

The current development debate starts from the post-World War II modernization theory and has become the dominant global development discourse (Cowen & Shenton, 2005). Modernization established a unique linear way in which the development of societies takes place, and is focused on the expansion of income, wealth and technological advance (Telfer & Sharpley, 2007). Based on Rostow's theory (1960), societies create preconditions to change their status from a "traditional society" to a "high consumption society," which therefore creates a dichotomy between traditional (backward, poor, underdeveloped) and modern (advanced, rich, developed). Butler (1980) developed a widely applied model that defined a linear way of tourism development based on two variables: time and tourist flow. In this model, tourism became an agent of modernization; an industry that could reach spaces that have not been modernized by traditional industrialization, and therefore have kept cultural and natural diversity preserved (Fazito, 2015).

Modernization has been criticized for producing inequalities and environmental destruction. As a result of its evolution after the 1970s, neoliberalism exacerbated the ability of modernization to produce such inequalities and environmental degradation (Castells, 2003; Escobar, 1995; Peck, 2004). The neoliberal recipe for development became known as the "Washington Consensus" (Westra, 2010). Deregulation, privatization, and withdrawal of the state from the provision of social services have been implemented everywhere in the world. Despite the criticism, neoliberalism has become a hegemonic discourse of development and best practice (Harvey, 2005).

Butler's model was criticized for lacking a political economy critique of tourism. Britton (1982) recommended alternative forms of tourism, other than mass tourism. In the 1980s, environmental concerns emerged as an important issue in global politics, inspired by earlier ideas that natural resources are not limitless and should be preserved (Meadows et al., 1972). The different levels of environmental thinking vary from a radical perspective, such as deep ecology, whereby nature is not considered to be a resource but rather having inherent value and, therefore, humans should prostrate before its sanctity (Acott et al., 1998; Ness, 1986), to more "flexible" perspectives which aim to balance economic development and environmental preservation. This idea of equilibrium underpins the concept of sustainable development, which was diffused to the whole world and became the most important environmental discourse in global politics (WCED, 1987). Equilibrium deals with "needs' and "limits," and considers available resources for present and future generations. In tourism studies and practice, it is described as "sustainable tourism" (Butler, 1999).

Many authors argue that sustainable development achieved global acceptance and political support because it is a vague and inaccurate concept (Mebratu, 1998; Reid, 1995), and ended up dominated by the dominant principle of economic growth, reinforcing the modernization paradigm (Sharpley, 2020). Sustainable tourism has also been criticized for not being operational or achievable (Butler, 1999; Sharpley, 2020) due to fostering an addiction to growth (Higgins-Desbiolles, 2018), and for focusing on future generations, thus avoiding the urgency of acting within this generation (Reid, 1995). The sustainable talk became fundamental to legitimizing conservative discourses championed by political groups that are more interested in expanding their wealth than protecting both people from social inequalities and also nature from destruction. However, it does so dangerously disguised as alternative discourse (Fazito et al., 2016).

In the 1990s, a critical discourse of development emerged to imagine an era of a more harmonic life. Known as post-development, it incorporates and appreciates the uncertain and spontaneous side of development, with a focus on human emancipation. Post-development understands development to be a contextualized process and is conscious of the destructive side of development and the violence left in the wake of neoliberalism. It proposes that current social sciences have neglected important voices, such as those of indigenous groups, the unproductive, the local, the non-European culture, and the traditional/backward. Therefore, post-development defends that an enriched social analysis would benefit from epistemologies produced from the South (Gibson-Graham, 2005; Santos, 2004). The role of tourism in such an enriched view of development is that of active leisure, meaning playful leisure with consciousness and context awareness, which is also capable of producing culture creation and autonomy, in contrast to the typical alienation of passive leisure (Marcellino, 2006). Such an enriched view of tourism contrasts with business tourism, when travel is motivated by working obligations instead of free time.

In this sense, the main role of tourism in human development would be to generate "conscientization" (Freire, 1970, 1973), foster learning, provide environmental and heritage education, produce cultural encounters, and build up peaceful relationships rather than expanding income and wealth, as dominant modernization neoliberalism defends (Fazito et al., 2017). It requires a more active role of the host communities in governing tourism development in their regions (Murphy, 2013). Specifically, in nature-based destinations, the role of tourism also can lead to an enriched view of ecotourism, which can minimize impacts, build environmental awareness, provide financial benefits for conservation and empowerment of local people, with respect to the local culture and human rights and democratic movements (Honey, 2008). Moreover, different works evidence the benefits of the interdependent relationships between biodiversity conservation, community livelihood improvements, and tourism development (Nyaupane & Poudel, 2011; Stone & Nyaupane, 2016).

The dominant discourse of neoliberal modernization and the alternative discourses of deep ecology, sustainable tourism, and post-development will be used to examine our results. The next section describes the case study.

4. The case study region

In Brazil, protected areas are classified in three categories: strict protection, sustainable use, and indigenous and *quilombola*[1] areas (Brazil, 2000). The study region is home to protected areas from all three categories, which also coexist together with farms, plantations, settlements, rivers with high potential for hydropower and some existing and dozens of potential mining sites.

In this context, a myriad of socio-environmental conflicts has already taken place in the region. These conflicts are related to conflicting interests among private companies, local communities, outsiders, public administrators at different levels, different protected area categories, and a variety of tourism stakeholders over the local resources and their uses, for example small-scale and industrial farming, hydropower, mining, and tourism (Ferreira & Almeida, 2014). However, the conflicts that emerge from mining projects in nature-based tourism contexts have more scientific relevance due to a lack of previous studies, reports and complaints of illegal activity in the region. This greater relevance is also due to rumours of new mining projects, which have deep environmental and social impacts. For instance, the mainstream media (Globo, 2020) recently showed that new spots were being deforested to make way for mining in the town of Cavalcante located in the Brazilian central-west state of Goiás.

Our study area is located within the Pouso Alto sustainable use Protected Area (APA Pouso Alto), which is situated within the UNESCO Cerrado Biosphere Reserve. The most visited site is Chapada dos Veadeiros National Park (CVNP), followed by the Kalunga territory, which is the largest quilombola area in the country, and located in the municipality of Cavalcante (see Figure 1).

Cavalcante has 9,829 inhabitants, of which 49.5% dwell in rural areas and 50.5% in urban areas. Communities formed by less than ten to more than a hundred families live in the ten districts spread throughout the rural area. Cavalcante has the lowest Human Development Index in the region (0.584) (IBGE, 2017).

Mining declined in Cavalcante after the 1980s. However, more recently, the number of mining research surveys[2] has strongly increased in the state of Goiás (Gonçalves & Mendonça, 2015) and more specifically in Cavalcante (e.g. manganese, tin, iron and gold), as a result of the evolution of prices of ore on the global market. Cavalcante is a poor and unequal municipality. Its economy is based on both the primary sector and tourism, but there is growing interest among mining companies to explore its mineral deposits. Moreover, Cavalcante has a highly informal labour market. Impacts of the region's mining history are still present on the local landscape. Images of a tailing decontamination project in the Lavapés River in the centre of Cavalcante and natural

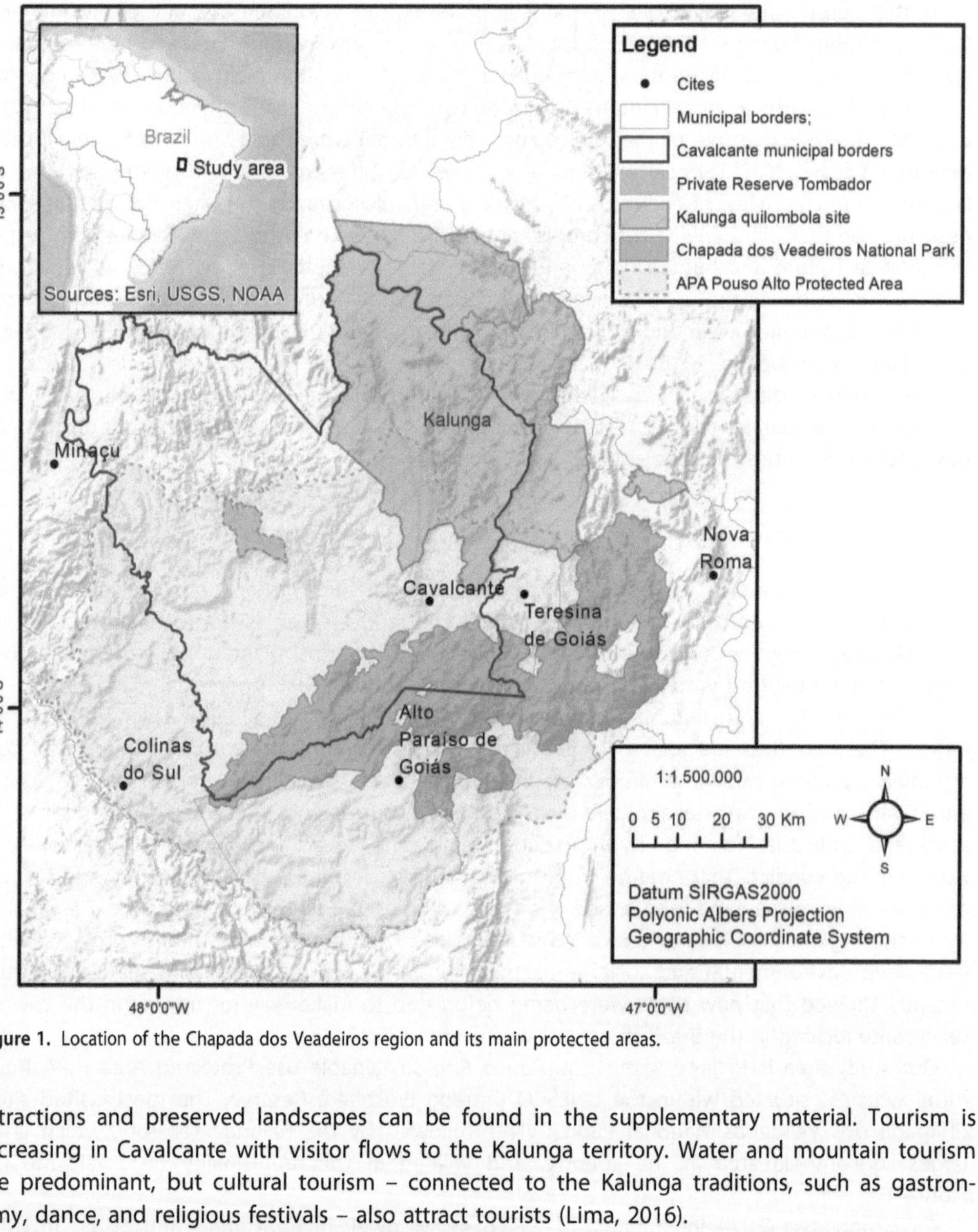

Figure 1. Location of the Chapada dos Veadeiros region and its main protected areas.

attractions and preserved landscapes can be found in the supplementary material. Tourism is increasing in Cavalcante with visitor flows to the Kalunga territory. Water and mountain tourism are predominant, but cultural tourism – connected to the Kalunga traditions, such as gastronomy, dance, and religious festivals – also attract tourists (Lima, 2016).

Since 2010, the growing interest of mining investors has aggravated socio-environmental conflicts throughout the region. We explore the dynamics of tourism development discourses in this context of conflicts and assess the policy outcomes that have resulted from this process. In the next section we describe the methodological path adopted to achieve this objective.

5. Research approach

A single case study was analysed to respond to the research problem. Case studies enable researchers to inspect a contextualized situation in detail and share the results with the academic community, thereby contributing to a greater level of understanding a given field (Flyvbjerg, 2006). A Foucauldian discourse analysis was developed to unveil the tourism development discourses that emerge from the Cavalcante case study and demonstrate its dynamics in the policy arenas. Foucauldian discourse analysis seeks a deeper understanding of a given empirical reality and its respective complexity (Wilson & Hollinshead, 2015) by incorporating the hidden power-play behind the development rhetoric and revealing details that traditional methodologies cannot (Hewitt, 2009).

According to Hewitt (2009), discourse analysis focuses on the role of language in the process of social construction of meanings and representations. Discourse analysis approaches vary in structure, from a complete lack of structure (Flyvbjerg, 1998; and Foucault's work) to some structure to guide the process (Dryzek, 2005; Hajer, 1995; Wilson & Hollinshead, 2015).

Here, we adapted the conceptual framework from Fazito et al. (2016) and the recommended steps for conducting such an analysis from Hajer (2006). These frameworks seek to reveal which rationalities legitimatize the competing discourses that influence public actions. The conceptual framework proposed by Fazito et al. (2016) aims to identify the fragments of tourism and development discourse that emerge in the process of establishing a large mining project on a tourist site, and how they gain a narrative aspect which, when clustered together, form storylines that compete in policy arenas. Storylines consist of a combination of distinct elements in the form of statements that summarize complex narratives, for example, metaphors, and are used as shortcuts to argumentation (Hajer, 1995). When exposing their perspective of reality, storylines attract stakeholders, who form coalitions of support. Thus, storylines are the cement that keep these coalitions together. Storylines are legitimized by rationalities in policy arenas. Rationalities represent knowledge and using them to legitimatize policy decisions represents power. From a Foucauldian perspective, power and knowledge are two inseparable elements (Cheong & Miller, 2000; Foucault, 1982).

Five rationalities emerged from the discourse fragments identified in documents and interviews, which could be compared to similar findings in previous research. They are: (1) *economic*, i.e. a focus on income and wealth expansion, and job creation; (2) *deep ecology*, i.e. a belief in the intrinsic value of nature; (3) *scientific*, i.e. based on scientific data, research findings, or endorsement from scientists; (4) *communicative*, i.e. based on the acceptance of a storyline by the involved actors; (5) *critical*, i.e. empirical or lay observations that promote resistance against dominant discourses (see also Fazito et al., 2016; Hajer, 1995; Healey, 1999; Rydin, 2003; Scott, 2008) . Rationalities underpin the competition among storylines in policy arenas and the storyline that attracts more support wins and generates a policy outcome. When the policy outcomes generate concrete interventions that affect people's lives, they take the form of an episode. Episodes are relevant units of analysis because they summarize a sequence of actions (Dijk, 1981), and materialize the prevalence of a winning storyline over another.

Following recommendations from Hajer (2006), data collection started in June 2017 with desk research, followed by fieldwork, which consisted of two site visits to develop a global view of the conflicts in the region, identify key stakeholders, register photographs, and carry out non-participant observation. In two additional field trips, in May and August 2018, semi-structured in-depth interviews with key stakeholders were conducted. The interviewees were selected from different areas, such as researchers, experts, environmental activists, members of the tourism and mining industries, affected communities and governmental authorities. Interviewees were sampled from the desk research, previous field visits, and snowball sampling. Documentary material was analysed in order to triangulate data, consisting of mostly legal and official documents e.g. the Chico Mendes Institute for Biodiversity Conservation process of expansion of the CVNP, meeting minutes (from Goiás State Environmental Council), management plans (e.g. Cavalcante's Tourism plan of 2011–2014), the Situational Diagnosis of the Municipality of Cavalcante (2012), and the project to establish the São Félix State Park (Funatura, 2013).

A total of forty-six interviews were conducted with thirty-nine people, some of whom were interviewed twice. Given the context of conflict, with rumours of illegal acts and corruption, the interviewees were anonymised. Although some of the interviewees represented more than one specific group (we detail this information when quoting them in the text), they were classified into six predominant categories: (1) leisure tourism industry, (2) modernization tourism industry, (3) mining industry, (4) members of Cavalcante local communities, (5) governmental actors, and (6) non-governmental organization (NGO) representatives. Mining was perceived differently by the tourism entrepreneurs, corroborating the previously discussed Dwyer et al. (2016) findings. Some of them saw an opportunity to increase the tourism flow and the city's economy, following the business needs of the mining activity, which would benefit their own business. For others, mining represented a threat to nature, and therefore to their business. Thus, we divided the tourism entrepreneurs into two groups, namely "tourist modernization" and "leisure tourism."

We proceeded to triangulate the collected data and to discuss with the development theories previously described. Based on this analysis, we identified four storylines and a timeline with the main episodes related to tourism development in the context of mining growth in Cavalcante. In the next section, we present the storylines identified with respect to the tourism development dynamics in Cavalcante.

6. Tourism development storyline dynamics

In this section, the meanings and discourse fragments of tourism development that emerged from data collection are presented in quotes. Four main storylines were identified: (1) alternative tourism; (2) sustainable tourism; (3) tourism modernization; and (4) community approach. The discourse dynamics of tourism and mining in Cavalcante are presented in a three-stage chronology: 1) the origins of tourism in Cavalcante – 1950s to 1990s; 2) the institutionalization of tourism in Cavalcante – 2000s to 2010; and 3) the return of mining – 2010 to date. For each stage, we identify the *prevalent storyline*, the *rationalities* that are employed to legitimate the prevalent storyline in the policy arenas, the local, national and global *context*, the main *episodes* that result from the policymaking process, the *discourse coalitions* among stakeholders and their role in the policymaking process, the "*challenging storyline*" that displaces the prevalent storyline, and finally the main *catalyst* that underpins the displacement of the prevalent storyline and influences the next stage. Table 1 summarizes the results, which are further detailed in the next subsections.

6.1. Origins of tourism in Cavalcante – 1950s to 1990s

The first tourists to visit Chapada dos Veadeiros dates back to the 1950s and 1960s, following the construction of Brasília (the capital of Brazil) in the recently established Federal District, both of which are located 320 Km from the area, and with the establishment of the CVNP in 1961. However, it was only in the 1980s, after the road from Brasília to Cavalcante was paved, that the region received a more permanent flow of tourists (Gentile et al., 2016; Sampaio, 2013). Thus, the volume of tourism was incipient with no formal leisure accommodation in Cavalcante. According to both representatives of the tourism industry and non-governmental organizations, these first visitors were highly motivated by traditional religious festivals and the pursuit of a closer connection with nature. Tourism in the neighbouring town of Alto Paraíso de Goiás, where the main gate to the CVNP is located, impacted the flow of visitors to Cavalcante.

We named the *prevalent storyline* "Alternative Tourism." It was formed by discourse fragments raised from community members and people that visited the area still in the 1980s such as *"a place to find inner peace"* (interviewee 15, 26 May 2018), *"tourists live in big cities, but want to go to the countryside, to a quiet place"* (interviewee 32, 11 August 2018). They argued that tourists to the area wanted to flee from the busy urban life to find peace in a quiet place. A community member, who lived in Cavalcante in the 1980s, stated:

Table 1. Tourism development storyline dynamics.

Timeline (period)	Prevalent storyline/ *Rationalities*	Discourse coalitions	Context	Challenging storyline/ *Rationalities*	Episodes	Catalysts
Origins of Tourism in Cavalcante (1950s to 1990s)	Alternative tourism • *Deep ecology*	First visitors, exoteric people, environment activists	Declining mining sector; Increased environmental awareness	Sustainable Tourism • *Economic* • *Scientific*	• Establishment of the CVNP – 1961 • Road to Cavalcante is paved – 1981 • Establishment of the UNESCO Cerrado Biosphere Reserve – 1994	International support and the environmental agenda of the 1990s
Institutionalization phase (2000 to 2010)	Sustainable Tourism • *Economic* • *Scientific*	Governmental agencies, UNESCO, local and international NGOs, leisure tourism industry	First leisure accommodations; Growing number of visitors; Manganese mining	Tourism Modernization • *Economic*	• Establishment of the APA Pouso Alto and expansion of the CVNP – 2001 • Establishment of the Lavapés Municipal Park – 2002 • Municipal tourism regulations are established	Perception of tourism as an industry, focused on an increased tourism flow, and capable of increasing wealth and income in remote areas
The Return of Mining (2010- Present)	Tourism modernization • *Economic*	Modernization tourism industry, local and state governments, SEBRAE and the local tourism council	Interest among mining investors; Consolidation of tourism; Global economic and political crisis; Cavalcante Sustainable Tourism Watch Institute	Community Approach • *Critical* • *Communicative*	• Cavalcante tourism development plan is developed 2011-2014 • Failed attempt to establish the Rio São Félix State Park – 2013 • The road to CVNP is paved (road GO-239) – 2014 • Expansion of the CVNP – 2017 • Resumption of mining – 2018 • Social movements against mining are established	Growing consciousness of the negative impacts of mining and "tourism modernization"

> Tourists were looking for ways to achieve inner peace, therefore… [they sought] less crowded places… There are indeed sacred places in the region. The Prata River Spring [for example] is a sacred place (Interviewee 15, 26 May 2018).

The "alternative tourism" storyline is underpinned by a "deep ecology" *rationality*, in which nature has an intrinsic value. The pursuit of harmony between humans and nature, a more spiritualized experience far from the urban areas as well as the holistic and sacred perception of places (Acott et al., 1998; Ness, 1986), as stated in the passages above, illustrate how the deep ecology rationality underpinned this storyline. Therefore, the *discourse coalition* that supported this storyline was mostly formed by admirers of nature, exoteric people, and environmental activists, who first visited the region as tourists.

The *context* of the 1980s and 1990s was a declining mining sector, with minerals valued at very low prices, thus making most mineral exploration in this historic mining region economically unfeasible. The main gold mine located in the urban area of Cavalcante had to close in the late 1990s. Meanwhile, there was growing environmental awareness in the Cerrado region, as a result of global political debates. Brazil was strongly influenced by these discussions, evidenced by the environmentally sound new Constitution of 1988 and the Rio de Janeiro UN ECO92 summit. In the 1990s, actions were taken to "professionalize" the tourism industry and promote economic growth alongside environmental protection, which was already promoted in sustainable development discourses that became highly accepted at the global level. These ideas influenced the national tourism policies from that period, with the creation of PROECOTUR – a programme to finance tourism projects in the Amazon region – and the Parques do Brasil programme, whose objectives were to take advantage of the tourism potential to generate economic growth and social benefits for the areas surrounding the national parks (Beni, 2006). Thus, "sustainable tourism" was the *challenging storyline* that displaced "alternative tourism" in the 1990s.

The main *episodes* that marked this stage were the establishment of the CVNP, in 1961, which defined the region as an important ecosystem to be preserved and fostered the first visitors to the region (Brazil, 1961; Sampaio, 2013); the paving of the road to Cavalcante, which enabled a more permanent flow of visitors (Gentile et al., 2016); and creation of the UNESCO Cerrado Biosphere Reserve in 1994 (UNESCO, 2011). The international NGOs and the UNESCO were highly influenced by the sustainable discourse, and their ideas have underpinned the mentioned episodes, with the aim of producing economic development through tourism alongside environmental protection. Local agencies inspired by these institutions developed a set of actions to change tourism from a spontaneous movement of alternative people to a structured economic development and environmental protection legal framework (Barbosa, 2008; Gentile et al., 2016; Sampaio, 2013). Thus, the *catalyst element* that defined the establishment of "sustainable tourism" as the prevalent storyline was international support and the environmental agenda of the 1990s, which created a suitable environment for the institutionalization of tourism development.

6.2. Institutionalization phase – 2000 to 2010s

"Sustainable tourism" is the *prevalent storyline* in this phase. The balance between economic growth and environmental protection that characterize the sustainable tourism discourse can be found in fragments raised by people with different backgrounds and interests, such as: *"tourism is the best solution to promote conservation and contributes to a diversified economy"* (interviewee 3 – tourism modernization entrepreneur, 24 May 2018*)*, and *"we should organize so that this [the landscapes] can be kept for future generations"* (interviewee 36 - community member and representative of an NGO, 12 August 2018). Documents also incorporate the sustainable discourse. For instance, State Decree n° 5.419/2001, to establish the APA Pouso Alto, states the intention to "promote sustainable development" and to "preserve flora, fauna, water sources, geology and landscaping in the region" (Goiás, 2001). This storyline encompasses the global discourse of sustainable development and its concern with future generations (WCED, 1987). As previously discussed, it has also been criticized for being too vague (Mebratu, 1998) and difficult to operationalize (Butler, 1999; Sharpley, 2020). The passages below demonstrate this:

> Nature preservation is important for our survival… in all aspects. We live in a world in which we still depend on the economy, the flow of money. So, how can we preserve [nature] and make money out of it? Well, the response – natural and obvious – is tourism. Tourism joins [nature] preservation and economy [growth]. (Interviewee 3 – a tourist modernization entrepreneur, 24 May 2018).

> I am not against [mining], but studies should be conducted, to see if it is feasible, because everything has impacts, and will generate impacts. (Interviewee 25 – an NGO representative, 9 August 2018).

The passages above illustrate the description and the criticism of the sustainable tourism storyline. It employs a scientific *rationality* ("studies should be conducted") to foster nature conservation and an economic *rationality* ("flow of money") to legitimize economic growth. Interviewees from different groups were all for sustainable tourism, showing that this storyline can attract people with different interests. The *coalition* that supported this storyline was led by the Goiás State Secretary of the Environment and Water Resources, the administration of Cavalcante, the World Wildlife Fund, and local NGOs, which were behind the implementation of this agenda (Barbosa, 2008). It attracted tourist-related businesses to the region, which were run by outsiders, mostly from São Paulo, as stressed by representatives of the tourism industry.

This phase was *contextualized* by a dominance of global sustainable development values, which underpinned the expansion of the protected areas. The Cerrado biome attracted increased global interest with the nominations of the CVNP and Emas National Park to the World Heritage List (UNESCO, 2019). Although there was no tourist flow control, there is a perception that a growing number of visitors began to flow into the region. Evidence is that tourist guide courses were offered to the host communities of Cavalcante to qualify the workers to receive these tourists (Barbosa, 2008). In 2002, the trained tourist guides organized the first tourist guide association of Cavalcante, which is still active today. The atmosphere of collaboration and opportunities for sustainable tourism has generated numerous tourism development actions.

The important *episodes* of this phase refer to the establishment of the APA Pouso Alto (Ferreira & Almeida, 2014; Goiás, 2001), the establishment of the Lavapés Municipal Park (Decree 24/2002), and three municipal regulations (e.g. law 862/2001) that were approved by the town councillors: creation of the Cavalcante Municipal Tourism Council to hear the stakeholders' opinions about local tourism; the establishment of the Local Municipal Tourism Fund to enable local public actions for tourism development; and a law to regulate the use of tourist attractions (Cavalcante, 2001). A fundamental actor that emerged in this period was the Brazilian Support Service for Micro and Small Enterprises (SEBRAE). SEBRAE started a tourism development project for the towns of Alto Paraíso and Cavalcante with the main goal of increasing tourism flow in the low season (SEBRAE, 2006).

The *challenging storyline* that emerged to displace "sustainable tourism" was "tourism modernization," which has been previously described as a linear way to modernize societies in isolated areas, where traditional industrialization has not been achieved (Fazito, 2015; Telfer & Sharpley, 2007). Sustainable tourism storyline ended as part of the "neoliberal conservation toolkit" and did almost nothing to address the local community disparities (Hunt et al., 2015). The process through which "tourism modernization" displaces "sustainable tourism" has to do with the tourism's addiction to growth (Higgins-Desbiolles, 2018) and to the previously described conceptual flexibility of the sustainable discourse, in which environmental protection played a major role when mining (traditional industrialization) was not feasible (Fazito et al., 2016; Sharpley, 2020). However, as soon as mining began to attract interest, a tourism modernization discourse emerged that contributed, along with mining, to its possibilities for wealth and income expansion. This example illustrates the claim by Fazito et al. (2016) that sustainable tourism has a "chameleon flexibility," which can become conservative, but is dangerously disguised in an alternative discourse. Tourism modernization benefits from the flow of visitors and the income generated by the establishment of mining projects (Dwyer et al., 2016). A tourist modernization entrepreneur who is also a governmental actor states:

> The population thought it [manganese mining] was good … because it created jobs, right? There were well developed regions, because of them [mining companies] they did the roads' maintenance … and created a lot of jobs and so forth, you know? (Interviewee 30, 11 August 2018).

The passage above illustrates the contribution of mining to improving tourist infrastructure, exemplifying the perspective that the relationship between tourism and mining is more complementary than counterproductive (Similä & Jokinen, 2018). The *catalyst* element that enabled "sustainable tourism" to be displaced by "tourism modernization" was the possibility for tourism businesses to benefit from mining, which implied a perception of tourism as an industry, with a focus placed solely on its ability to expand wealth and income, and on an increased and permanent tourism flow (Butler, 1980). The next section explores the dynamics of tourism development discourses in this context of conflict.

6.3. The return of mining – after 2010s

This is the consolidation of tourism stage and the "Tourism modernization" is the *prevalent storyline*. It is formed by discourse fragments drawn from documents, such as *"increasing the flow of tourists"*, *"strengthening competitiveness"*, *"creating jobs and income"* (SEBRAE, 2006) and the Cavalcante Tourism Development Plan 2011-2014 (Cavalcante, 2011), which defines its objective as *"to expand the flow of tourists and increase the competitiveness of Cavalcante in the national market."* Mining emerges after 2010 as an opportunity for modernization tourism entrepreneurs and the local government to increase income and wealth. For the government, mining increases tax collection and investments, and for the tourism industry, mining generates a whole chain of activities that use tourism facilities, such as hotels and restaurants, and improve infrastructure. Although it does not sound correct, since mining is an industry related to environmental degradation, the possibility of changing the motivation of tourists from leisure to business has created a sense of opportunity among the modernization tourism entrepreneurs. Statements, such as *"I am not against mining"* (interviewee 21, 7 August 2018) and *"The only thing that matter is employment and income for the people of Cavalcante"* (interviewee 30, 22 May 2018) were made by these entrepreneurs. However, mining is a very sensitive topic in the region. People chose words carefully to talk about mining. A more supportive statement comes always with some conditions, to minimize the impacts of mining. For example, a very active stakeholder, who represents the modernization tourism industry and is a member of an NGO states:

> About mining… if we are here, using cell phones, watches… we benefit from mining. But I think it must be done in the right way, impacts must be controlled, and part of the wealth must remain with the people. But what we see is that sometimes it does not occur (Interviewee 21, 7 August 2018).

The *context* of this phase is marked by the global economic crisis that has affected Brazil since 2013, which has also caused a political breakdown that has led to a more authoritarian regime. As a result, pro-environment and social inclusion policies that marked the previous phases have been challenged (Sauer et al., 2020). The second part of the Cavalcante Sustainable Tourism Watch Institute was implemented between 2011 and 2012. The institute was created to organize a legal framework. However, different from this initial intention, it has changed over time to fostering sustainable tourism, and then simply reporting quantitative data about the tourism economy (Sampaio, 2013). The 2011–2014 Cavalcante Tourism Development Plan was launched as a result of the partnership between the municipal and state governments, SEBRAE and the local tourism council. It focused on improving tourism infrastructure, professional qualifications, promotion, and destination management (Cavalcante, 2011; Sobrinho et al., 2015). Such an institutional framework demonstrates how the *coalition* toward the "tourism modernization" storyline quickly dominated the development policy arenas of Cavalcante.

Some *episodes* reveal the predominance of "tourism modernization" in this stage. The highway GO-239 was paved in 2015, which improved access to the CVNP. According to members of the local administration, this has led to an increase in the number of visitors to the park and to Cavalcante. However, the paving of the road has also increased local fauna roadkill rates (Ribeiro, 2016). Tourism modernization has been criticized for the same reasons as the broader modernization discourses: it contributes to environmental degradation and ecological insecurity, as illustrated in the example above, and produces inequalities (Escobar, 1995; Loperena, 2017). Although it was not registered in the official mining data (DNPM, 2013), manganese exploitation was active in 2012, as some interviews revealed, and technical reports have highlighted that mining was a relevant threat to tourism in Cavalcante (Brazil, 2012; Cavalcante, 2012). This may explain why another important episode is the failed attempt to establish the Rio São Félix State Park near to the district of São José (Funatura, 2013), considering it was close to manganese deposits (Figure 2). Most of the people who attended a public meeting in 2013 rejected the proposed park, along with its advertised future job opportunities and economic development. However, leisure tourism entrepreneurs reported that the local government was sponsored by the mining industry, thereby acting against establishing the park. This fact created an atmosphere of lack of trust in institutional mediation, which has been a fundamental issue in the mediation of such conflicts, as previously discussed (Almeida et al., 2017; Siakwah et al., 2019).

Finally, in 2017 the CVNP expansion project was concluded (Brazil, 2017). However, it left the ore deposit sites out of its borders, explicitly demonstrating the growing importance of the national and international mining lobby in the region (Brazil, 2017). Although mining within the APA Pouso Alto was not yet regulated in its management plan (Goiás, 2016), Cavalcante became the second target of mineral research survey requests in the entire state (Figure 2 shows the polygons of soil research request).

Mining had a strong impact on the development of leisure tourism in Cavalcante, as it causes river siltation, water pollution, and has altered the landscape and increased the risk of accidents with so many heavy trucks on the road. These situations evidence the difficult relationship between mining and leisure-based ecotourism (Honey, 2008; Similä & Jokinen, 2018). Community members also reported impacts from mining on water sources and the precarious situation of the workers (including children). These impacts have already caused public institutions to intervene. For instance, the public prosecutor's office selected Cavalcante to establish a social project to fight the high rates of child labour (MPT-GO, 2020). In the last field visit (August 2018), we observed that manganese was explored in the northern districts of Cavalcante, which created intense traffic of trucks. Transportation is the element of manganese production with the highest cost, given that manganese is a voluminous and heavy ore (Drummond, 2005). Trucks use the same road to access the most visited tourist attractions in Cavalcante (see Figure 2). A member of a local community who is also an NGO representative argues:

> Trucks leave São José with tons of manganese and pass through the road that goes to Santa Barbara [waterfall]. ... when there is a high flow of trucks, it destroys the road. (...) ... Tourism is strongly undermined [by mining], I'm not in favour of mining companies! (Interviewee 27, 9 August 2018).

Differently from the modernization tourism entrepreneurs, the leisure tourism entrepreneurs spoke loudly and clearly against mining. One of them argued:

> I think ... tourism and mining are adversaries ... like oil and water, they don't mix. I believe tourism is much more viable because tourism will not generate impacts ... but of course, you must know how to manage it (Interviewee 29, 10 August 2018).

Although the mining operation had clearly restarted, no one among our interviewees (even governmental actors) was able to provide us with the environmental licences that legal

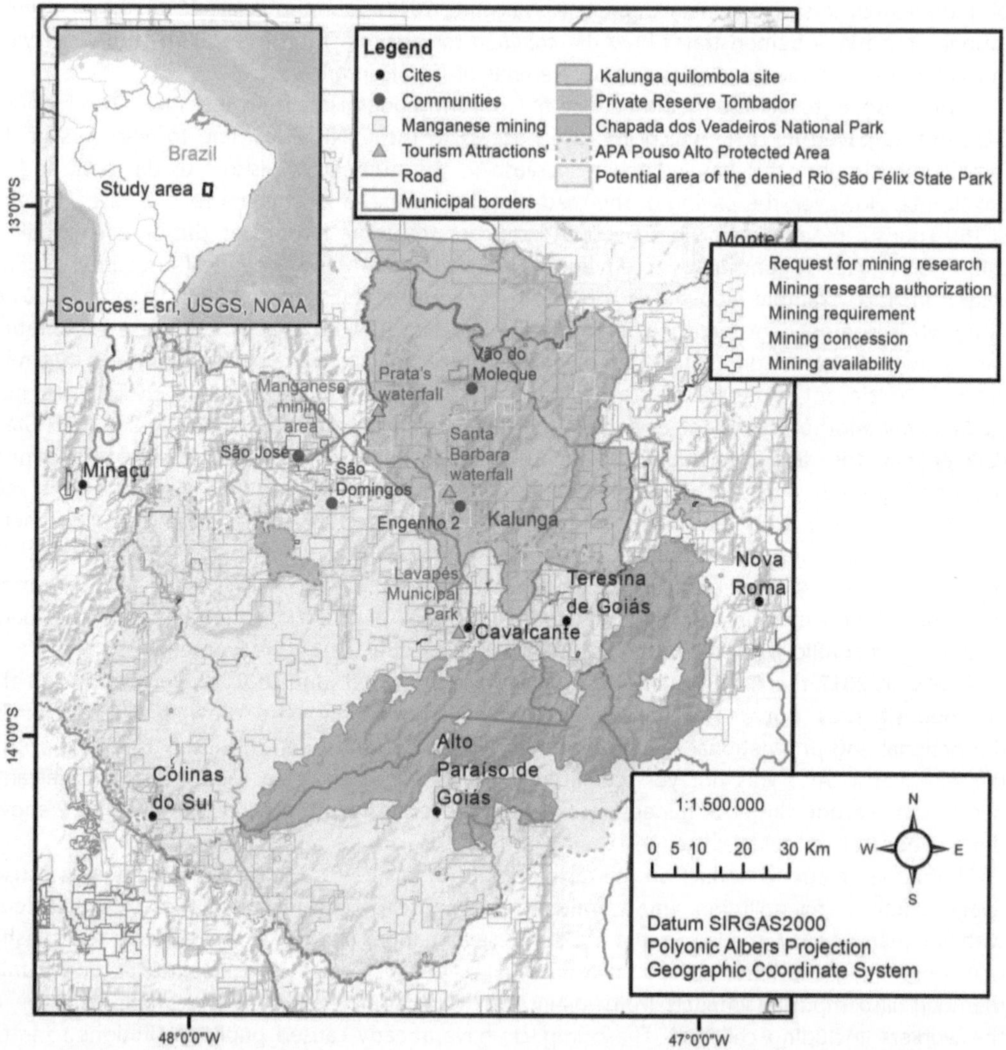

Figure 2. Location of the main tourist attractions, the mining processes areas, the road used by manganese mining trucks and tourists, and the area of the denied Rio São Félix State Park.

processes in Brazil require. Leisure tourism entrepreneurs and community members affirm that there was no public participation in the decision-making processes that enabled the mining operation to take place, similar to the process of poor public participation and corruption evidenced by Leonard and Lebogang (2018) in South Africa and lack of dialogue between public sphere, community leaders and populations in Goiás (Ferreira & Almeida, 2014).

Likewise, the negative impacts of "tourism modernization" in the neighbouring city of Alto Paraíso, such as mass tourism, real estate speculation, animal roadkill and inflation, have generated growing awareness among members of the leisure tourism industry and local communities of Cavalcante that, like mining, the tourism industry's addiction to growth (Higgins-Desbioles, 2018) can be very harmful to the local communities and the environment. This realization has led to the establishment of a growing *challenging storyline*, which we named "community approach." A local community member declared: "*We want sustainable tourism, as long as it respects our traditions, cultures, and historic sites*" (Interviewee 23, 8 August 2018). Thus, the *catalysts* to establishing the resistance movement were a growing consciousness of the failure of "tourism modernization" in Alto Paraíso and the negative impacts generated by the return of mining in Cavalcante.

A local movement against mining emerged in Cavalcante and received the support of the National Movement for Popular Sovereignty in Mining. They organized two public meetings, in partnership with local students, in 2017 to discuss mining experiences in Cavalcante and Brazil as a whole, the violence that the communities affected by mining suffer, and alternatives to the dominant discourses that cause such violence (MAM, 2017). The growing awareness among community members of implementing tourism and mining modernization projects and their impacts created a moment of conscientization in the affected communities. For instance, a community member and representative of an NGO explained why he was against large mining projects in the region:

> there is no point in employing ten people and seeing [environmental] degradation that will affect one hundred… I cannot agree with that… (Interviewee 36, 12 August 2018).

This is how we understand "conscientization" (Freire, 1970, 1973): the result of an increased capability to be aware and criticize the context in which a person is inserted. The "community approach" storyline values local reality and contexts, listens to the voices of the affected communities, and does not try to hide the destructive side of the development process, thereby promoting awareness and resistance, as preconized by the champions of post-development (Gibson-Graham, 2005).

Following Little's (2001) concept, the two sides of the conflict are: the government actors, the mining and tourism entrepreneurs that profit from the mining projects and support it through the "tourism modernization" storyline – focused on the dominance of economic interests over other interests (Porter & Salazar, 2005) – and both the communities affected by the mine and the leisure tourism entrepreneurs that depend on the preservation of nature for their businesses to flourish. The latter form the discourse coalition that supports the "community approach" storyline, thus evidencing how the engagement of tourism stakeholders with local communities could resist the implantation of large mines (Hales & Larkin, 2018).

The discourse fragments identified that form this critical storyline among the communities affected by mining were: *"there should be more active participation"* (interviewee 20 - community member, 7 August 2018), *"get more people engaged"* (interviewee 3, 24 May 2018), *"the community should be more engaged, shouldn't it?"* and *"when community members are engaged, communities gain value"* (interviewee 30, 22 May 2018). These passages evoke the ratification of public policies by residents, which imply the value of a communicative *rationality*, where decisions have to be accepted and supported by the people involved. They also defend that local communities should benefit from the decisions that govern the development of their societies (Murphy, 2013). Thus, criticism of the dominant modernization discourses is legitimized by local experiences and local wisdom, i.e. the main features of what we previously named critical rationality. A community member who is also a local NGO worker argues:

> … what kind of tourism do the people want? Like CVC[3]? I completely disagree. We support community-based tourism, but it will not happen. The local government has not supported us since 2004 (Interviewee 5, 22 May 2018).

Despite the imbalance of power between the local communities and the mining entrepreneurs, who received governmental support, the "community approach" storyline has been attracting the support of intellectuals, activists and journalists, who shared the indignation of the local communities seeking to break with the dominance of local modernization power, as argued by the above-mentioned interviewee, and empower a discourse coalition to challenge mining and tourism modernization. Evidence can be seen in the national news (Globo, 2020) and the presence of these people in the public meetings we attended during fieldwork. Finally, a focus on active tourism and ecotourism emerges as an opportunity to promote collaboration, peace, and enhance the positive relationship between nature conservation and livelihood improvements (Causevic & Lynch, 2013; Honey, 2008; Marcellino, 2006; Nyaupane & Poudel, 2011; Stone &

Nyaupane, 2016). Thus, a post-development perspective has started to emerge in the region with a perspective of tourism that can reduce the violence and dispossession derived from neoliberalism and contribute to the production of contextualized epistemologies of the South (Devine & Ojeda, 2017; Gibson-Graham, 2005; Santos, 2004).

7. Conclusion

The dynamics of tourism development discourses in Cavalcante were presented in a three-stage timeline, with the "alternative tourism" storyline, backed by a deep ecology rationality, being displaced by "sustainable tourism" in its origins, which generated a phase of institutionalization and organization of tourism in a context of declining mining activity, mostly focused on nature-based leisure tourism. During the institutionalization phase, several protected areas were established, based on a combination of possible economic gains (economic rationality) and nature conservation (scientific rationality). After 2010, when mining became attractive for investors again, an economic rationality prevailed to legitimize the "tourism modernization" storyline, which is defined as a linear process focused on the expansion of wealth and income (Telfer & Sharpley, 2007). Thus, it has demonstrated the inaccuracy (Mebratu, 1998) and the "chameleon flexibility" that characterizes the sustainable discourse (Fazito et al., 2016), and causes sustainable tourism to be difficult to achieve (Butler, 1999; Sharpley, 2020). However, the negative impacts of both tourism modernization and mining have been awakening a resistance impetus among the local communities, based on the mandatory acceptance among them of the decisions that govern their development (communicative rationality) and on the local contextualized experiences (critical rationality) that underpin the "community approach" storyline. Similarities were identified between this storyline and the post-development perspective, which incorporates the voices and local experiences that are normally neglected in the development practice and theory (Gibson-Graham, 2005).

Subsequently, a socio-environmental conflict was established (Little, 2001). On the one side there was a coalition of local governmental actors, tourism modernization and mining industry and, on the other side, communities affected by the negative impacts of mining, including the leisure tourism industry (e.g. stakeholders more inclined to the ecotourism principles). Economic interests have prevailed over other interests (Porter & Salazar, 2005), hence local and state governments have taken side with the mining entrepreneurs and the "tourism modernization" coalition. This has led to a lack of trust in their role as conflict mediators. Institutional trust is crucial to mediating such conflicts (Almeida et al., 2017).

The "sustainable tourism" storyline, rather than being understood as native sovereignty, which could help minimize the conflict and bring peace (Causevic & Lynch, 2013; Devine & Ojeda, 2017), became an industry analogous to extractivism and addicted to growth (Higgins-Desbiolles, 2018). Moreover, it has resulted in ecological insecurity among indigenous communities (Loperena, 2017), and served as part of the "neoliberal conservation toolkit" (Hunt et al., 2015). The conservative "tourism modernization" storyline emerged from the sustainable discourses, thus causing confusion among stakeholders.

This article contributes both to the interdependent relations among nature conservation and livelihood improvements and to the critical tourism literature, by giving a discursive treatment to a political economy development context (Nyaupane & Poudel, 2011; Stone & Nyaupane, 2016; Wilson & Hollinshead, 2015), and by producing epistemologies from the South (Santos, 2004). It also contributes to the under-researched issue of mining impacts on specific protected environments and tourism sites (Leonard & Lebogang, 2018) and demonstrates that leisure tourism loses out while modernization tourism benefits from a boom in mining (Dwyer et al., 2016), thus calling for a need to separate leisure and modernization tourism in public planning processes. The case study of Cavalcante can also serve as a learning site and promote knowledge exchange with regards to tourism development in similar contexts, as recommended for biosphere reserves (UNESCO, 2008).

The main limitations of this research have to do with the fact that it took place in a conflict context, where people were afraid to talk about mining, and the lack of official data on tourism flow and its impacts. Therefore, we had to use a set of qualitative data collection technics to validate our data. Future research could explore some specific dynamics of contextualized tourism development, such as the conflicts inside the tourism industry and issues of power underpinning policymaking in local policy arenas. It could also address other tourism conflicts in the region, namely agribusiness and small-scale hydropower projects.

Notes

1. Former escaped slave settlements, and historical symbols of resistance against oppression.
2. Exploration of Brazilian subsoil is under the responsibility of the Brazilian Mining Agency (Brazil, 2018), which regulates the polygons that are authorized to receive soil surveys. The procedure starts with a request for geological research by the mining company. If the ore exploration is viable, a set of environmental licenses and other mitigating actions are required for the mine to operate.
3. CVC is the largest tour operator in Brazil. It is notorious for promoting mass tourism.

Acknowledgement

We gratefully acknowledge the support of the Brazilian agency CNPQ, Centre for Excellence in Tourism - University of Brasília (UnB), UnB Cerrado Research Centre, Tourism and Sustainability Studies Lab. (LETS/UnB). We are also grateful to Y. Salmona for the maps and for contributions from J. L. A. Franco, L. C. S. Pena, and H. Günther. We are grateful to the contributions received at the MMV9 International Conference, and to the anonymous reviewers for their valuable comments. We also express our special thanks to all the communities of Chapada dos Veadeiros, who contributed directly or indirectly to this study. This paper is part of N. R. Marques's master thesis on Tourism.

Funding

This work was supported by Conselho Nacional de Desenvolvimento Científico e Tecnológico – CNPq.

ORCID

Nayara Marques http://orcid.org/0000-0002-6255-3845
Mozart Fazito http://orcid.org/0000-0002-7528-6619
André Cunha http://orcid.org/0000-0003-4928-2494

References

Acott, T. G., Trobe, H. L. L., & Howard, S. H. (1998). An evaluation of deep ecotourism and shallow ecotourism. *Journal of Sustainable Tourism*, *6*(3), 238–253. https://doi.org/10.1080/09669589808667314

Almeida, J., Costa, C., & da Silva, F. (2017). A framework for conflict analysis in spatial planning for tourism. *Tourism Management Perspectives*, *24*, 94–106. https://doi.org/10.1016/j.tmp.2017.07.021

Barbosa, A. G. (2008). *As Estratégias de Conservação da Biodiversidade na Chapada dos Veadeiros: Conflitos e Oportunidades* [Masters dissertation, Universidade de Brasília]. UnB Repository. https://repositorio.unb.br/handle/10482/1040

Beni, M. (2006). *Política e Planejamento do Turismo no Brasil*. Aleph.

Bianchi, R. (2009). The "critical turn" in tourism studies: A radical critique. *Tourism Geographies, 11*(4), 484–504. https://doi.org/10.1080/14616680903262653

Brazil. (1961). *Decree n° 49,875*. Brazil. http://www.planalto.gov.br/ccivil_03/decreto/antigos/d49875.htm#:~:text=DECRETO%20No%2049.875%2C%20DE,Goi%C3%A1s%20e%20d%C3%A1%20outras%20provid%C3%AAncias

Brazil. (2000). *Decree n° 9,985*. Brazil. http://www.planalto.gov.br/ccivil_03/leis/l9985.htm#:~:text=L9985&text=LEI%20No%209.985%2C%20DE%2018%20DE%20JULHO%20DE%202000.&text=Regulamenta%20o%20art.,Natureza%20e%20d%C3%A1%20outras%20provid%C3%AAncias

Brazil. (2012). Sustainable tourism development plan-Chapada dos Veadeiros. Ministry of Tourism. Brasília.

Brazil. (2017). *Decree from 5/7/2017 about the ICMBIO process 02070.000116/2011-10*. http://www.planalto.gov.br/ccivil_03/_ato2015-2018/2017/Dsn/Dsn14471.htm#:~:text=DECRETO%20DE%205%20DE%20JUNHO,da%20Alian%C3%A7a%2C%20Estado%20de%20Goi%C3%A1s

Brazil. (2018). Decree n° 9,406. Brazilian Mining Regulation. Brazil.

Britton, S. (1982). The political economy of tourism in the third world. *Annals of Tourism Research, 9*(3), 331–358. https://doi.org/10.1016/0160-7383(82)90018-4

Butler, R. W. (1980). The concept of a tourism area cycle of evolution: Implications for management resources. *The Canadian Geographer/Le Géographe Canadien, 24*(1), 5–12. https://doi.org/10.1111/j.1541-0064.1980.tb00970.x

Butler, R. W. (1999). Sustainable tourism: A state-of-the-art review. *Tourism Geographies, 1*(1), 7–25. https://doi.org/10.1080/14616689908721291

Castells, M. (2003). The rise of the Fourth World. In A. G. McGrew & D. Held (Eds.), *The global transformations reader: An introduction to the globalization debate* (pp. 430–439). Polity Press.

Causevic, S., & Lynch, P. (2013). Political (in)stability and its influence on tourism development. *Tourism Management, 34*, 145–157. https://doi.org/10.1016/j.tourman.2012.04.006

Cavalcante. (2001). *Municipal Law 862/2001*. https://www.cavalcante.go.leg.br/leis/legislacao-municipal/2001/lei-862-2001-regulam-atrativos-turist/lei-862-2001-regulam-atrativos-turisticos.1/view

Cavalcante. (2011). *Cavalcante tourism development plan 2011-2014*. Cavalcante.

Cavalcante (2012). *Situational diagnosis of the municipality of Cavalcante*. Cavalcante.

Cheong, S., & Miller, M. (2000). Power and Tourism: A foucauldian observation. *Annals of Tourism Research, 27*(2), 371–390. https://doi.org/10.1016/S0160-7383(99)00065-1

Cole, D. (2004). Exploring the sustainability of mining heritage tourism. *Journal of Sustainable Tourism, 12*(6), 480–494. https://doi.org/10.1080/09669580408667250

Cole, S. (2014). Tourism and water: From stakeholders to right holders, and what tourism business need to do. *Journal of Sustainable Tourism, 22*(1), 89–106. https://doi.org/10.1080/09669582.2013.776062

Cowen, M., & Shenton, R. (2005). *Doctrines of development*. Routledge.

Devine, J., & Ojeda, D. (2017). Violence and dispossession in tourism development: A critical geographical approach. *Journal of Sustainable Tourism, 25*(5), 605–617. https://doi.org/10.1080/09669582.2017.1293401

Dijk, T. A. V. (1981). Episodes as units of discourse analysis. In D. Tannen (Ed.), *Analyzing discourse: Text and talk* (pp. 177–195). Georgetown University Press.

DNPM. (2013). *Desempenho do setor mineral GO-DF*. https://www.gov.br/anm/pt-br/centrais-de-conteudo/dnpm/publicacoes-economia-mineral/arquivos/desempenho_mineral_go_df_2013.pdf/view

Dredge, D. (2010). Place change and tourism development conflict: Evaluating public interest. *Tourism Management, 31*(1), 104–112. https://doi.org/10.1016/j.tourman.2009.01.004

Drummond, J. A. L. (2005). O manganês do Amapá: O seu papel no desenvolvimento regional e nacional do Brasil (1957-1998). *Desenvolvimento e Meio Ambiente, 11*(12), 51–69. https://doi.org/10.5380/dma.v11i0.7817

Dryzek, J. (2005). *The politics of Earth: Environmental discourses*. Oxford University Press.

Dwyer, L., Pham, T., Jago, L., Bailey, G., & Marshall, J. (2016). Modeling the impact of Australia's mining boom on tourism: A classic case of Dutch disease. *Journal of Travel Research, 55*(2), 233–245. https://doi.org/10.1177/0047287514541007

Escobar, A. (1995). *Encountering development: The making and unmaking of the third world*. Princeton University Press.

Fazito, M. (2015). Modernização Turística: O papel do turismo nos discursos dominantes de desenvolvimento. In S. Figueiredo, F. Azevedo, & W. Nóbrega (Eds.), Perspectivas Contemporâneas de Análise em Turismo (pp. 108–126). NAEA.

Fazito, M., Rodrigues, B., Nascimento, E., & Pena, L. (2017). O Papel do Turismo no Desenvolvimento Humano. *Papers Do NAEA, 372*, 1–21.

Fazito, M., Scott, M., & Russell, P. (2016). The dynamics of tourism discourses and policy in Brazil. *Annals of Tourism Research, 57*, 1–17. https://doi.org/10.1016/j.annals.2015.11.013

Ferreira, L. C., & Almeida, M. G. (2014). Usos e conflitos na APA do Pouso Alto (GO): Uma abordagem sobre a percepção dos atores locais de Colinas do Sul e Cavalcante. *Ateliê Geográfico, Goiânia, 8*(1), 215–230.

Fletcher, C., Pforr, C., & Brueckner, M. (2016). Factors influencing Indigenous engagement in tourism development: International perspective. *Journal of Sustainable Tourism, 24*(8-9), 1100–1121. https://doi.org/10.1080/09669582.2016.1173045

Flyvbjerg, B. (2006). Five misunderstandings about case-study research. *Qualitative Inquiry, 12*(2), 219–245. https://doi.org/10.1177/1077800405284363

Flyvbjerg, B. (1998). *Rationality and power: Democracy in practice*. University of Chicago Press.

Foucault, M. (1982). *Microfísica do Poder*. Graal.

Freire, P. (1970). *Pedagogia do Oprimido*. Paz e Terra.

Freire, P. (1973). *Education for critical cousciousness*. Sheed and Ward.

Funatura. (2013). Proposta de criação do Parque Estadual Rio São Félix – Documento para Consulta Pública. Funatura e Semarh, Goiás.

Gentile, C., Franco, J. L. A., & Sayago, D. A. V. (2016). Um Modelo de Capacitação Rumo à Sustentabilidade: Os guias de Alto Paraíso de Goiás – Chapada dos Veadeiros (GO). *Fronteiras: Journal of Social, Technological and Environmental Science, 5*(1), 168. https://doi.org/10.21664/2238-8869.2016v5i1.p168-185

Gibson-Graham, J. K. (2005). Surplus possibilities: Post-development and community economies. *Singapore Journal of Tropical Geography, 26*(1), 4–26. https://doi.org/10.1111/j.0129-7619.2005.00198.x

Globo. (2020). *JA 2a Edição: Operação flagra novos pontos de desmatamento e minerações irregulares em área kalunga*. Retrieved July 2, 2020, from https://globoplay.globo.com/v/8647183/?utm_source=whatsapp&utm_medium=share-bar

Goiás. (2001). *Decree n° 5.419/2001*. http://www.gabinetecivil.goias.gov.br/decretos/numerados/2001/decreto_5419.html

Goiás. (2016). *APA Pouso Alto Management Plan*. http://www.sgc.goias.gov.br/upload/arquivos/2016-06/resumo-executivo.pdf

Gonçalves, R. J. F. A., & Mendonça, M. R. (2015). Expansão dos grandes empreendimentos de mineração e territórios em disputa no cerrado goiano (Goiás/Brasil). *Sociedade e Território, Natal, 27*(l), 206–228.

Hajer, M. (1995). *The politics of environmental discourse: Ecological modernization and the policy process*. Oxford University Press.

Hajer, M. (2006). Doing Discourse Analysis: Coalitions, practices, meanings. In M. Brink & T. Metze (Eds.), Words Matter in Policy and Planning: Discourse theory and method in the social sciences (pp. 65–74). Labor Grafimedia.

Hales, R., & Larkin, I. (2018). Successful action in the public sphere: The case of a sustainable tourism-led community protest against coal seam gas mining in Australia. *Journal of Sustainable Tourism, 26*(6), 927–941. https://doi.org/10.1080/09669582.2017.1423316

Harvey, D. (2005). *A brief history of neoliberalism*. Oxford University Press.

Healey, P. (1999). Institution analysis, communicative planning, and shaping places. *Journal of Planning Education and Research, 19*(2), 111–121. https://doi.org/10.1177/0739456X9901900201

Hewitt, S. (2009). Discourse analysis and public policy research. *Centre for Rural Economy Discussion Paper, 24*, 1–16.

Higgins-Desbiolles, F. (2018). Sustainable tourism: Sustaining tourism or something more? *Tourism Management Perspectives, 25*, 157–160. https://doi.org/10.1016/j.tmp.2017.11.017

Hoffman, M. (2010). The impact of conservation on the status of the world's vertebrates. *Science, 330*, 1503.

Honey, M. (2008). *Ecotourism and sustainable development: Who owns paradise?* Island Press.

Hunt, C., Durham, W., Driscoll, L., & Honey, M. (2015). Can ecotourism deliver real economic, social, and environmental benefits? A study of the Osa Peninsula, Costa Rica. *Journal of Sustainable Tourism, 23*(3), 339–357. https://doi.org/10.1080/09669582.2014.965176

IBGE. (2017). *Cidades – Cavalcante*. https://cidades.ibge.gov.br/brasil/go/cavalcante/panorama

Klink, C. A., & Machado, R. B. (2005). Conservation of the Brazilian Cerrado. *Conservation Biology, 19*(3), 707–713. https://doi.org/10.1111/j.1523-1739.2005.00702.x

Leonard, L. (2016). Mining and/or tourism development for job creation and sustainability in Dullstroom. *Local Economy: The Journal of the Local Economy Policy Unit, 31* (1-2), 249–263. https://doi.org/10.1177/0269094215621875

Leonard, L., & Lebogang, T. (2018). Exploring the impacts of mining on tourism growth and local sustainability: The case of Mapungubwe Heritage Site, Limpopo, South Africa. *Sustainable Development, 26*(3), 206–216. https://doi.org/10.1002/sd.1695

Lima, I. B. (2016). Ecotourism community enterprises and ethnodevelopment: Modelling the Kalunga empowerment possibilities in the Brazilian savannah. *Brazilian Journal of Science and Technology, 3*, 1.

Little, P. E. (2001). Os conflitos socioambientais: Um campo de estudo e ação política. In M. Bursztyn (Org.), *A difícil sustentabilidade: Política energética e conflitos ambientais* (pp. 107–122.). Garamond.

Loperena, C. (2017). Honduras is open for business: Extractivist tourism as sustainable development in the wake of disaster? *Journal of Sustainable Tourism, 25*(5), 618–633. https://doi.org/10.1080/09669582.2016.1231808

MAM. (2017). *MAM realiza Assembleia Popular da Mineração em Cavalcante*. https://mamnacional.org.br/2017/11/17/mam-realiza-assembleia-popular-da-mineracao-em-cavalcante/

Marcellino, N. (2006). *Estudos do Lazer: Uma introdução* (4th ed.). Autores Associados.

Markwick, M. (2000). Golf Tourism Development, Stakeholders, Differing Discourses and Alternative Agendas: The case of Malta. *Tourism Management, 21*, 515–524. https://doi.org/10.1016/S0261-5177(99)00107-7

Meadows, D., Randers, J., & Behrens, W. (1972). *The limits to growth: A report for the Club of Rome's project on the predicament of mankind.* Potomac.

Mebratu, D. (1998). Sustainability and sustainable development: Historical and conceptual review. *Environmental Impact Assessment Review, 18*(6), 493–520. https://doi.org/10.1016/S0195-9255(98)00019-5

MPT-GO. (2020). *Município de Cavalcante assina TAC para combater trabalho infantil.* http://www.prt18.mpt.mp.br/2-uncategorised/184-municipio-de-cavalcante-assina-tac-para-combater-trabalho-infantil

Murphy, P. (2013). *Tourism: A community approach.* Routledge.

Myers, N., Mittermeier, R., Mittermeier, C., Fonseca, G., & Kent, J. (2000). Biodiversity hotspots for conservation priorities. *Nature, 403*(6772), 853–858. https://doi.org/10.1038/35002501

Ness, A. (1986). Intrinsic value: Will the defenders of nature please rise? In M. Soulé (Ed.), *Conservation biology: The science of scarcity and diversity*, pp. 504–515. Michigan.

Nyaupane, G. P., & Poudel, S. (2011). Linkages among biodiversity, livelihood, and tourism. *Annals of Tourism Research, 38*(4), 1344–1366. https://doi.org/10.1016/j.annals.2011.03.006

Peck, J. (2004). Geography and public policy: Constructions of neoliberalism. *Progress in Human Geography, 28*(3), 392–405. https://doi.org/10.1191/0309132504ph492pr

Porter, B., & Salazar, N. (2005). Heritage tourism, conflict, and the public interest: An introduction. *International Journal of Heritage Studies, 11*(5), 361–370. https://doi.org/10.1080/13527250500337397

Porto-Gonçalvez, C. W. (2019). *Dos Cerrados e de suas riquezas: De saberes vernaculares e de conhecimento científico.* FASE e CPT.

Qian, J., Jiewen Wei, Rob Law. (2017). Review of critical discourse analysis in tourism studies. *International Journal of Tourism Research 20*(3-4). https://doi.org/10.1002/jtr.2202

Reid, D. (1995). *Sustainable development: An introductory guide.* Earthscan.

Reis, T. (2017). *Climate challenges and opportunities in the Brazilian Cerrado.* IPAM, COP23.

Ribeiro, T. R. S. (2016). Influências da pavimentação de rodovias em índices de atropelamento de fauna: o caso da rodovia GO-239 em Alto Paraíso de Goiás. [Masters dissertation, Universidade de Brasília]. UnB Repository. https://repositorio.unb.br/handle/10482/23100

Rostow, W. (1960). *The stages of economic growth: A non-comunist manifesto.* Cambridge University Press.

Rydin, Y. (2003). *Conflict, consensus, and rationality in environmental planning: An institutional discourse approach.* Oxford University Press.

Saarinen, J., Rogerson, C. M., & Hall, M. (2017). Geographies of tourism development and planning. *Tourism Geographies, 19*(3), 307–317. https://doi.org/10.1080/14616688.2017.1307442

Sampaio, A. C. M. A. (2013). *Cultura da participação e o saber-fazer do turismo: Estudo de caso observatório para o turismo sustentável de Cavalcante – GO* [Masters dissertation, Universidade de Brasília]. UnB Repository. https://repositorio.unb.br/handle/10482/13807

Santos, B. S. (2004). The WSF: Toward a counter-hegemonic globalization. In A. Sen, A. Anand, A. Escobar, & P. Waterman (Eds.), *Titulo do livro* (pp. 235–245). The Viveka Foundation.

Sauer, S., Leite, A. Z., & Tubino, N. L. G. (2020). Agenda Política da Terra no Governo Bolsonaro. *Revista da Anpege, 16*(29), 285–318.

Scott, M. (2008). Managing rural change and competing rationalities: Insights from conflicting rural storylines and local policy making in Ireland. *Planning Theory & Practice, 9*(1), 9–32. https://doi.org/10.1080/14649350701843689

SEBRAE. (2006). *Atendimento ao turista ganha reforço na Chapada dos Veadeiros.* http://www.sebrae.com.br/sites/asn/uf/GO/atendimento-ao-turista-ganha-reforco-na-chapada-dos-veadeiros,00cb36fdbd45641OVgnVCM1000003b74010aRCRDb

Sharpley, R. (2020). Tourism, sustainable development and the theoretical divide: 20 years on. *Journal of Sustainable Tourism, 28*(11), 1932–1946. https://doi.org/10.1080/09669582.2020.1779732

Siakwah, P., Musavengane, R., & Leonard, L. (2019). Tourism governance and attainment of the sustainable development goals in Africa. *Tourism Planning & Development, 17*(4), 355–383.

Similä, J., & Jokinen, J. (2018). Governing conflicts between mining and tourism in the Arctic. *Arctic Review on Law and Politics, 9*(0), 148–173. https://doi.org/10.23865/arctic.v9.1068

Sobrinho, F. L. A., Alves, I. C. S., & Vieira, P. L. M. (2015). Uma análise geográfica do plano de desenvolvimento turístico de Cavalcante-Goiás. *Boletim de Geografia, Maringá, 33*(3), 31–45.

Stone, M. P., & Nyaupane, G. P. (2016). Protected areas, tourism and community livelihoods linkages: A comprehensive analysis approach. *Journal of Sustainable Tourism, 24*(5), 673–693. https://doi.org/10.1080/09669582.2015.1072207

Strassburg, B. B. N., Brooks, T., Feltran-Barbieri, R., Iribarrem, A., Crouzeilles, R., Loyola, R., Latawiec, A. E., Oliveira Filho, F. J. B., Scaramuzza, C. A. d. M., Scarano, F. R., Soares-Filho, B., & Balmford, A. (2017). Moment of truth for the Cerrado hotspot. *Nature Ecology & Evolution, 1*(4), 0099. https://doi.org/10.1038/s41559-017-0099

Telfer, D., & Sharpley, R. (2007). *Tourism and development in the developing world.* Routledge.

UNESCO. (2008). *Madrid action plan for biosphere reserves (2008—2013)*. http://www.unesco.org/new/en/natural-sciences/environment/ecological-sciences/man-and-biosphere-programme/strategy-and-action-plan/madrid-action-plan/

UNESCO. (2011). *Biosphere reserve information: Brazil – Cerrado*. http://www.unesco.org/mabdb/br/brdir/directory/biores.asp?mode=all&code=BRA+02

UNESCO. (2019). *Cerrado protected areas: Chapada dos Veadeiros and Emas National Parks*. https://whc.unesco.org/en/list/1035/documents

WCED. (1987). *Our common future* (G. H. Brundtland, Ed.). Oxford University Press.

Westra, R. (2010). *Confronting global neoliberalism: Third world resistance and development strategies* (R. Westra, Ed.). Clarity Press.

Wilson, E., & Hollinshead, K. (2015). Qualitative tourism research: Opportunities in the emergent soft sciences. *Annals of Tourism Research*, *54*, 30–47. https://doi.org/10.1016/j.annals.2015.06.001

Connecting landscape-scale ecological restoration and tourism: stakeholder perspectives in the great plains of North America

Connor Clark and Gyan P. Nyaupane

ABSTRACT

This grounded theory research seeks to understand stakeholder attitudes towards tourism development and ecological restoration around a landscape-scale ecological restoration project in Montana, USA. Due to the limited number of landscape-scale ecological restoration projects across the globe, research on stakeholder attitudes and the relationship of tourism with such projects is missing. Only recently have researchers begun to take an interest in the relationship between tourism and rewilding, a subfield of ecological restoration that involves the restoration of wildlife. The results derived from interviews with various stakeholder groups, including nonprofit conservation groups, a national wildlife refuge, the local ranching community, a county commission, and local residents working in the tourism industry demonstrate important, emerging themes, such as contrasting worldviews, ecological paradigm, landscape-scale land management, and lack of shared community vision. As local ranchers are particularly threatened by landscape-scale ecological restoration efforts, integrated threat theory is used to describe intergroup conflict in a tourism and ecological context. A revised tourism area life cycle (TALC) model using a tourismscape framework and core-periphery concept is proposed to understand the interrelationships between regional tourism and landscape-scale ecological restoration areas. The importance of bottom-up solutions, community-driven initiatives, and stakeholder engagement in the planning process is emphasized.

Introduction

While scholarly literature surrounding the environment often reports on the negative impacts resulting from human activities, certain conservation practices, such as landscape-scale ecological restoration (ER), help to reverse the effects of habitat and biodiversity loss (Moore et al., 1999). Tourism and ecological studies are mostly focused on understanding the negative consequences of tourism activities on species and habitat, or the communities who live around protected areas (PAs) (Dodds & Butler, 2019; Stankey & Cole, 1985; Walpole & Goodwin, 2001). Scholars have also examined attitudes towards tourism development and biodiversity conservation in communities near traditional PAs (Ghoddousi et al., 2018; Mbaiwa & Stronza, 2011; Stone & Nyaupane, 2018) where rural livelihoods are threatened or improved by conservation and tourism practices. More recently, there is a growing body of literature that supports tourism's role in conservation (Liu et al., 2012; Nyaupane & Poudel, 2011; Poudel et al., 2016).

A plethora of studies have also looked at attitudes towards ER activities, such as restoring rivers (Åberg & Tapsell, 2013) and reintroducing large carnivore species (Caruso & Jiménez, 2013), but few have looked at stakeholder attitudes towards tourism development in an ER context. Despite the growing research interest on the concept of rewilding, a subfield of ER that focuses on the restoration of wildlife (Anderson et al., 2019; Gillson et al., 2011), its connection with tourism development has only been brought up very recently (Hall, 2019). Because this line of research is so new, there is a paucity of academic assessment of stakeholder attitudes towards tourism development within landscape-scale ER projects. The purpose of this study, therefore, is to increase our understanding of how stakeholders perceive tourism development and ER in and around landscape-scale ER areas.

Initially, the conservation practice of protecting natural areas was for the benefit of the railway industry and for attracting tourists to iconic landscapes, such as Yosemite, Yellowstone, Banff, and Grand Canyon national parks (Jamal & Stronza, 2009). However, since the Great Plains of North America lacks tall mountains or deep canyons, it was overlooked during the initial park creation movement (Flores, 2016). This juncture between tourism and PAs is widely documented, and research shows that there are both positive and negative social, economic, and environmental outcomes resulting from tourism development within and near PAs (Dodds & Butler, 2019; Stronza, 2007). By employing grounded theory, this study seeks to fill this void of understanding regarding stakeholder views towards tourism development and ER efforts at a landscape-scale ER project in north central Montana, USA, led by the nonprofit, American Prairie Reserve (APR). Stakeholder theory holds that attitudes will vary among stakeholder groups, and these attitudes should be heard and included beginning from the initial planning stages of tourism development (Jamal & Stronza, 2009; Mbaiwa & Stronza, 2011) and ER projects (Clewell & Aronson, 2013).

This study makes several contributions to tourism research. First, this study contributes to the current knowledge gap regarding tourism's impact and role with landscape-scale ER projects that is only beginning to be addressed. Moreover, this study contributes to tourism literature by presenting a revised TALC model that incorporates spatio-temporal dimensions which combines both resident attitudes over time and tourism development across core and peripheral areas within tourism and ecological landscapes. Finally, this research contributes to the current understanding of community perceptions of threat brought on by tourism development and ER by highlighting the complex social attitudes towards these topics. By filling this research void, destination planners, conservation organizations, and communities near ongoing, landscape-scale ER projects can better prepare for the social, economic, and environmental responses that can result from these efforts.

Literature review

Previous research analyzing the nexus of tourism development, stakeholder attitudes, and the natural environment have typically focused on social, economic, and conservation-related issues as they pertain to communities near PAs (Jamal & Stronza, 2009; Ghoddousi et al., 2018; Mbaiwa & Stronza, 2011; Nyaupane & Poudel, 2011). In these studies, the challenges and benefits of tourism development in and around traditional PAs to surrounding communities and their natural environment are highlighted. Few studies, however, have looked at the interactions between tourism development and ER efforts (Hall, 2019), and how such interactions impact local stakeholders (Koninx, 2019; Leduc & Von Essen, 2019; Pellis, 2019). This is likely due to the limited number of landscape-scale ER projects that can be studied and the recency of this form of ecological work. Similarly, there is a paucity of research that analyzes stakeholder views towards tourism development across a regional scale between core and peripheral areas. Hence, by highlighting stakeholder attitudes near one of the few landscape-scale ER projects that exist, little known social intricacies behind this recent practice can be better understood and a framework for tourism development in this context can be developed. The following literature review provides context by summarizing existing tourism development models and frameworks to understand residents' attitudes towards tourism development and ER.

Tourism development models

Various tourism development models, conceptual frameworks, and attitudinal scales have been evolving to understand tourism development and its relationship with stakeholders since the 1970s. The most widely cited literature on tourism development is Butler's (1980) Tourism Area Life Cycle (TALC) model (Hall, 2006). According to the TALC model, destinations function as tourism products and will pass through various stages along its life cycle, including exploration, involvement, development, and consolidation, after which a critical point is reached and three trajectories unfold: 1. Stagnation 2. Decline, or 3. Rejuvenation (Butler, 2006). Other scholars, such as Hovinen (2002) and Johnston (2006), feel that a maturity stage needs to be inserted into this framework, as many destinations have stable and constant tourist flows. Notwithstanding, destinations repeatedly show that a critical point exists (Clark & Nyaupane, 2020), and once a destination arrives at this critical point, local sentiments towards tourism can radically decline (Pechlaner et al., 2019). Doxey (1975) accounts for these declining sentiments in the Irritation Index, which posits that local residents initially view tourism with euphoria but then pass through stages of apathy, irritation, and eventually antagonism. Both the TALC model and the Irritation Index have made a lasting impact on tourism research as many studies have used these frameworks as a reference point to further build on (Clark & Nyaupane, 2020; Hovinen, 2002; Hunt & Stronza, 2014; Johnston, 2006). Although these frameworks are frequently cited in academic literature, some scholars find them to be incomplete.

Hunt and Stronza (2014) argue that both the TALC model and Irritation Index, among other frameworks, need to include a pre-tourism phase that accounts for destinations lacking prior experience with tourism and are less aware of its impacts. We similarly feel that the TALC model needs to be adapted in the context of ER in peripheral areas because the model treats each destination independently and ignores the fact that many destinations have either interdependency or hierarchy with other destinations, as indicated by the core-periphery concept (Murphy & Andressen, 1988; Salvatore et al., 2018).

Tourism development can bring about economic development for communities. In rural, peripheral communities, nature-based tourism may compete with agricultural livelihoods, such as herding, ranching, farming, and resource extraction industries (Davenport, 2018; Hall & Boyd, 2005; Nyaupane & Poudel, 2011; Stone & Nyaupane, 2016). Under such circumstances, these scholars warn that tourism should not be viewed as a replacement to existing industries but as complementary to them. More importantly, enhancing peripheral, rural areas through tourism largely depends on "core" areas where tourism flows, facilities, and services are concentrated (Salvatore et al., 2018). Thus, the core-periphery concept (Fennell, 1996) can be crucial for tourism research to recognize the spatial and economic relationships between destinations and sectors, which helps to form a regional tourism development plan (Yu-Ming, 2002). Core and peripheral areas, however, are often wrought with inequality and uneven power structures (Saarinen et al., 2017). Scholars additionally urge that nature-based tourism in peripheral areas should coincide with regional development plans and should balance regional development objectives with natural asset preservation (Hall & Boyd, 2005; Peeters et al., 2018). Furthermore, Chaperon and Bramwell (2013) illustrate how agency and dependency for peripheral communities involved in tourism are intimately interwoven as they often depend on core areas for tourists but can still maintain control over how tourism develops.

Considering the spatial scale of tourism dynamics between core and peripheral areas, using the tourismscape framework is useful. This framework is based on the economic-centered cluster concept introduced by Porter (1998) where a company's health is interdependent on the presence of competitors across a region. In tourism, clusters within a tourismscape represent attractions that form the backbone of tourism in a region (Gunn, 1997). Van Der Duim (2007) emphasizes the importance of viewing tourism as part of a tourismscape where a complex process of ordering between people, objects, organizations, and spaces occurs. Additionally, the primary components of a tourismscape are service communities, attractions, and the linkage corridors between them (Gunn, 1997). Linkage corridors connecting service communities to attraction clusters are an essential element of tourism in peripheral areas, as tourism is not possible without access (Hall & Boyd, 2005). To ensure the success of nature-based tourism development across a region, local stakeholders from each community need to retain control in order to guide such developments (Gunn, 1997; Johnson & Payne, 2005). In this study, we look at a tourismscape through an ecological lens that spans across core PAs (national parks) that attract tourism and have developed tourism services in gateway communities to peripheral conservation areas (ER sites, undeveloped or semi-developed parks) lacking more developed tourism infrastructure.

Successful planning and implementation of tourism development entails recognition, participation, and negotiations among all stakeholder groups (Breakey et al., 2019; Jamal & Stronza, 2009; Pechlaner et al., 2019; Poudel et al., 2016). Excluding stakeholder groups from discussions and the planning process can seriously thwart tourism development objectives and will most likely lead to conflict (Breakey et al., 2019). For this purpose, planning and policy making should be an inclusive and regional endeavor that avoids the fatal growth paradigm and embraces a regional development paradigm instead (Peeters et al., 2018). The growth paradigm has led many destinations into a state of overtourism (Dodds & Butler, 2019), which has started a larger conversation about decision-making, bottom-up solutions, and the rural/urban divide (Pechlaner et al., 2019). According to Hunt and Stronza (2014), stakeholder involvement in decision-making leads to increased positive attitudes towards tourism development, especially when local stakeholders possess managerial and ownership roles. In essence, those who are involved in tourism or benefit from it are more likely to view it favorably, which is the main tenet of social exchange theory (Andereck et al., 2005).

This myriad literature demonstrates various models and theories, such as the TALC model, Irritation Index, and social exchange theory, that are used to explain the development process and the resident attitudes toward tourism development. However, these models are limited to independent, single destinations and cannot fully explain where the destinations are spatially interconnected to form a landscape-scale tourismscape. The literature further shows how stakeholder attitudes are closely linked with the level of involvement in decision-making, the proper balance of growth and community well-being, and the effectiveness of conflict interventions. This same thought process for stakeholder engagement in the tourism planning and development process is equally critical for successful ER projects and will be detailed in the following section.

Ecological restoration

ER is the process of helping ecosystems recover their natural functions after having been disturbed, degraded, or destroyed (Clewell & Aronson, 2013; Suding et al., 2015). Many scholars see ER as a necessary process to reverse human-caused environmental damage and to mitigate the environmental impacts of climate change (Harris et al., 2006). In the field of restoration ecology, looking to history can be useful for finding appropriate benchmarks or guidelines for ER work (Jackson & Hobbs, 2009), although it can never serve as a clear template (Suding et al., 2015). In North and South America, Australia, and Oceania, historical benchmarks for restoration efforts typically correspond with pre-European settlement conditions, whereas in Europe, Asia, and Africa, historical benchmarks are more difficult to rely on (Jackson & Hobbs, 2009). In these cases, scholars refer to paleo ecological records, such as tree rings, rodent middens, and sediments from lakes, peatlands, estuaries, and oceans to help form goals for a restoration project where historical conditions are not easily discerned (Jackson & Hobbs, 2009).

In certain parts of the world, rural to urban migration trends among young people present an opportunity for ER interventions (Navarro et al., 2015). However, this migration trend also results in a loss of local socio-ecological knowledge and stewardship of landscapes (Allen, 2015). Looking at the broader economic impacts of ER, Zhang et al. (2011) find that ER facilitates the redistribution of industries with tourism as an alternative to extractive industries. ER is also essential for conservation goals as PAs alone are often not big or diverse enough to achieve these goals (Mattsson et al., 2020; Nabhan et al., 2014). For this reason, peripheral areas that often rely on agricultural production are crucial for biodiversity conservation, and regional cooperation between public and private actors across a landscape is vital (Nabhan et al., 2014).

In recent decades, the concept of 'rewilding', which typically refers to the wildlife restoration component of ER, has further demonstrated the need for regional cooperation as it has gained traction by practitioners and researchers (Arts et al., 2016). A number of scholars view rewilding as a subfield of ER (Anderson et al., 2019), and according to Navarro et al. (2015), rewilding is just one of the tools used to accomplish the goals of ER. Nevertheless, the topic is beginning to interest scholars as evidenced by recent research that looks at rewilded areas as potential tourist attractions (Hall, 2019; Pellis, 2019). Scholars refer to the three C's of rewilding as: cores (PAs); corridors; and carnivores (Arts et al., 2016). These three elements are critical for ecological landscapes as core PAs are part of a much broader ecological region (Chester, 2015) that ER can help to reconnect (Bauduin et al., 2018). Reintroducing large carnivores, however, is usually a polarizing endeavor since livestock producers bear the cost of these actions through livestock depredations and other related stresses (Rashford et al., 2010). Large herbivores can also be problematic to agricultural livelihoods because of concerns over disease transmission, crop damage, and competition for grazing resources (Kilpatrick et al., 2009; Mbaiwa & Stronza, 2011; Morrisette, 2000; Stone & Nyaupane, 2018).

Since attitudes towards ER efforts, such as animal reintroductions, can be mixed (Caruso & Jiménez, 2013; Koninx, 2019; Leduc & Von Essen, 2019; Pellis, 2019; Smith et al., 2003), it is paramount that all relevant stakeholders be involved in such endeavors (Byrd, 2007; Marshal et al., 2007). Ideally, ER efforts should start within the community and have the support of all stakeholders (Clewell & Aronson, 2013). Such are the underpinnings for stakeholder theory, which posits that stakeholders must be involved in the entire process, starting from the initial planning stages all the way through the execution stages (Jamal & Stronza, 2009). Stakeholder involvement in ER projects is key as these projects require substantial financial investment and high levels of public support for successful completion (Wohl et al., 2005).

When considering stakeholder attitudes towards ER projects, there can be a wide variety of outlooks on what constitutes the successful completion of such a project (Jähnig et al., 2011). These scholars illustrate how much of the general public appreciates the aesthetic values of plants and animals while conservationists and biologists place considerable value on the restoration of ecological processes. On the other hand, agricultural producers whose livelihoods can be affected by ER efforts, might strongly oppose them (Rashford et al., 2010). Despite these mixed attitudes, the implications of wildlife-based tourism as an ecosystem service can equate to tens of millions of dollars in annual revenue for communities near ER projects (Sage, 2017). According to Pellis (2019), since rewilding proposals can be met with local opposition, rewilding practitioners often incorporate economic benefits in the form of ecotourism to provide local employment opportunities, which can bring its own set of social tensions.

Research on tourism development in conjunction with ER is quite limited (Hall, 2019). Some research has looked at the financial, political, and sociocultural challenges organizations face when working to restore biodiversity (Bailey, 2015; Davenport, 2018; Pellis, 2019), but commonly lack a deeper analysis of stakeholder attitudes. Other research has evaluated attitudes towards smaller restoration projects in terms of land base (Åberg & Tapsell, 2013), but these projects have not been at a landscape-scale where many communities could potentially be impacted and do not consider tourism development. Considering the potential benefits of tourism to ER efforts, some scholars claim that tourism can act as an agent to conserve natural and cultural landscapes (Burger, 2000; Murphy, 1986). For these reasons, a deeper analysis of stakeholder attitudes near a landscape-scale ER project is needed, and since there is a paucity of literature on this emerging issue, a grounded theory approach can help gain more insight.

Methods

Grounded theory

Grounded theory was considered the most effective approach for understanding stakeholder attitudes (Kensbock & Jennings, 2011) towards tourism development and ER because it is well suited for exploratory research. The goal of grounded theory research is to develop a theory or framework from the data that explains events and relationships based on the lived experiences of individuals (Charmaz, 2006; Glasser & Strauss, 1967; Hardy, 2005; Matteucci & Gnoth, 2017). Rather than fitting data into a preconceived theory or set of codes, grounded theorists follow leads that they define in their data (Charmaz, 2006). Grounded theory research works in a reverse fashion in comparison to traditional research where researchers begin by selecting a theory and then proceed to form hypotheses, collect and analyze data, and make conclusions about the theory (Nyaupane & Poudel, 2012). Instead, data collection and analysis go hand-in-hand with grounded theory as theory is grounded or generated from the participants who experience the studied phenomenon (Charmaz, 2000).

The purpose of grounded theory is to develop middle-ranged theories or frameworks based on the uniqueness of a specific social setting and then apply such theories or frameworks more broadly outside that setting (Hardy, 2005; Matteucci & Gnoth, 2017). Although grounded theory research does not begin a study with an existing theory in mind, this type of research should not be done in isolation to existing theories (Glasser & Strauss, 1967). Notwithstanding, some scholars warn that prior knowledge can be a barrier to entering a study with an open mind (Connell & Lowe, 1997). Grounded theory allows for the research design to evolve in response to emerging themes (Daengbuppha et al., 2006), and the derived theory should be based off of the needs of local participants experiencing the phenomenon (Matteucci & Gnoth, 2017). Our study uses the constructivist grounded theory method which is value-laden, has a realist ontology and a relativism epistemology, and holds that multiple realities exist (Charmaz, 2008). This theoretical approach aims to illustrate the complexity of such multiple realities, views, and actions (Creswell et al., 2007).

Data analysis in grounded theory research consists of systematic steps using open coding, where categories are formed from the data; axial coding, where specific categories are positioned within a model; and selective coding, where categories within the model are interrelated around a central explanatory concept (Creswell, 2007; Creswell & Creswell, 2018; Strauss & Corbin, 1998). The data analysis process begins before all data are collected in order to build theory and drive further data collection from the appropriate contacts by means of theoretical sampling (Matteucci & Gnoth, 2017). Theoretical sampling does not follow a pre-planned approach for data collection but is a form of purposive sampling where individuals from different groups are contacted based off their unique knowledge and exposure to a topic in order to magnify the differences and similarities between responses (Daengbuppha et al., 2006; Nunkoo & Ramkissoon, 2016). Developed theories must do more than provide a description of events or relationships; they must help explain and predict such events or relationships to provide a helpful guide to action (Strauss & Corbin, 1998). The job of the grounded theorists is not to just describe the world as they see it, but to change the world for the better through the construction of their theories (Matteucci & Gnoth, 2017). Given this explanatory, predictive, and impactful charge of grounded theory in social science research, and having been privy to the conflicts surrounding tourism development, natural resource management, and livelihoods in the study area and around the world, we opted to use grounded theory to help us navigate through these issues and to conceptualize the poorly understood impacts of ER and tourism development on local stakeholders.

Study area

This study focused on the communities and stakeholders surrounding and within a landscape-scale ER project led by APR. Most of the study participants live and work in the relative vicinity of the two largest towns in the region: Lewistown and Malta (Figure 1). APR seeks to assemble a 3.5-million-acre wildlife reserve in north central Montana by strategically purchasing ranches from willing sellers adjacent to the Charles M. Russell National Wildlife Refuge (CMR) and the Upper Missouri River Breaks National Monument (Bailey, 2015). Once a property comes under the stewardship of APR, the organization carries out a series of actions to restore any missing

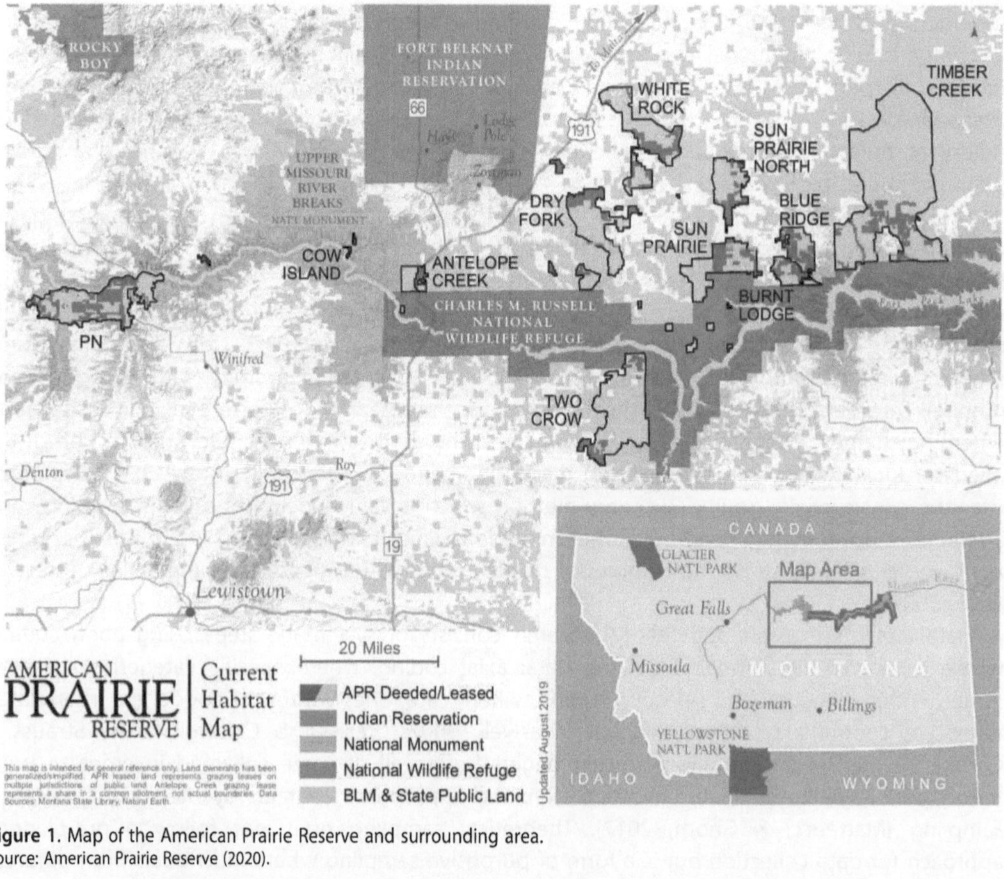

Figure 1. Map of the American Prairie Reserve and surrounding area.
Source: American Prairie Reserve (2020).

ecological processes. The nonprofit currently owns or has the public land leases to over 400,000 acres of land (Figure 1). Another pillar of APR's mission is to facilitate the development of a nature-based economy centered on tourism. To achieve their mission, APR has constructed developed camping areas and other lodging options, such as huts and cabins, for visitors.

Data collection and analysis

Data for this study were obtained through in-person and phone interviews with 21 participants from core stakeholder groups who possess in-depth knowledge about tourism development and ER or conservation issues in the study area. Data were also obtained through secondary sources, such as APR's website and the Bureau of Land Management's (BLM) publicly available documents. Study participants included individuals from the local ranching community, federal land management agencies, nonprofit conservation organizations, the local tourism sector, local government agencies, and in-state and out-of-state tourists. Participants were selected through a theoretical sampling method in order to build on theoretical concepts arising from interviews (Strauss & Corbin, 1998). We initially recruited participants with a purposive sampling strategy by directly contacting nonprofits, county commissions, and PAs in the study area and interviewing their leaders and representatives. When important issues were raised during interviews, we used a snowball sampling strategy to contact individuals representing other stakeholder groups, such as local ranchers, with an intimate knowledge of such issues and invited them to participate in the study. To interview tourists, however, a convenience sampling method was used to approach tourists at their campsites in designated campgrounds as they were in between activities. Interview questions were tailored to each participant for the purpose of uncovering their views towards the region becoming a nature-based tourism destination featuring abundant wildlife populations. Interviews lasted between 35 minutes and two hours. All interviews except for one were audio recorded for accuracy purposes. One tourist was not recorded due to weather conditions. In this instance, field notes were used during the interview.

Data collection and analysis were simultaneously conducted to look for underlying themes and patterns (Charmaz, 2000). Once the 21 participants were interviewed, the authors felt that the data had been saturated and that additional interviews would yield little new information. As the interviews were completed, the audio recordings and field notes were transcribed and analyzed through a content analysis using Maxqda software. The coding process began by first conducting open coding where many themes were identified from certain words, phrases, or combination of words. Following the process of open coding, data were then brought through an axial coding process to group the large group of themes into a smaller group of 17 categories. Finally, selective coding was used to merge categories into four overarching concepts (Nunkoo & Ramkissoon, 2016). The selective coding process was guided by grounded theory to identify these four overarching concepts that emerged from the data.

Results

The content analysis and coding process revealed four overarching themes arising from participant interviews. These themes were contrasting world views, landscape-scale land management, ecological paradigm, and lack of shared community vision. The interviews clearly demonstrated a wide spectrum of attitudes towards tourism development and ER in this region of Montana. The four themes inform a spatio-temporal tourism development model for landscape-scale regional tourism (Figure 2) that we elaborate on in the discussion section of this paper.

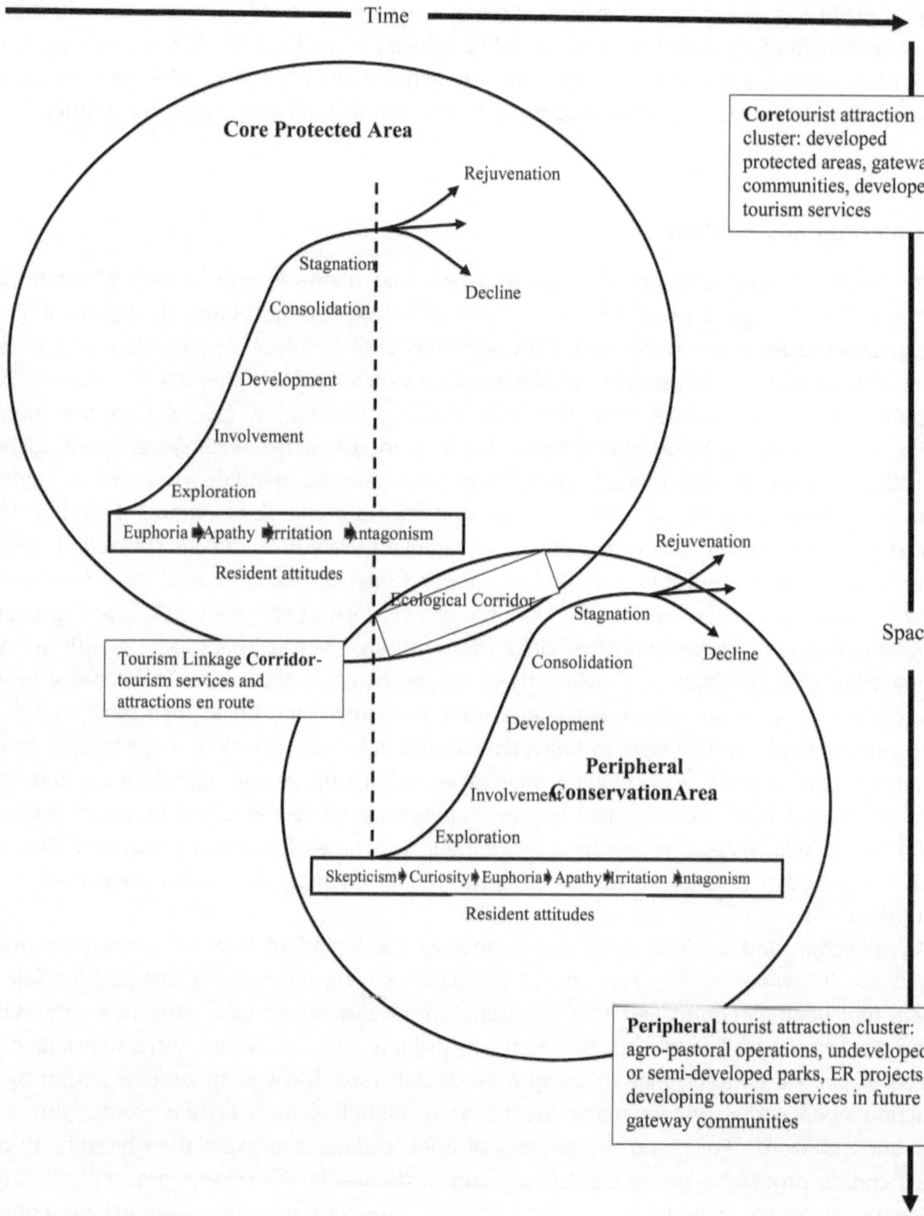

Figure 2. Spatio-temporal TALC model for landscape-scale tourismscapes.

Contrasting world views

There is ample evidence of highly contrasting worldviews between stakeholder groups in the study area. The widest range of attitudes were represented by the local ranchers and a political representative at one end and APR staff on the other. One APR manager illustrated this contrast by stating:

> Our vision for the landscape is one where biodiversity is the variable we are trying to maximize. Currently, ranchers, the variable they're trying to maximize is protein off-take. So, they want their cows as fat as possible. Those are different world views. One's not better. One's not worse.

A significant portion of the interviews with individuals from these two stakeholder groups was spent either validating or explaining their points of views. Ranchers also expressed their mistrust in the environmental groups. One rancher put it this way:

> Right at this very moment, APR is not, has not presented itself in a way in eastern Montana that most people locally see as an asset. They see it as a threat.

As a separate example, another rancher expressed their frustration and mistrust by stating,

> And at this point, I don't think that there's anything they [APR] can do to gain the trust of the people in Phillips County that don't, like, don't support their efforts. I can't imagine what they could do to get my trust back. They've just proven themselves over and over again to be untrustworthy.

Not only were contrasting world views apparent between ranchers and APR, but also between domestic and out-of-state tourists. Some in-state tourists shared many of the same sentiments as the local ranchers given their upbringing in rural Montana. They lamented how their community continues to shrink as people move out of the area and feel that the vision to create a large nature reserve and increase tourism exacerbates this declining population trend. Out-of-state tourists, on the other hand, came to visit the area for the very purpose of seeing the in-progress ER project that they had heard about or read about in outdoor magazines. Interestingly, some of the out-of-state tourists were from a part of the U.S. that is also seeing a population decline and have seen how tourism helped revive areas undergoing economic and population declines. An out-of-state tourist described how in one town,

> ... people got together, and they built up some bike trails. Within 10 years, they became some of the most popular and highly rated bike trails in the country ... And in the early days there were a lot of land owners that said, "I don't want these damn bicycles going across my property". All that kind of thing. It's not buffalo, it's bicycles, right?

This quote illustrates the width of outlooks towards tourism development and ER among stakeholders in the study area. The contrasting world views held by individuals in this region present challenges for developing a shared vision on how to protect and manage the larger landscape in a manner that appeases all stakeholder groups.

Landscape-scale land management

The theme of landscape-scale land management emerged during interviews, as land within the study area consisted of a patchwork of private, federal and state (public), and indigenous lands. Much of the rangeland used by ranchers and APR is public land managed by the BLM, the largest federal land management organization in the U.S. Interviews with domestic tourists highlighted the concern over access to public land, primarily for hunting and fishing. One in-state tourist expressed outrage that billionaires from Texas had bought several large properties in the region for their own private hunting use and closed off access to the public land parcels within their ranches. This in-state tourist stated,

> They come in and they just start buying land, buying up land, blocking people out, and pissing people off. And then they wonder why the locals have dug in their heels.

Since public land is owned by the public, public access is mandated by the BLM (Chavez, 1987). However, when an individual's private property surrounds a parcel of public land, private land owners have the right to close off access through their lands, rendering that parcel of public land inaccessible to the rest of the public. Such actions are condemned by local hunters, as demonstrated by this in-state tourist.

Grazing practices, rights, regulations, and privileges are a very important component of landscape-scale land management in the region. Grazing on federal lands can be a controversial topic for ranchers when they feel that a federal regulation does not take their best interests into account. For example, with the creation of the CMR refuge in 1976 came more stringent grazing regulations for the ranchers whose cattle had been grazing in this area. A manager from the wildlife refuge explained how legally they could have halted grazing immediately but,

> it was decided not to do that at first because they felt like it wasn't politically feasible. … [We're] trying to be socially responsible and not kicking a bunch of people off that have been using that land for multiple generations to make their living.

Due to the mistrust that can exist between ranchers and the federal government, the CMR refuge is striving to balance conservation values with accommodating local livelihoods by allowing the ranching families who had existing grazing leases on the refuge to continue grazing their cattle as long as they or their direct descendants choose this profession.

Data obtained from interviews with APR staff and secondary sources revealed their innovative approach to manage landscape-scale conservation and ER through a public-private partnership. As a nonprofit, APR can purchase private property as well as the leases to public land and make decisions to maximize biodiversity across their landholdings. Interviews revealed challenges, however, for APR to reintroduce wildlife species as these decisions must be made by state and federal fish, wildlife, and parks agencies who must satisfy all their constituents, including ranchers who want to reduce wildlife conflict with their operations. Another challenge for APR in conducting landscape-scale ER was evident in their 2018 request to the BLM to replace cattle with bison on the majority of their public land leases. Stiff public opposition quickly resulted, and APR drastically scaled back their request in 2019 in response to this opposition.

Ecological paradigm

Stakeholder attitudes towards ER and conservation widely varied. The mere premise that conservation groups are working 'save' the grasslands is perceived as insulting to many ranchers who feel that they and their forbearers have been good stewards of these lands for generations. One rancher describes this scenario:

> … this land wouldn't be the desirable location that it is … if the landowners that are on it now and ranching haven't done something right to get it to that point. And it, quite frankly, angers me that they think that they're going to save it, 'cause I don't know what they're saving it from.

Ranchers were also concerned about the return of large wildlife species, such as bison, grizzly bear, or wolves that can directly harm agricultural operations through depredations, disease transfer, property damage, personal safety, or consumption of resources intended for livestock. For participants in this region, the anxiety surrounding bison is especially prevalent given APR's goal to restore them in abundance. To address these concerns, APR developed an innovative tool called the Wild Sky initiative that compensates ranchers by the number of photos of large carnivore species caught on trail cameras placed on their ranches. One participating rancher lauded the initiative and felt that the compensation was helpful for their ranch operation. APR, however, can only afford to work with a limited number of ranches but plans to expand the program in the future.

In response to environmental groups' efforts in the area, local grassroot organizations have formed that work to conserve and restore grasslands in a way that does not disturb traditional agricultural livelihoods. One such group, the Rancher's Stewardship Alliance, has even received grant funding to restore 10,000 acres of grassland. This clearly demonstrates how all stakeholders value the protection and restoration of the grasslands. Yet, when it comes to restoring bison, local ranchers see bison as a symbol of the threat to their way of life. Just as Anderson et al. (2019) pointed out, rewilding and ecological restoration are not one in the same.

The ecological paradigm of ranchers presents challenges for executing the vision to re-create a large prairie ecosystem based on a pre-European historical reference. Another nonprofit environmental group in the area, The Nature Conservancy (TNC), uses a conservation model that is received more positively by private land owners because it seeks to preserve existing grasslands and provides incentives for ranchers to cooperate with their vision. One local rancher praised the TNC for their grass bank model where 24 different ranching families lease sections of a TNC ranch for grazing livestock at a reduced rate in exchange for complying with certain grassland conservation measures.

Lack of shared community vision

Lack of shared community vision was the strongest theme emerging from interviews. While all stakeholder groups reiterated the importance of forming partnerships to secure a joint vision for this region, many felt that this had yet to happen. Stakeholder cooperation was seen as vital for this region where the economy is continually evolving and the population has been declining for many decades. Study participants also articulated that there is presently not a joint vision among groups but instead conflict. For many study participants, creating a popular tourism attraction is not desired as living in a state with highly-visited national parks gives them an awareness of the economic, social, and environmental drawbacks inherent in a tourism economy. Locals expressed skepticism that tourists would even want to visit a remote area where few amenities exist. Local residents also did not like the thought of large tourist crowds, given the negative impacts of hunters who have left trash and cattle gates open, traveled at high speeds through the community, and caused traffic congestion. One county commissioner expressed disdain about tourism crowds in Montana's national parks that supports Pellis' (2019) finding that tourism brings its own set of social tensions:

> We have Yellowstone and we have Glacier, and I very seldom go to those areas because the last thing I want to do is rub elbows with a zillion tourists from somewhere. I mean, suddenly, it's not- that's Disneyland. That's not what I live here for.

The desire for controlled tourism growth was observed across all stakeholder groups, even among the out-of-state tourists who support APR. Interestingly, some rancher participants displayed curiosity about the potential for tourism development during interviews and even discussed their ideas to accommodate tourists. Other participants shared how the rancher community grassroot organizations they lead or belong to are envisioning a private land model for ER, conservation, and tourism development that allows for tourists to experience Montana's plains in a ranch setting. An out-of-state tourist even cited tourism development in the region as a method to combat the overtourism phenomenon in Montana's national parks by dispersing tourists to the peripheral regions:

> We should bring tourism out here. This would relieve pressure on the national parks. People are hungry to see wilderness and wildlife, but you can't even get a parking spot on Logan Pass after 7 o'clock in the morning.

While stakeholder groups in the region appear to be in conflict, there is still common ground to be found. All participants agree on the importance of strengthening local communities and preserving the grasslands through collaboration. An APR employee living on one of APR's landholdings made it clear that their vision requires the wellbeing of neighboring ranches: "We want to be successful, so we want our neighbors to be successful as well so that we can work together". Similarly, one local rancher posed questions for how all parties can be satisfied:

> I mean, in the end it's, like, can you protect the grassland, can you sustain these rural communities, can you protect family ranching? Can you do all that, and can you still have as great a component of wildlife that could be expected to, you know, coexist with human beings out there?

The goals to preserve an endangered ecosystem and strengthen rural communities are possible, according to participants, but how to accomplish these goals remains unestablished. Through increased collaboration and communication, a local tourism outfitter believes this can be done. This outfitter offered a reminder that, "we're not all that different after all. If we would just talk to each other and sit down and not pin each other against each other."

Discussion

The attitudes towards tourism development and ER held by study participants from various stakeholder groups exhibit a broad spectrum of viewpoints and challenge existing tourism development and attitudinal models. The diversity of attitudes corresponds with previous findings concerning stakeholder attitudes as the interests, values, and goals of different groups widely diverge (Dodds & Butler, 2019; Navarro et al., 2015). The local residents' objections to tourism development adds complexity to existing attitudinal scales in tourism literature which assumes that attitudes will initially be positive towards the addition of tourism and then decline as tourism development progresses, as indicted by the Irritation Index (Butler, 2006; Doxey, 1975). Stakeholder groups are highly aware of the impacts of tourism through observing the overtourism phenomenon in Montana's popular national parks (core areas) and know very well that tourism in their state is seasonal because of harsh winters and short summers. Some of the participants have also already witnessed undesirable behavior from visiting hunters. This educated outlook on tourism's negative impacts creates a need for adopting a new tourism development model that accounts for spatial and ecological relations between core and peripheral areas across a regional landscape, as well as the temporal aspect of changing resident attitudes.

We propose a spatio-temporal tourism development model for landscape-scale regional tourism that embeds Butler's (1980, 2006) TALC model and Doxey's (1975) temporal Irritation Index in a core-periphery, ecological, and spatial dynamic (Figure 2). The model shows that when peripheral conservation areas outside of core PAs are in a pre-exploration stage, the sentiments of skepticism and curiosity preclude euphoria. Skepticism stands at the front of the process because of a disbelief in the potential or ability for tourism to create a viable alternative to existing industries. According to the model, peripheral attraction clusters can develop once tourism access is facilitated through linkage corridors to core attraction clusters. In this fashion, a landscape-scale tourismscape resembles landscape-scale conservation and ER where peripheral habitat is connected to core PAs through ecological corridors, so wildlife and ecosystems can have sufficient space to thrive (Arts et al., 2016). In a tourismscape, a successful linkage corridor from core to periphery should provide essential services and other tourism opportunities while in route (Gunn, 1997). For the area of focus in this study, the tourism opportunities span both a natural and cultural landscape that can offer tourists a wide variety of opportunities where ranching, indigenous, and rural communities can benefit provided they are empowered to take part in the planning and development process.

By utilizing a grounded theory approach, data derived from interviews help to further conceptualize the contrasting views held by stakeholder groups and contribute to scientific understanding of intergroup conflicts in a tourism and ecological context. Ranchers' perceived threats by environmental groups' efforts very closely resemble the postulation of integrated threat theory (ITT), a theory used in social psychology research used to analyze how negative attitudes form between interacting groups who are socially and culturally different (Monterrubio, 2016). According to the tenets of ITT, the four major types of threats that can be perceived in intergroup interactions are: realistic, involving the wellbeing of group members; symbolic, associated with beliefs, values, and overall world view; intergroup anxiety; and negative stereotypes (Stephan & Stephan, 1996). These four types of threats have been used to predict negative attitudes among 'ingroups' towards 'outgroups', such as immigrants, ethnic minorities, and nationalities, and may be social, political, or economic in nature or involve competition over limited resources (Ward & Berno, 2011). Since ingroups tend to have strong cultural identification

(Monterrubio, 2016), these tenants of ITT are useful for explaining the phenomenon occurring around ER projects in rural areas.

Many local residents have a strong ingroup identification as generational ranchers and farmers and see environmental groups, such as APR, as outsiders who threaten their way of life. The potential threat to ranchers of landscape-scale ER can come in several forms, including realistic threats to their economic well-being through reduced ranching jobs and symbolic threats to their worldviews regarding human-environment relations (Stephan & Stephan, 1996; Ward & Berno, 2011). Given the potential gravity of such threats, it comes as little surprise that local opposition to APR is so palpable. The threat local stakeholders perceive fuels their skepticism to the tourism component of APR's vision and precludes the more positive sentiments of curiosity and euphoria, as the Figure 2 illustrates.

By incorporating tourism development within a regional development plan, tourism can be more successful (Hall & Boyd, 2005; Peeters et al., 2018). The spatio-temporal TALC model for landscape-scale tourismcapes can help diverse stakeholder groups plan for sustainable tourism development over a larger landscape across core PAs with developed tourist attractions to peripheral conservation areas where ER projects are occurring and tourist services are limited. Tourism can not only act as an agent to help stakeholders collaborate on better connecting the region ecologically (Murphy, 1986) but also economically and culturally (Burger, 2000) through offering tourism services in peripheral areas. This framework helps increase understanding of the relationship between ER and tourism, an area of research that has not received much attention (Hall, 2019), and can be used by PA authorities to promote tourism opportunities beyond crowded national parks.

In Montana, the tourismscape contains developed tourism attractions at Yellowstone and Glacier national parks, and developing attractions in peripheral areas, such as APR's campgrounds and lodgings. Diverse stakeholders, such as nonprofits, rural communities, rancher communities, grassroots organizations, and indigenous communities reside and work across this region and must be engaged and involved in regional development planning. These stakeholders all have disparate goals that may lead to conflict and social tension (Pellis, 2019) in the absence of collaboration. The spatio-temporal TALC model for landscape-scale tourismcapes can facilitate an inclusive approach to meet the needs of diverse stakeholders through collaborative planning and management of the area.

The bifurcation of attitudes within groups adds complexity to the planning process and must also be considered. For tourists, the divergence in attitudes can be seen between the in-state and out-of-state tourists. The attitudes and values between these two groups largely result from their preference in recreational activities. While in-state tourists value access to consumptive activities, such as hunting and fishing opportunities, out-of-state tourists tend to value non-consumptive recreation, including birding, wildlife viewing, biking, and hiking. Tourism providers, both private and public, need to safely and sustainably provide opportunities for these different groups of visitors to increase visitor satisfaction, which can also have the result of augmenting support for ER projects (Burger, 2000). To understand the existing differences within stakeholder groups, future research can conduct quantitative and spatial studies among larger samples to evaluate how spatial (residents from core to periphery spectrum) and demographic attributes impact their attitudes.

In bracing for a return to an era of overtourism, tourism destinations must involve local stakeholders and make appropriate plans to ensure tourism development fits within the area's carrying capacity and regional development goals (Peeters et al., 2018). The more stakeholders feel involved in the planning and implementation process of tourism development, the more favorable views they will have towards continued developments (Byrd, 2007; Hunt & Stronza, 2014). In a region where tourism visitation is low and still undeveloped, making plans for limited growth is paramount.

In order to establish a shared vision for the region, stakeholder groups must collaborate, communicate, interact, and sufficiently address each other's major concerns (Nabhan et al., 2014). This is the environmental groups' primary challenge. While the TNC's innovative work with their grass bank provides benefits to ranchers, APR's Wild Sky initiative does as well. Expanding these initiatives will provide more incentives for ranchers to collaborate with environmental groups. According to the literature, successful ER initiatives should come from within the community and have their full support (Clewell & Aronson, 2013). APR's vision to restore an ecosystem and drive tourism to a peripheral region of Montana is fraught with controversy yet embodies an innovate model for landscape-scale ER. Their nonprofit status allows for flexibility in management and development alternatives that can and should involve all stakeholders to ensure successful regional development.

Conclusion

This study fills in a void of understanding about stakeholder attitudes towards tourism development in a landscape-scale ER context, a topic previously overlooked by tourism researchers (Hall, 2019). The results demonstrate the complex socio-ecological environment in which ER is taking place in north central Montana. The complexity and diversity of views is likely to be seen in other parts of the world as the divide between rural and urban communities continues to grow (Pechlaner et al, 2019). This study contributes to conceptual understanding of stakeholder attitudes and tourism development on a regional scale by introducing the spatio-temporal TALC model for landscape-scale tourismscapes. This proposed model applies Porter's (1998) economic-centered, regional approach into an ecosystem-centered approach to tourism development and uses Butler's (1980) TALC model and Doxey's (1975) Irritation Index to account for core-periphery interactions across landscape-scale tourismscapes. Future research can test this model in other contexts to make comparisons and additional revisions.

By using a constructivist grounded theory approach, this study also contributes to the analysis of contrasting stakeholder views by incorporating elements of integrated threat theory (ITT). ITT holds that intergroup differences can lead to perceptions of threat to an in-group, thus creating prejudice towards an out-group (Stephan & Stephan, 1996). In rural areas where traditional vocations of ranching and farming have been the dominant industry, landscape-scale ER projects that encourage tourism have been perceived as a major threat to the livelihoods, culture, and values of local communities. To navigate such threats, tourism development and ER efforts need to be guided by communities who are involved in decision making and collaborate with federal, state, local, nonprofit, and community institutions to increase support for projects and ensure their success (Clewell & Aronson, 2013; Johnson & Payne, 2005).

Finally, this study contributes to the literature on sustainable tourism development. In complex socio-ecological systems where attitudes and recreational activities widely vary, accommodating and involving different groups of stakeholders is critical for operating within an area's carrying capacity and augmenting support for ER initiatives that invite tourism development. As efforts to restore lost ecosystem services, wildlife species and habitat, and ER projects increase in numbers and geographic scale, closely collaborating with local communities must always be a top priority.

Disclosure statement

No potential conflict of interest was reported by the authors.

References

Åberg, E., & Tapsell, S. (2013). Revisiting the River Skerne: The long-term social benefits of river rehabilitation. *Landscape and Urban Planning, 113*, 94–103. https://doi.org/10.1016/j.landurbplan.2013.01.009

Allen, K. (2015). Trade-offs in nature tourism: Contrasting parcel-level decisions with landscape conservation planning. *Ecology and Society, 20*(1), 21. https://doi.org/10.5751/ES-07058-200121

Andereck, K., Knopf, R., Valentine, K., & Vogt, C. (2005). Residents' perceptions of community tourism impacts. *Annals of Tourism Research, 32*(4), 1056–1076. https://doi.org/10.1016/j.annals.2005.03.001

Anderson, R. M., Buitenwerf, R., Driessen, C., Genes, L., Lorimer, J., & Svenning, J. C. (2019). Introducing rewilding to restoration to expand the conservation effort: A response to Hayward et al. *Biodiversity and Conservation, 28*(13), 3691–3693. https://doi.org/10.1007/s10531-019-01845-1

Arts, K., Fischer, A., & Van der Wal, R. (2016). Boundaries of the wolf and the wild: A conceptual examination of the relationship between rewilding and animal reintroduction. *Restoration Ecology, 24*(1), 27–34. https://doi.org/10.1111/rec.12309

Bailey, R. (2015). Where the private buffalo roam and the private antelope play. *Reason, 47*(5), 18–19.

Bauduin, S., Mcintire, E., St-Laurent, M., & Cumming, S. (2018). Compensatory conservation measures for an endangered caribou population under climate change. *Scientific Reports, 8*(1), 1–10. https://doi.org/10.1038/s41598-018-34822-9

Breakey, N., Curcija, M., & Driml, S. (2019). Development of a conflict management model as a tool for improved project outcomes in community-based tourism. *Tourism Management, 70*, 341–354.

Burger, J. (2000). Landscapes, tourism, and conservation. *The Science of the Total Environment, 249*(1-3), 39–49. https://doi.org/10.1016/S0048-9697(99)00509-4

Butler, R. (1980). The concept of a tourism area cycle of evolution: Implications for the management of resources. *The Canadian Geographer/Le Géographe Canadien, 24*(1), 5–12. https://doi.org/10.1111/j.1541-0064.1980.tb00970.x

Butler, R. (2006). *The tourism area life cycle. Vol. 1: Applications and modifications*. Channel View Publications.

Byrd, E. (2007). Stakeholders in sustainable tourism development and their roles: Applying stakeholder theory to sustainable tourism development. *Tourism Review of Review, 62*(2), 6–13. https://doi.org/10.1108/16605370780000309

Caruso, F., & Jiménez, I. (2013). Tourism, local pride, and attitudes towards the reintroduction of a large predator, the jaguar *panthera onca* in Corrientes. *Endangered Species Research, 21*(3), 263–272. https://doi.org/10.3354/esr00519

Chaperon, S., & Bramwell, B. (2013). Dependency and agency in peripheral tourism development. *Annals of Tourism Research, 40*(1), 132–154. https://doi.org/10.1016/j.annals.2012.08.003

Charmaz, K. (2000). Grounded theory: Objectivist and constructivist methods. In N. Denzin & Y. Lincoln (Eds.), *Handbook of qualitative method* (2nd ed., pp. 509–535).Sage Publications.

Charmaz, K. (2006). *Constructing grounded theory: A practical guide through qualitative analysis*. Sage Publications.

Charmaz, K. (2008). Constructionism and the grounded theory. In J. A. Holstein & J. F. Gubrian (Eds.), *Handbook of constructionist research* (pp. 397–412). The Guilford Press.

Chavez, M. (1987). Public access to landlocked public lands. *Stanford Law Review, 39*(6), 1373–1401. https://doi.org/10.2307/1228850

Chester, C. (2015). Yellowstone to Yukon: Transborder conservation across a vast international landscape. *Environmental Science & Policy, 49*, 75–84. https://doi.org/10.1016/j.envsci.2014.08.009

Clark, C., & Nyaupane, G. (2020). Overtourism: An Analysis of its coverage in the media by using framing theory. *Tourism Review International. 24* (2/3) https://doi.org/10.3727/154427220X15845838896314

Clewell, A., & Aronson, J. (2013). *Ecological restoration: principles, values, and structure of an emerging profession* (2nd ed.). Island. Science and Practice of Ecological Restoration.

Connell, J., & Lowe, A. (1997). Generating grounded theory from qualitative data: The application of inductive methods in tourism and hospitality management research. *Progress in Tourism and Hospitality Research, 3*(2), 165–173. https://doi.org/10.1002/(SICI)1099-1603(199706)3:2<165::AID-PTH87>3.0.CO;2-I

Creswell, J. (2007). *Qualitative inquiry and research design: Choosing among five traditions* (2nd ed.). Sage Publications.

Creswell, J., & Creswell, J. (2018). *Research design: Qualitative, quantitative, and mixed methods approaches* (5th ed.). Sage Publications.

Creswell, J., Hanson, W., Clark, V., & Morales, A. (2007). Qualitative research designs: Selection and implementation. *The Counseling Psychologist, 35*(2), 236–264. https://doi.org/10.1177/0011000006287390

Daengbuppha, J., Hemmington, N., & Wilkes, K. (2006). Using grounded theory to model visitor experiences at heritage sites. *Qualitative Market Research: An International Journal, 9*(4), 367–388. https://doi.org/10.1108/13522750610689096

Davenport, J. (2018). Making the buffalo commons new again: Rangeland restoration andbison reintroduction in the Montana highline. *Great Plains Quarterly, 38*(2), 199–225. https://doi.org/10.1353/gpq.2018.0024

Dodds, R., & Butler, R. (Eds.). (2019). *Overtourism: Issues, realities and solutions*. De Gruyter Oldenbourg. https://www.degruyter.com/view/product/505514

Doxey, G. (1975). *A causation theory of visitor–resident irritants: Methodology and research inferences* [Paper presentation]. 6th Annual Conference Proceedings of the Travel Research Association, San Diego, CA (pp. 195–198).

Fennell, D. (1996). A tourist space-time budget in the Shetland Islands. *Annals of Tourism Research, 23*(4), 811– 829. https://doi.org/10.1016/0160-7383(96)00008-4

Flores, D. (2016). *American Serengeti: The last big animals of the Great Plains*. University Press of Kansas.

Ghoddousi, S., Pintassilgo, P., Mendes, J., Ghoddousi, A., & Sequeira, B. (2018). Tourism and nature conservation: A case study in Golestan National Park, Iran. *Tourism Management Perspectives, 26*, 20–27. https://doi.org/10.1016/j.tmp.2017.12.006

Gillson, L., Ladle, R., & Araújo, M. (2011). Baselines, patterns and process. In R. Ladle & R. Whittaker (Eds.), *Conservation biogeography* (pp. 31–44). Wiley-Blackwell.

Glasser, B., & Strauss, A. (1967). *The discovery of grounded theory: Strategies for qualitative research*. Aldine.

Gunn, C. (1997). *Vacationscape: Developing tourist areas* (3rd ed.). Taylor and Francis.

Hall, M. (2006). Introduction. In R. Butler (Ed.), *The tourism area life cycle, vol. I: Applications and modifications* (pp. xv–xviii). Channel View Publications.

Hall, M. (2019). Tourism and rewilding: An introduction – definition, issues and review. *Journal of Ecotourism, 18*(4), 297–308. https://doi.org/10.1080/14724049.2019.1689988

Hall, M., & Boyd, S. (Eds.). (2005). *Tourism and nature-based tourism in peripheral areas: Development or disaster*. Channel View Publications.

Hardy, A. (2005). Using grounded theory to explore stakeholder perceptions of tourism. *Journal of Tourism and Cultural Change, 3*(2), 108–133. https://doi.org/10.1080/09669580508668490

Harris, J., Hobbs, R., Higgs, E., & Aronson, J. (2006). Ecological restoration and global climate change. *Restoration Ecology, 14*(2), 170–176. https://doi.org/10.1111/j.1526-100X.2006.00136.x

Hovinen, G. (2002). Revisiting the destination lifecycle model. *Annals of Tourism Research, 29*(1), 209–230. https://doi.org/10.1016/S0160-7383(01)00036-6 https://doi.org/10.1016/S0160-7383(01)00036-6

Hunt, C., & Stronza, A. (2014). Stage-based tourism models and resident attitudes towards tourism in an emerging destination in the developing world. *Journal of Sustainable Tourism, 22*(2), 279–298. https://doi.org/10.1080/09669582.2013.815761

Jackson, S., & Hobbs, R. (2009). Ecological restoration in the light of ecological history. *Science (New York, N.Y.), 325*(5940), 567–569. https://doi.org/10.1126/science.1172977

Jähnig, S., Lorenz, A., Hering, D., Antons, C., Sundermann, A., Jedicke, E., & Haase, P. (2011). River restoration success: A question of perception. *Ecological Applications : A Publication of the Ecological Society of America, 21*(6), 2007–2015. https://doi.org/10.1890/10-0618.1

Jamal, T., & Stronza, A. (2009). Collaboration theory and tourism practice in protected areas: Stakeholders, structuring and sustainability. *Journal of Sustainable Tourism, 17*(2), 169–189. https://doi.org/10.1080/09669580802495741

Johnson, M., & Payne, R. (2005). Ecotourism and Regional Transformation in Northwestern Ontario. In M. Hall & S. Boyd (Eds.), *Tourism and nature-based tourism in peripheral areas: Development or disaster?* (pp. 21–35). Channel View Publications.

Johnston, C. (2006). Shoring the foundations of the TALC in tropical island destinations: Kona, Hawai'i. In R. Butler (Ed.), *The tourism area life cycle, vol. 1: Applications and modifications* (pp. 198–221). Channel View Publications.

Kensbock, S., & Jennings, G. (2011). Pursuing: A grounded theory of tourism entrepreneur's understanding and praxis of sustainable tourism. *Asia Pacific Journal of Tourism Research, 16*(5), 489–504. https://doi.org/10.1080/10941665.2011.597574

Kilpatrick, A., Gillin, C., & Daszak, P. (2009). Wildlife-livestock conflict: The risk ofpathogen transmission from bison to cattle outside Yellowstone National Park. *Journal of Applied Ecology, 46*(2), 476–485. https://doi.org/10.1111/j.1365-2664.2008.01602.x

Koninx, F. (2019). Ecotourism and rewilding: The case of Swedish Lapland. *Journal of Ecotourism, 18*(4), 332–347. https://doi.org/ https://doi.org/10.1080/14724049.2018.1538227

Leduc, J., & Von Essen, E. (2019). The compromises of rewilding in Swedish Laponia: Implications for nature reconciliation. *The Journal of Transdisciplinary Environmental Studies, 17*(1), 38–54.

Liu, W., Vogt, C., Luo, J., He, G., Frank, K., & Liu, J. (2012). Drivers and socioeconomic impacts of tourism participation in protected areas. *PLoS One, 7*(4), e35420. https://doi.org/10.1371/journal.pone.0035420

Marshal, K., White, R., & Fischer, A. (2007). Conflicts between humans over wildlife management: On the diversity of stakeholder attitudes and implications for conflict management. *Biodiversity and Conservation*, *16*(11), 3129–3136. https://doi.org/10.1007/s10531-007-9167-5

Matteucci, X., & Gnoth, J. (2017). Elaborating on grounded theory in tourism research. *Annals of Tourism Research*, *65*(C), 49–59. https://doi.org/10.1016/j.annals.2017.05.003

Mattsson, B., Devries, J., Dubovsky, J., Semmens, D., Thogmartin, W., Derbridge, J., & Lopez-Hoffman, L. (2020). Linking landscape-scale conservation to regional and continental outcomes for a migratory species. *Scientific Reports*, *10*(1), 4968–4968. https://doi.org/10.1038/s41598-020-61058-3

Mbaiwa, J., & Stronza, A. (2011). Changes in resident attitudes towards tourism development and conservation in the Okavango Delta, Botswana. Journal of Environmental Management , *92*(8), 1950–1959. https://doi.org/10.1016/j.jenvman.2011.03.009

Monterrubio, C. (2016). The impact of spring break behaviour: An integrated threat theory analysis of residents' prejudice. *Tourism Management*, *54*, 418–427. https://doi.org/10.1016/j.tourman.2015.12.004

Moore, M., Covington, W., & Fulé, P. (1999). Reference conditions and ecological restoration: A southwestern ponderosa pine perspective. *Ecological Applications*, *9*(4), 1266–1277. https://doi.org/10.1890/1051-0761(1999)009[1266:RCAERA]2.0.CO;2

Morrisette, P. (2000). Is there room for free-roaming bison in greater Yellowstone? *Ecology Law Quarterly*, *27*, 467–1383.

Murphy, P. (1986). Tourism as an agent for landscape conservation: An assessment. *Science of the Total Environment*, *55*(C), 387–395. https://doi.org/10.1016/0048-9697(86)90196-8

Murphy, P. E., & Andressen, B. (1988). Tourism development on Vancouver Island: An assessment of the core-periphery model. *The Professional Geographer*, *40*(1), 32–42. https://doi.org/10.1111/j.0033-0124.1988.00032.x

Nabhan, G., Knight, R., & Charnley, S. (2014). The biodiversity that protected areas can't capture: How private ranch, forest, and tribal lands sustain biodiversity. In S. Charnley, T. Sheridan, & G. Nabhan (Eds), *Stitching the west back together: Conservation of working landscapes* (pp. 33–48). University of Chicago Press.

Navarro, L., Pereira, H., Proença, V., & Kaplan, J. (2015). Maintaining disturbance-dependent habitats. In L. Navarro & H. Periera (Eds.), *Rewilding European landscapes* (pp.143–170). Springer International.

Nunkoo, R., & Ramkissoon, H. (2016). Stakeholders' views of enclave tourism: A grounded theory approach. *Journal of Hospitality & Tourism Research*, *40*(5), 557–558. https://doi.org/10.1177/1096348013503997

Nyaupane, G., & Poudel, S. (2011). Linkages among biodiversity, livelihood, and tourism. *Annals of Tourism Research*, *38*(4), 1344–1366. https://doi.org/10.1016/j.annals.2011.03.006

Nyaupane, G., & Poudel, S. (2012). Application of appreciative inquiry in tourism research in rural communities. *Tourism Management*, *33*(4), 978–987. https://doi.org/10.1016/j.tourman.2011.10.009

Pechlaner, H., Innerhofer, E., & Erschbamer, G. (2019). *Overtourism: Tourism management and solutions*. Routledge: New York.

Peeters, P., GöSsling, S., Klijs, J., Milano, C., Novelli, M., Dijkmans, C., Eijgelaar, E., Hartman, S., Heslinga, J., Isaac, R., Mitas, O., Moretti, S., Nawijn, J., Papp, B., &., & Postma, A. (2018). *Research for TRAN Committee - Overtourism: Impact and possible policy responses*. European Parliament, Policy Department for Structural and Cohesion Policies.

Pellis, A. (2019). Reality effects of conflict avoidance in rewilding and ecotourism practices – the case of Western Iberia. *Journal of Ecotourism*, *18*(4), 316–331. https://doi.org/10.1080/14724049.2019.1579824

Porter, M. (1998). Clusters and the new economics of competition. *Harvard Business Review*, *76*(6), 77–90.

Poudel, S., Nyaupane, G., & Budruk, M. (2016). Stakeholders' perspectives of sustainable tourism development: A new approach to measuring outcomes. *Journal of Travel Research*, *55*(4), 465–480. https://doi.org/10.1177/0047287514563166

Rashford, B., Foulke, T., & Taylor, D. (2010). Ranch-level economic impacts of predation in a range livestock system. *Rangelands*, *32*(3), 21–26. https://doi.org/10.2458/azu_rangelands_v32i3_foulke

Saarinen, J., Rogerson, C. M., & Hall, C. M. (2017). Geographies of tourism development and planning. *Tourism Geographies*, *19*(3), 307–317. https://doi.org/10.1080/14616688.2017.1307442

Sage, J. (2017). *Analyzing economic and social opportunities and challenges related to bison conservation in northeast Montana*. Institute for Tourism and Recreation Research Publications, 353. https://scholarworks.umt.edu/itrr_pubs/353

Salvatore, R., Chiodo, E., & Fantini, A. (2018). Tourism transition in peripheral rural areas: Theories, issues and strategies. *Annals of Tourism Research*, *68*, 41–51. https://doi.org/10.1016/j.annals.2017.11.003

Smith, D., Peterson, R., & Houston, D. (2003). Yellowstone after wolves. *BioScience*, *53*(4), 330–340. https://doi.org/10.1641/0006-3568002820030029053005B0330:YAW005D2.0.CO;2

Stankey, G., & Cole, D. (1985). *The Limits of Acceptable Change (LAC) system for wilderness planning*. USDA Forest Service Intermountain Research Station.

Stephan, W., & Stephan, C. (1996). An integrated threat theory of prejudice. *International Journal of Psychology*, *31*(3–4), 1635.

Stone, M., & Nyaupane, G. (2016). Protected areas, tourism and community livelihoods linkages: A comprehensive analysis approach. *Journal of Sustainable Tourism*, *24*(5), 673–693. https://doi.org/10.1080/09669582.2015.1072207

Stone, M., & Nyaupane, G. (2018). Protected areas, wildlife-based community tourism and community livelihoods dynamics: Spiraling up and down of community capitals. *Journal of Sustainable Tourism*, *26*(2), 307–324. https://doi.org/10.1080/09669582.2017.1349774

Strauss, A., & Corbin, J. (1998). *Basics of qualitative research: Techniques and procedures for developing grounded theory* (2nd ed.). Sage Publications.

Stronza, A. (2007). The economic promise of ecotourism for conservation. *Journal of Ecotourism*, *6*(3), 210–190. https://doi.org/10.2167/joe177.0

Suding, K., Higgs, E., Palmer, M., Callicott, J., Anderson, C., Baker, M., Gutrich, J., Hondula, K., LaFevor, M., Larson, B., Randall, A., Ruhl, J., & Schwartz, K. (2015). Committing to ecological restoration. *Science (New York, N.Y.)*, *348*(6235), 638–640. https://doi.org/10.1126/science.aaa4216

Van Der Duim, R. (2007). Tourismscapes: An actor-network perspective. *Annals of Tourism Research*, *34*(4), 961–976. https://doi.org/10.1016/j.annals.2007.05.008

Walpole, M., & Goodwin, H. (2001). Local attitudes towards conservation and tourism around Komodo National Park. *Environmental Conservation*, *28*(2), 160–166. https://doi.org/10.1017/S0376892901000169

Ward, C., & Berno, T. (2011). Beyond social exchange theory: Attitudes toward tourists. *Annals of Tourism Research*, *38*(4), 1556–1569. https://doi.org/10.1016/j.annals.2011.02.005

Wohl, E., Angermeier, P., Paul, L., Bledsoe, B., Kondolf, G., MacDonnell, L., Merritt, D., Palmer, M., Poff, N., & Tarboton, D. (2005). River restoration. *Water Resources Research*, *41*(10), W10301. https://doi.org/10.1029/2005WR003985

Yu-Ming, W. (2002). Core-periphery geographic theory in regional tourism planning. *Economic Geography*, *3*, 372–375.

Zhang, J., Fu, M., Hassani, F., Zeng, H., Geng, Y., & Bai, Z. (2011). Land use-based landscape planning and restoration in mine closure areas. *Environmental Management*, *47*(5), 739–750. https://doi.org/10.1007/s00267-011-9638-z

OPEN ACCESS

Environmentality, green grabbing, and neoliberal conservation: the ambiguous role of ecotourism in the Green Life privatised nature reserve, Sumatra, Indonesia

Chantal Elizabeth Wieckardt, Stasja Koot and Nadya Karimasari

ABSTRACT

Against the background of neoliberal conservation and green grabbing, this paper investigates an ecotourism initiative through the notion of 'multiple environmentalities' (neoliberal, disciplinary, sovereign and truth), which concerns the governance of people for environmental causes. We apply this to the inhabitants of Batu Katak, Sumatra, Indonesia, where the Green Life Project has been established next to Gunung Leuser National Park as a Private Protected Area. An important governance strategy for Green Life is the establishment of volunteer ecotourism to generate funds and to include the local community. We argue that ecotourism can play a crucial role in legitimising the activities and increasing the power of nature conservation organisations, which, in the case of Green Life, is visible in the growth of coercive means to secure protected areas. If ecotourism is the kernel of neoliberal environmentality at Green Life, the core issue is the prevalence of truth environmentality, in which nature is often valued higher than people. This valuation led to the strengthening of disciplinary and sovereign environmentality. Arguably, ecotourism at Green Life will not establish the intended sustainable relations with the community, and this paper functions as an important lesson for practitioners to scrutinise the consequences of their governance strategies.

Introduction

Over the last few decades, Private Protected Areas (PPAs) have increasingly been recognised as an important strategy to conserve nature, including by big international NGOs such as the International Union for Conservation of Nature (IUCN, see Stolton et al., 2014). Despite this increasing recognition, the definition of what exactly constitutes a PPA and what its effectiveness is in terms of biodiversity conservation continues to be subject to debate. Additionally, the total global area that is privately protected remains uncharted (Holmes, 2013). Even though PPAs (insofar recognised as such) make a significant contribution to protect nature in the 'traditional developing world' (Holmes, 2013), for example in Africa (e.g. Ramutsindela et al., 2011;

Spierenburg & Wels, 2006) or Latin America (e.g. Brondo, 2013; Holmes, 2014; Pearce, 2015), PPAs are relatively uncommon in Southeast Asia (Stolton et al., 2014). This paper therefore explores the impact of the privatisation of land for nature conservation within the Southeast Asian context, namely by analysing the Green Life Project on Sumatra, Indonesia, at the border of Gunung Leuser National Park (GLNP).[1] As has been the case more broadly in Southeast Asia, contentions between local inhabitants and conservation agencies have arisen over the establishment of protected areas. These are often related to the disruption of already existing local livelihoods. Conservation agencies often perceive local inhabitants as a threat to the protected area as these people contest the parks' legitimacy and its unilateral imposition of boundaries which disregards their entitlement to access land and natural resources (Li, 2008).

In the case study of this paper, the Green Life Project (see below), civil society actors—a Czech and an Indonesian NGO—are the new owners and managers of the land, instead of the government or the private sector (e.g. an ecotourism company). These NGOs use the strategy of privatising forest land to ostensibly prevent large-scale deforestation in the North Sumatra province (Green Life, 2020a), mostly by the oil palm and paper industry, but also as the result of a demand for tropical wood and rubber (Green Life, 2020b). As such, the Green Life Project presents a clear case of what Fairhead et al. (2012, p. 238) named "green grabbing", which is "the appropriation of land and resources for environmental ends" and which "is an emerging process of deep and growing significance". Globally, green grabbing has already shown the problematic consequences that land acquisition for conservation may have on local communities (Duffy, 2008; Fairhead et al., 2012; Holmes, 2014; Hutton et al., 2005; Koot & Van Beek, 2017; Mowforth & Munt, 2003). In this paper, however, we move beyond the consequences of green grabbing. Instead, we address the local responses *to* green grabbing, as well as to the governance strategies environmental NGOs can use to create particular self-regulating 'subjects' (cf. Agrawal, 2005; Deutsch, 2020) in order to create a particular 'environment' or 'nature' on the ground. Our focus is on the ambiguous role of ecotourism in the governance strategies deployed by environmental NGOs through the notion of 'environmentality', more specifically we use Fletcher's (2010) framework of 'multiple environmentalities' (explained in more detail below). We thus contribute to the literature on tourism and conservation governance strategies in a context where green grabbing and neoliberal conservation tactics, namely privatisation and ecotourism, are aimed against large-scale land grabs by commercial multinationals.

In this paper, we show that for Green Life ecotourism functions as a crucial livelihood diversification strategy to govern the inhabitants of Batu Katak, a rural village located adjacent to the Green Life Project's office and reserves, but often without creating the intended support for nature conservation. In fact, ecotourism provides for a discourse to cover up the perceived negative impacts of the project on the livelihoods of the local community, who show a variety of, at times conflicting, responses to the project's conservation practices. Although most inhabitants from the local community cooperate regarding the project's ecotourism initiative, this cooperation is ambiguous as these inhabitants also undermine aspects of the larger Green Life Project through the continuation of banned livelihoods and the use of violence and threats towards the project's management and employees. Most of these responses, however, remain largely unnoticed by the Green Life management. We argue that ecotourism plays a crucial role in legitimising Green Life's activities including the increase of coercive means employed by Green Life to secure its protected areas. Most likely, such coercive tactics will not establish sustainable links with the community.

In what follows, we first provide the necessary background on ecotourism and the neoliberalisation of nature, which is followed by the conceptualisation of 'environmentality'. Next, we provide an overview of the case study, after which we describe our methodology and data analysis. We then present our empirical findings, first by describing how Green Life has constrained the local community of Batu Katak, and second by showing how the inhabitants from Batu Katak respond to ecotourism initiatives and benefits, and third by exposing how the inhabitants from Batu Katak respond to Green Life more generally. In the discussion that follows, we relate our findings to green grabbing and neoliberal conservation, and the local responses specifically to the multiple environmentalities. Last, in the conclusion, we wrap up and iterate our main argument.

Ecotourism and the neoliberalisation of nature

The establishment of ecotourism is firmly embedded within a larger global movement of neoliberalising nature. In fact, "green grabbing operates [...] in ways that are enabled, structured and shaped by the neo-liberal logics of commodification and appropriation" (Fairhead et al., 2012, p. 248). Neoliberalisation, in this context, means "a political-economic project which advocates, first, a strong free-market economic system, [...] and, second, the use of the market as a model for other areas of political and social life" (Farrell, 2015, p. 256). As such, it is "a political ideology that aims to subject political, social, and ecological affairs to capitalist market dynamics" (Büscher et al., 2012, p. 5). These definitions allow us to emphasise the ideology, practices and governmentality that attempt to reproduce capitalist market dynamics in the public and social spheres, creating an "on-going process of *neoliberalisation*" (Farrell, 2015, p. 256), in which various crises are addressed, including the global biodiversity crisis (Büscher et al., 2012; Fletcher et al., 2014).

In this ideology, (eco)tourism has always played a crucial role as a market-based incentive that further spreads neoliberal capitalist mechanisms and values, and that creates 'environmental subjects' (Agrawal, 2005; Carrier & West, 2004; Duffy, 2006, 2008; Fletcher, 2010, 2011). As such, ecotourism promises "to promote environmental consciousness for Western consumers" (Igoe & Brockington, 2007, p. 434), but also among the local hosts: under neoliberalisation, ecotourism should generate income for the local groups in and around protected areas, based on the rationale that it will make local people support conservation because they now receive benefits from this (Brockington et al., 2008; Duffy, 2008, 2013; Fletcher, 2011). Thus, the *financial* value of a commodified nature is centralised, instead of intrinsic values, changing local inhabitants' attitudes towards and connection with nature (Serhadli, 2020). This is also subject in our focus on ecotourism, since we analyse how the local population of the rural village Batu Katak responds to the project, thus contributing to the above debates by placing the emphasis "on agency *within* rather than on analysing the broader structural political economy *of* neoliberal conservation" (Matose, 2014, p. 67). Generally, such agency received little attention in the literature on neoliberal conservation (Matose, 2014), and it is thus highly relevant to explore how people are governed, and how they respond to this, which we do by applying the concept of 'environmentality'.

Multiple environmentalities

We investigate Green Life, and particularly its ecotourism activities, through the notion of 'environmentality'. Environmentality finds its origin in Foucault's work on 'governmentality', which today has become a generic term for "various particular modes of [...] 'conducting conduct'" (Fletcher, 2010, p. 173). As such, governmentality often serves to indicate the governing of human behaviour through a variety of modes. Since the late 1990s, the neologism environmentality has gained a foothold in academic literature, especially with the works of Luke (1995, 1999) and Darier (1999). Environmentality aims to understand the ways in which governmentality can be applied to analyses of the environment and environmental governance. In such an analysis, the environment is an object of knowledge, about which certain 'truths' can be constructed. These 'truths' necessitate the regulation, management and governance of the environment (Rutherford, 2007).

In order to better understand the different modes of environmental governing, Fletcher (2010) developed a 'multiple environmentalities' framework in which four types of environmentality can be discerned. First, *disciplinary environmentality* aims to create 'environmental subjects' who internalise particular norms and values, e.g. through environmental education. Second, *neoliberal environmentality* aims to prevent environmental degradation "through the creation of incentive structures intended to influence individuals' use of natural resources by altering the cost-benefit ratio of resource extraction" (Fletcher, 2010, p. 176; cf. Deutsch, 2020). Third, *sovereign environmentality* aspires to conserve nature through "top-down creation and enforcement of regulations" (Fletcher, 2010, p. 178), for example through the establishment of protected areas—and patrolling these areas. Last, *truth environmentality* dictates appropriate behaviour based on certain conceptions of the "fundamental nature of life and the universe" (Fletcher, 2010, p. 176; cf. Montes et al., 2020). These modes and rationalities of governance are not solely top-down, but rather are renegotiated by those who are 'governed', especially in nature conservation projects that bring with them a set of ideas and ideologies that often differ from those of local actors.

Case study

The Green Life Project borders GLNP in the Bohorok district of North Sumatra, Indonesia. Starting off as several nature and wildlife reserves initiated since 1934, GLNP was established in 1980 as an amalgamation of these reserves. After several expansions, the park comprised 838,872 hectares in 2014, and it is estimated that today around 690,000 people live in or adjacent to GLNP. In 2011, UNESCO enlisted GLNP as part of the Tropical Rainforest Heritage of Sumatra, a Natural World Heritage in-danger, based on increasing poaching activities, illegal logging, agricultural encroachment, road construction, pressure on surrounding forests to change them into crop plantations, oil palm plantations and refugees from Aceh Province (Purwanto, 2016). Historically, conflicts between nature conservation and local inhabitants have existed within and around GLNP (cf. Li, 2008), predominantly because conservation limits livelihood possibilities, e.g. to benefit from export-oriented agriculture (McCarthy, 2006).

Green Life aims to prevent large-scale deforestation by the oil palm and paper industry in the North Sumatra area (Green Life, 2020a, 2020b), and to curb "the illegal encroachment of poachers, loggers and plantations" (Green Life, 2020a). They claim that GLNP is not properly protected, since "it consists almost entirely of government forest land, one third of which can be legally logged or cleared for agricultural use" (Green Life, 2020c). Their response was to establish a PPA as "a network of properly monitored, privately owned reserves on the border of the National Park", which they consider the best way to preserve nature, to provide an undisturbed habitat for wildlife, and to deter poachers (Green Life, 2020d).

Currently, Green Life consists of three geographical areas that they have named Reserve I (87.3 hectares), Reserve II (9 hectares) and Reserve III (11 hectares), totalling 107.3 hectares (Green Life, 2020a) (see Figure 1). The light green areas on Figure 1 indicate the current size of the Green Life reserves, whereas the dark green areas and the yellow lines indicate the project's future vision for expansion. The river flowing from north to south is the Berkail River, the one flowing south of Reserves I and II is the Sembelang River and past Reserve III flows the Sekelam River. These rivers originate in GLNP.

Since it is forbidden by law for foreigners to own land in Indonesia (Green Life, 2020a; USAID, 2010), the three smaller reserves that together comprise the Green Life reserves are "owned by [the] Indonesian conservationist and chairman of the NGO Yayasan Hutan Untuk Anak" (Green Life, 2020a). Subsequently, these lands are rented to the Czech NGO Forest for Children for 50 years, with the option to renew the contracts. Together, both NGOs manage these lands, and they have the ambition to ultimately own and preserve 700 hectares of rainforest (Green Life, 2020f). Green Life buys the land from local landowners, claiming to give them a very good price for land of little value.

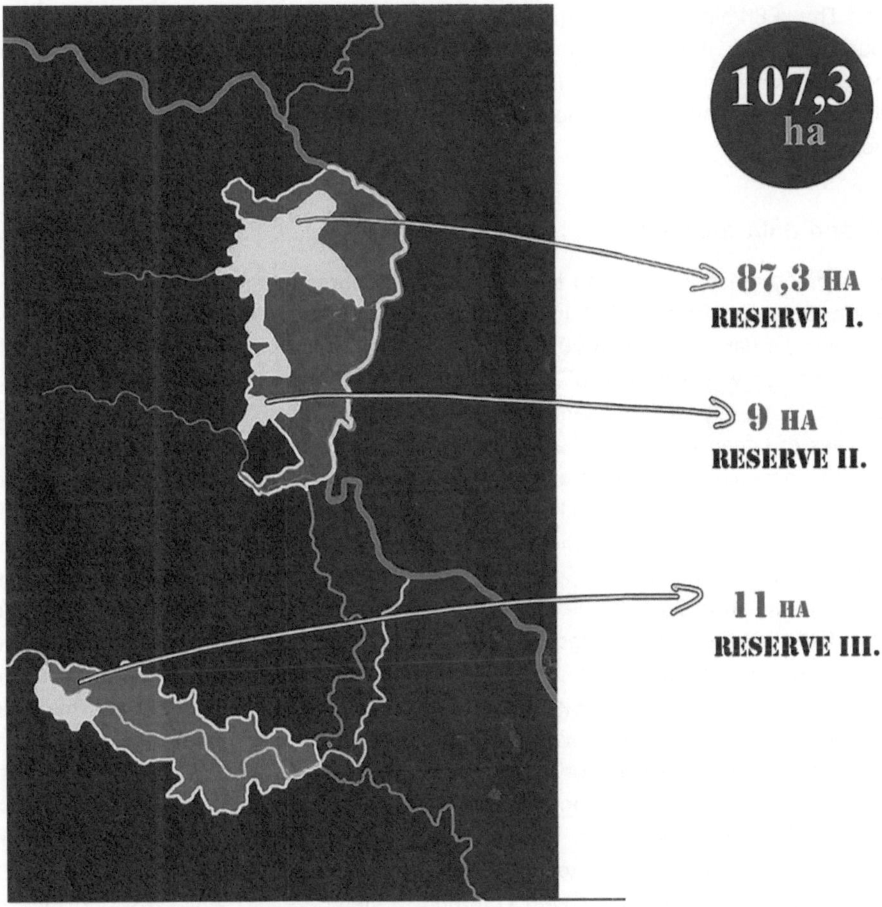

Figure 1. The Green Life reserves, *Source*: Green Life, 2020e.

The 'locals' Green Life is referring to emanate from Batu Jong Jong, a village comprising of several sub-villages, of which Batu Katak is of specific interest to the project due to its accessibility by road and its location close to one of the entrances to GLNP. Batu Katak is situated along the riverbank of the Berkail River about two kilometres southwest of Green Life's Reserves I and II. Many inhabitants have long been, or still are, dependent upon small-scale agriculture, especially rubber and subsistence fishing. About half the population owns land and those who do not often work on other people's land, either cutting rubber or transporting blocks of rubber by motorbike. However, as a result of rubber price reductions, several farmers have decided to grow oil palm trees instead. These are less labour-intensive and hence demand for labour has decreased, leaving especially young men searching for jobs.

Since the arrival of Green Life in the area in 2009, nature conservation organisations and ecotourism initiatives have sprouted in Batu Katak. Green Life has not established any collaborations with these organisations at the time of fieldwork, but it "welcomes the move towards ecotourism by the Batu Katak community" (Green Life, 2020g), as is also shown through the promotion of Batu Katak as an ecotourism destination on the website ecotourismbatukatak.com.[2] Though written from the perspective of the Batu Katak community, this website carries the Green Life logo, promotes the Tiger House (described further below) and sells a range of tours, including a Tiger Trekking tour with Green Life (Ecotourism Batu Katak, 2015). In order to coordinate the increased (eco)tourism activities in Batu Katak, a local tourism organisation, Lembaga Pariwisata Batu Katak (LPBK) has been founded by some members of the Batu Katak community, separately from

Green Life.[3] The LPBK aims to "assist the local community with an alternative livelihood program via ecotourism, and to prevent illegal practices in the National Park" (Raw Wildlife Encounters, 2020). The LPBK has 21 members and is involved both with local and international tourism, playing an important role in the distribution of guiding jobs for the different organisations active in the village.

Methods and data analysis

The paper is predominantly based on two months of ethnographic fieldwork in Batu Katak (with a population of approximately 250 inhabitants, see Sumatra Ecoventures, 2019) by the first author between January and March 2018, during which she conducted 22 semi-structured interviews, most of them with the help of a local interpreter. These semi-structured interviews were conducted with a variety of inhabitants, including farmers, (often self-proclaimed) tour guides, and employees of Green Life, as well as with the Indonesian manager of the project. Earlier, in February 2017, the second and third author together conducted an open, exploratory interview about Green Life with both the Czech and Indonesian managers, after they had heard about Green Life's existence and were triggered to investigate the project due to its distinctive approach for nature conservation in Southeast Asia. The Czech manager of the project has not been interviewed again in 2018, as he was not responsive to the researcher's interview requests at that time. Interviewees were selected by purposive sampling and by using a snowball method, until answers became repetitive at its core and a certain level of 'thick description' (Geertz, 1973) had been achieved in combination with observations and informal conversations. The selection of interviewees aimed to include a diversity of perspectives present within Batu Katak, revealed to the first author during participatory observation. Participatory observation has been key to build up trust with the community more broadly, and to get a sense of the key issues at stake. Importantly, participatory observation complemented the semi-structured interviews with informal conversations on more sensitive topics. In fact, often the most valuable stories were told to the first author while spending time with community members and while assisting informants during their day-to-day work, as well as during a guided tour with one of the farmers/guides into GLNP, and a one-day hike inside GLNP with the Tiger Commando (the Green Life anti-poaching unit) together with a small group of volunteers at Green Life. The semi-structured interviews and first-hand fieldnotes of participatory observation were systematically analysed using thematic content analysis; identifying common themes and patterns (including activities such as fishing, small-scale agriculture, but also perceptions of local people of the Green Life Project) as well as inconsistencies within and between informants' stories by annotating and coding the data, and segment them into categories (such as the continuation of banned livelihoods, the use of violence and threats). The categories, in turn, have been used to inform and structure the results section below.

Ever since the first interview in February 2017, the third author, who is a native speaker of Bahasa Indonesia, maintained contact with the Indonesian manager, which included telephone calls and social media messages. She also conducted interviews in GLNP headquarters in Medan with government officials of GLNP who commented about the Green Life Project, as part of her larger PhD research, for which she has conducted fieldwork in North Sumatra and Aceh Provinces between January and March 2017, and between October 2017 and January 2019. This research took place in the buffer zone of GLNP and focused on the dynamics of, and the tension between, smallholder agriculture and nature conservation.

Additionally, we have analysed various publicly available websites, most prominently the Green Life website, using Critical Discourse Analysis (CDA). In CDA the analytical focus is "*on the role of discourse in the (re)production and challenge of dominance* [which is] the exercise of social power by elites, institutions or groups, that results in social inequality, including political, cultural, class, ethnic, racial and gender inequality" (Van Dijk, 1993, pp. 249–250). It is essential to acknowledge that such dominance appeals to "the relations between power and discourse" (Van Dijk, 1993, p. 249; see also Fairclough, 2012). In our case, the information from the Green Life website has been an important addition to the ethnographic fieldwork because it showed some discrepancies with the fieldwork findings that are essential in our analysis.

Results

In this section, we first outline Green Life's connections with ecotourism and nature. Next, we show how certain activities constrain people's lives, and subsequently zoom in on the ecotourism projects. Last, we describe how people respond to these activities.

An important issue for Green Life is the prevention of illegal practices, for which they have set up the 'Tiger Commando', an anti-poaching unit. Furthermore, they monitor their reserves using camera traps (Green Life, 2020h). According to Green Life, the Tiger Commando has the authority to arrest people that enter Green Life's reserves without permission, and hand them over to GLNP rangers. Moreover, the project shares its camera trap data with GLNP rangers, who can subsequently arrest trespassers. In addition to the Tiger Commando, Green Life also established the 'Tiger House' that aims to "provide information to tourists and local people about conservation issues, sustainable forest use, and safe and appropriate behaviour in the forest" (Green Life, 2020i). They have established an English teaching programme for the children of Batu Katak, for which they hired a local woman as teacher. Ideally, the Tiger House is also used to inform inhabitants from Batu Katak and other villages about poaching and illegal wildlife trade, and to encourage them to inform Green Life if they have suspicions about poaching (Green Life, 2020i). Moreover, Green Life puts up posters and sign boards that list all forbidden activities inside GLNP and on Green Life's reserves.

To establish a good relationship with local inhabitants, connect to global audiences, and to raise funds, a crucial activity is ecotourism, which is presented by Green Life as an income-generating opportunity through guided tours, to ensure that "local people protect nature because it has become valuable to them" (fieldnotes, 13 January, 2018). Ecotourism is done with volunteers who are interested in nature conservation, thereby generating a steady source of income for the project, and through the involvement of the inhabitants of Batu Katak by supporting the establishment of community-based tourism to create employment.

Green Life's activities are based on several underlying assumptions about human-nature relationships. Although they perceive "nature and animals as equal to human beings" (Green Life, 2020j), in their perception there are two different types of humans: "an ethical human society" and "characterless and greedy people" (Green Life, 2020j). Humans and nature, especially wildlife, can no longer peacefully co-exist and thus nature needs to be separated from "the greedy, spiteful, stupid, or characterless people who, around us, commit crimes against nature" (Green Life, 2020j). In order to do so, the project initially focused on buying and redeeming land and managing these lands through the implementation of strict rules and regulations, barely engaging with the local community. This changed when the project tried to obtain a license to extend the range of the Tiger Commando and to gain authority to patrol against illegal activities inside GLNP. To receive this license, the Indonesian government requested Green Life to involve the surrounding communities. This resulted in them employing around 25 people, mostly men, from the surrounding villages: four people (two from Batu Katak) became a member of the Tiger Commando and the others became mostly irregularly employed in tourism.

In an interview with the Green Life management (14 February, 2017) the threat of large-scale oil palm companies was contradicted: it was explained that the specific tracts of land that Green Life buys are often not attractive for such large-scale commercial companies, as the hilly land is inaccessible for cars or motorbikes. In fact, various Green Life employees and enthusiasts, including the management, considered people living around GLNP as the biggest threat to nature, especially poachers from the neighbouring villages who "haven't been confronted yet" (Green Life, 2020b). In response, Green Life actively tries to govern local people's behaviour. This became particularly evident when the first author wanted to interview a member of the Green Life management, who refused and disregarded the local communities' voices, explaining that interviewing community members would be "useless" because they would "just tell you what you want to hear [...] You have to *teach* them [about nature conservation]" (fieldnotes, 26 January, 2018). Thus, a clear discrepancy became visible between public communication, such as the Green Life website, and the local discourse, particularly in relation to the inhabitants of the area.

Green Life's constraints to local livelihoods

Since its arrival in the area of Batu Katak, Green Life has been buying up small-scale rubber and oil palm plantations on the border of GLNP, increasingly reducing agricultural opportunities. Although Green Life states that the inhabitants from Batu Katak can buy land that is located closer to the village, one young tour guide claims that there is no land available there, while another tour guide and a farmer, as well as the Green Life management itself, claim that the price of land has risen near Batu Katak since the arrival of Green Life. Green Life thus strongly affects agricultural practices that are central to people's livelihoods.

Furthermore, several respondents from Batu Katak explained that they consider *access* to the Green Life reserves and GLNP a serious threat, since this has become increasingly restricted over time. Several farmers explained that accessing small-scale plantations on the border of GLNP has become difficult, as they are no longer allowed to cross Green Life's reserves to enter their own land. This happens intentionally, as the management explained that many locals use the river to access GLNP, and by cooperating with the government and placing camera traps, Green Life tries to prevent this (interview, 7 February, 2018). If someone gets caught on camera crossing the reserves, Green Life summons this person to their office to justify their 'trespassing'. Accessing GLNP for guided tours with tourists from the other NGOs in the village, or to collect leaves for traditional medicines, becomes increasingly difficult as well, according to one of the tour guides, as Green Life's reserves function as a strictly monitored border around GLNP. However, this was contradicted by one of the rangers of GLNP, who claims that locals can still "easily access" GLNP but are solely prevented from buying and logging on the land at the border of GLNP (interview, 13 February, 2018).

Hunting as a livelihood strategy is also actively discouraged by Green Life, who consider it poaching. Almost all farmers and tour guides we spoke to claim that most inhabitants from Batu Katak have stopped hunting wildlife, both for consumption and for the market, although they acknowledge that there are a few exceptions of people who continue to catch songbirds in order to sell them. However, according to Green Life 'poaching' also includes fishing, which is officially forbidden inside GLNP and no longer allowed by Green Life in or close to its reserves either. Yet, many inhabitants from Batu Katak are at least partly dependent on fishing for their livelihoods. Access to, especially, the Sembelang River has become a sensitive matter between Green Life and the community. Green Life actively patrols the area around this river, trying to enforce a strict no-fishing policy. People who are seen entering GLNP carrying fishing equipment are, according to the management, caught and brought back to the community to "pay a minimum of 5 million rupiah and they have to put little fish back in the river" (interview, 7 February, 2018). Whether Green Life has the authority to fine locals for fishing in GLNP remains unclear, although it seems as if Green Life is offering the fine as an 'alternative' to being handed over to GLNP rangers, who are supposed to treat fishing as "illegal" and thus "a criminal act" (interview, 7 February, 2018). In practice local employees of Green Life have a 'softer' approach to fishing, at least on Green Life's reserves; inhabitants first receive two warnings and will be informed about the consequences of fishing on the fish population (interview, 7 February, 2018). The rangers of GLNP are not as strict on fishing either, as one of them explained that most locals would only fish when they do not have enough money to buy fish. As long as local inhabitants fish with a fishing rod or a small net that is not left overnight, the rangers will not act. Only if locals use poison or electricity to fish, GLNP rangers will take action (interview, 13 February, 2018).

Tourism-based employment and Batu Katak

Green Life thrives on the contributions of so-called 'visitors' or 'volunteers' under their 'Visitors Project' (Green Life, 2020g), originating almost exclusively from the Czech Republic. These volunteers are a source of revenue for the project, help to maintain the camps and reserves, and shy away poachers by their presence in the forest. In order to support the local community, volunteers can also go on excursions with local guides from the LPBK in Batu Katak. As such, people working in tourism in Batu Katak do so mostly for Green Life's volunteers, not only as tour guides, but also as porters of food to Green Life's camps in the forests, as assistants to do shopping, or as drivers to the airport in Medan. New forms of (irregular) employment have thus emerged from Green Life's volunteer programme, diversifying the livelihood strategies of the inhabitants from Batu Katak. Working as a tour guide for one day can earn someone a wage that sometimes equals a week's work on a rubber plantation and is therefore a popular job. However, excursions are optional in Green Life's visitor's programme, limiting the stability of this type of employment.

As such, the employment opportunities offered by Green Life are a sensitive issue in the village. As Green Life's requests for tours are irregular, an often heard complaint is that Green Life is not involved with the local community. As a respondent from Batu Katak explained: "Tourists that are coming only go to Green Life. Green Life does not give jobs to the community. The tourists only work on the project and not on the community" (interview, 30 January, 2018). The employment opportunities are also a sensitive issue as some inhabitants have been able to gain more benefits out of the project than others, thus changing socio-economic structures. The extent to which inhabitants have been able to benefit from Green Life depends on their social and human skills and knowledge (e.g. speaking English), and on their financial situation and assets (e.g. possession of a car to do grocery shopping or airport pick-ups).

Another, often heard complaint is that Green Life does not work through the LPBK to employ tour guides for their volunteers, but rather asks one of the inhabitants of Batu Katak to distribute the guiding jobs. According to a GLNP ranger who was involved in the establishment of the LPBK, Green Life initially cooperated with the LPBK. However, after the government transferred the ranger to another area, he could no longer be involved in the LPBK, and the cooperation between Green Life and the LPBK came to a halt. The Batu Katak village head claims that Green Life consequently approached him to distribute the jobs, while one of the LPBK members also claims this. However, today a different man distributes Green Life's guiding jobs. He explained that Green Life wants to cooperate with the LPBK, but that it is currently not possible due to "problems within the organisation" and the fact that they are "not organised" (fieldnotes, 8 February, 2018). During our fieldwork, the LPBK indeed was not active in the village. Although most members of the LPBK confirm this, they still consider it rude that Green Life does not collaborate, and some members even argue that Green Life does not cooperate in order to "grow their own business", disrespecting the local culture (interview, 30 January 2018).

Local responses to the larger Green Life Project

Altogether, cooperation between the inhabitants of Batu Katak and Green Life has proven fragile. In fact, most locals who also occasionally act as a tour guide for Green Life or carry food to Green Life's camps, simultaneously continue livelihoods that are now banned, especially fishing (with the exception of those people who have obtained a more permanent position with Green Life). They explain that they now go fishing when the Green Life employees or volunteers are not around, or they fish at night, as a way to sustain their "tradition" (interviews, 12, 15, 24 and 26 February, 2018) and livelihoods. Moreover, although it is forbidden by law inside GLNP and by Green Life on their reserves, many also continue to fish with tuba roots (poison), describing it as "the way we fish" (fieldnotes, 7 March, 2018). It might be for this reason that Green Life has stepped up its activities and has put up signs around the village that aim to stop fishing with tuba roots.

A similar issue is the continuation of hunting. According to one of Green Life's longer-term volunteers, local people who continue hunting have been caught on Green Life's camera traps, yet the rangers of GLNP refuse to act on these images. Green Life's own employees do not arrest these people themselves, as they explain that they are afraid of retributions. This fear is grounded: some inhabitants from Batu Katak and the other sub-villages of Batu Jong Jong have responded to Green Life and its employees with violence and threats in the past, often anonymously. For instance, there have been anonymous phone calls and threats to burn down Green Life's camps on their reserves. Furthermore, camera traps have been stolen and damaged and some of Green Life's signposts have been demolished. It might be for this reason that Green Life employs coercive tactics today to protect their reserves, for example by finalising their application process to obtain a "Smart Patrol Licence" that allows the Tiger Commando to patrol inside GLNP, by sharing all data on their camera traps with the GLNP Administration in Medan (Green Life, 2020k), and by becoming a member of the International Ranger Federation (Green Life, 2020l).

Such responses by the community obviously undermine Green Life's conservation practices and goals and are interpreted by the Green Life management as evidence that local communities do not know (yet) how to live with, or protect, nature. This is emphasised by the Czech manager, who claims that local people

> are angry because a white foreigner comes here to buy the land and then does nothing with it. We use it for nature only. They don't understand and think that we are just here to make money. Because that is what land and nature means to them, money. (fieldnotes, 26 January, 2018)

This quote shows how the manager's and community members' perceptions of the value of land and nature differ. In contrast to what the manager seems to conclude, the valuation of land and nature by the local community seems to be more complex and is not necessarily valued in terms of money only, but also in terms of livelihoods and intrinsic value (cf. Montes et al., 2020; Serhadli, 2020). From such a perspective, establishing a nature reserve to which the local community does not have access might therefore seem illogical and wasteful to the local community itself. Even though most of the threats and complaints were also targeted at the Indonesian manager of the land, he takes a more moderate stance, explaining that the Green Life management believes that protecting nature and separating it from human beings is also in the (long-term) interest of the local community. Moreover, there are many different views and interests that Green Life needs to take into account, including those of the government and the locals. Balancing these views and interests is not an easy job: "It is difficult, you know. You need to be friendly with the locals and cooperate with the government. You cannot be too strict, or they won't like you". He therefore tries to support the village and wants to establish good relationships with the locals, but "it is not easy. [Some others in the management do] not understand why I spend all the money buying everybody food and drinks. [...] We have a different approach. [These others care] about the locals, but in a different way" (interview, 7 February, 2018). However, one of these other managers of Green Life bluntly stated: "I am not interested in the people, I am interested in the nature" (fieldnotes, 26 January, 2018).

Discussion

Neoliberal conservation versus corporate power?

Green Life is a project that uses neoliberal strategies to 'save nature', which is its prime concern. Its practices reveal 'conventional' neoliberal ideologies to save nature (Büscher et al., 2012; McAfee, 1999), specifically through privatisation and the implementation of market-based eco-tourism for community development. As is well-known today, such strategies do not always deliver the expected results and they contain problematic elements, including the creation of local elitism, the implementation of externally driven agendas, homogenisation of local groups, the perpetuation of neo-colonial relations and the prioritisation of nature over local people's interests (Agrawal & Gibson, 1999; Carrier & West, 2004; Dressler et al., 2010; Fletcher, 2011; Mowforth & Munt, 2003). At Green Life, all these problems have been exposed in different degrees. Nonetheless, the project has the intention to challenge the expansion of global corporate power, in particular oil palm and paper multinationals. This is part of the legitimisation of the project and its activities, by arguing that their 'good' privatised project is a response to 'bad' privatisation. From a conservation point of view then, Green Life's proposition makes much sense. What clearly needs more attention though, is how they relate to community development.

The appropriation of land areas for conservation purposes comes with changing power relations in which ownership, user rights and control over resources are transferred from generally poor people into the hands of the powerful (Fairhead et al., 2012). In the instance of the Green Life Project the land and its resources are changing ownership and control, into the hands of an individual, who leases it to an overseas nature conservation NGO. For the primary goal of nature conservation, privatisation thus plays a pivotal role, and so does ecotourism, which legitimises the project by fulfilling the requirement to involve community members, and which provides important revenues.

However, our results also question Green Life's lands' suitability for large-scale oil palm plantations and Green Life's arrival in the area has meant an increase of land prices. This is a more general problem globally where green grabbing takes place: there is less land left for others because it concentrates into the hands of a few, thus creating novel forms of commodification and valuation and new markets, actors and alliances (Fairhead et al., 2012). Thus, Green Life's own legitimisation does not hold. What *is* prevented by this project, are expansions and establishments of (new) small-scale plantations, instead of large multinational activities. From Green Life's point of view and with their conservation interests in mind, it makes sense to curb these small-scale plantations but to not communicate this publicly, because to contest large oil palm multinationals sounds much better in marketing and public discourse than constraining local people's livelihoods, particularly for volunteer tourists who provide critical funding for the project.

From the perspective of the inhabitants from Batu Katak, what matters most are the constraints and benefits that come with Green Life. However, the benefits that some inhabitants receive also mean that the socio-economic dynamics of the village change substantially. There is a combination of acquiescence and resistance (cf. Koot & Van Beek, 2017). Initially, Green Life had little interest in the local communities and wanted to create an area in which they would avoid local people. However, government pressure *forced* them to collaborate, which they did not do out of their own interests, despite a long tradition of community-based conservation in Indonesia (see for example Eghenter, 2002; Wadley, 2002). Moreover, the colonial style implementation of fortress conservation (Duffy, 2008; Hutton et al., 2005) instigated militarisation, most prominently through the Tiger Commando anti-poaching unit, but also the camera traps and sharing that information with GLNP rangers and government officials can have strong social implications such as creating fear and infringing privacy (Sandbrook et al., 2018). Overall, militarised approaches carry the danger to *exclude* local people from nature conservation and to create *negative* attitudes towards conservation (Duffy et al., 2019; Duffy et al., 2015), which is of course the opposite of what Green Life aims for. Such a negative attitude already starts to show in some of the responses by the local inhabitants, who use anonymous threats and violence, who steal camera traps and who demolish sign boards to express their discontent with some crucial elements of the larger Green Life Project.

Ecotourism and multiple environmentalities

Green Life's reliance on tourism is not uncommon globally, in particular when nature conservation and ecotourism are established close to or together with 'local communities' (Agrawal, 2005; Brockington et al., 2008; Dressler et al., 2010; Koot & Van Beek, 2017; Mostafanezhad et al., 2016; Sullivan, 2002; West, 2016). Ecotourism provides employment for the local communities, a process that resembles Marx's concept of 'primitive accumulation'; through primitive accumulation, wealth and power get concentrated in the hands of a few while drastically reducing livelihood opportunities for most others. Human labour power is transformed into a commodity and property becomes privatised, increasing dependency and inequality (Marx, 2013). As such, ecotourism "is a never-ending form of accumulation by dispossession" (West, 2016, p. 60; cf. Harvey, 2003). Buffer-zone protected areas often show a strong focus on participation and community development, but in reality end up implementing more coercive strategies, often based on (negative) stereotypes of local inhabitants (Neumann, 1997). Market-based development mechanisms such as ecotourism function within a limited understanding of community development and are set up to prioritise nature above people. In the case of Green Life, it masks their initial lack of interest in the local community, as well as some of its difficult aspects, such as the ethical questions regarding the prioritisation of nature above people (see also Dressler et al., 2010; Sullivan, 2002). Green Life attempts to alter local people's behaviour to act in such a way that they contribute to, or at least do not obstruct, its conservation ideology. These attempts are, all in their own way, informed by the four different types of environmentality we described above (namely disciplinary, neoliberal, sovereign and truth environmentality). The case also shows how these different types do not stand apart, but continuously influence each other (cf. Deutsch, 2020).

When looking at the project from a truth environmentality point of view, Green Life has shown rather explicit ideas about the separation of humans and nature, as if humans are a threat *to* nature, acting from 'outside' of it, instead of a part *of* nature. This ontological assumption is crucial in informing the project's disciplinary and sovereign environmentality strategies. Sovereign environmentality has been put to work by restricting local people's access to GLNP, the Green Life reserves and increasingly the rivers. Official government rules on fishing that are not strictly implemented by the rangers themselves are now enforced by Green Life and its Tiger Commando. Moreover, camera traps play an important role too, increasingly monitoring people's activities in the forest. Disciplinary environmentality techniques have been advocated by both NGOs: 'teaching' local communities about the project and nature conservation at the Tiger House and putting up sign boards and posters suggests that people need to be 'educated' in order to live harmoniously with their natural surroundings. This confirms the idea that local inhabitants are considered people who still need to 'learn' what 'proper' nature conservation is (Neumann, 1997), essentially framing them as a threat to nature.

Neoliberal environmentality has been put to work, most importantly, by establishing a volunteer ecotourism project which aims to change people's (mostly monetary) incentive structures to act in certain ways, resulting in a variety of socio-economic consequences. Indeed, the volunteer ecotourism programme at Green Life offers the inhabitants of Batu Katak the opportunity to diversify their livelihood strategies by offering alternative occupations in (eco)tourism, which is considered an important attribute to cope with and adapt to different stresses (Scoones, 2015). However, such alternative occupations have in many places delivered elusive results, both in terms of conservation as well as in terms of socio-economic development (Sene-Harper et al., 2019). Other mechanisms that strengthen the neoliberal, disciplinary and sovereign governance strategies employed by Green Life are, for example, fining whoever is caught fishing inside GLNP. This is not only a form of top-down regulation of rules and a form of punishment, but the fines are also economic incentive structures that aim to alter people's values and behaviour, in order to withhold them from fishing altogether. This also leads to what West (2016) calls the "dispossession of sovereignty", which entails a lot more than only material dispossession and a restriction of access to particular natural resources. As she explains, sovereignty is

> multifaceted with aspects connected to power or authority, the ability to self-govern (in the state-related political sense as well as in the sense of rights to make decisions about self, land, family and future), as freedom from domination and control, and as the ability to assert autonomy through daily practice and action. (West, 2016, p. 35)

The ecotourism discourse, together with the public discourse that Green Life prevents large oil palm multinationals, masks an exclusionary and protectionist conservation practice, in which the (monetary) benefits of volunteer ecotourism largely go to the project itself. Such a practice is not only counterintuitive to the idea of buffer-zones protected areas in which community participation and development are central, but consequently results in several, at times conflicting, responses by the inhabitants from Batu Katak. Such responses should be seen both in the light of resistance against losing access to particular resources (the forest and the rivers) (see also Holmes, 2007) and a way to survive. As such, Green Life is an example of what Giddens (1984) called an 'enabling constraint': structural properties in social systems can be considered as enabling and constraining at the same time. In the case of Green Life, however, the ecotourism benefits (enablements) do not seem to polish off the overall project's constraints and costs for the inhabitants of Batu Katak.

Conclusion

Green Life's volunteer ecotourism project has resulted in conflicting responses by the Batu Katak community: some people actively cooperate with the project's volunteer ecotourism initiative, while others, and sometimes the same people who support the ecotourism initiative, undermine the project through the continuation of banned livelihoods and the use of violence and threats towards the project's management and employees. Moreover, volunteer ecotourism is used to cover up the project's negative impact on local livelihoods and the challenges the project faces with regards to community involvement, while positively highlighting the larger project to obtain the necessary licenses and goodwill from government institutes. In addition, global corporate multinationals are also part of the discourse as threats to nature, which of course they often are, although it is ambiguous whether the physical terrain of the Green Life reserves would indeed be suitable for such multinationals.

Although the privatisation of land by civil society actors is a relatively new strategy to protect nature in Southeast Asia, the outcomes of the project on the ground are not completely new. Our argument is that the (volunteer) ecotourism discourse can play a crucial role in legitimising the activities and increasing power of nature conservation organisations, which, in the case of Green Life, is visible in the growth of coercive means to secure its protected areas. If ecotourism is the kernel of neoliberal environmentality at Green Life, the core issue is the prevalence of truth environmentality, in which nature is valued higher than people at times. This valuation has led to the strengthening of disciplinary and, especially, sovereign environmentality. However, as Matose (2016, p. 312) already found in southern Africa, "without resolving property rights in the form of rights and access for local people at the same time, such projects do not yield the intended outcomes". Arguably, in its current form, ecotourism at Green Life is unlikely to establish the intended sustainable relations with the community, necessary for long-term conservation success. This paper thus functions as an important lesson also for practitioners to scrutinise the consequences of their governance strategies.

Notes

1. Green Life is not the first PPA on Sumatra: we are aware of two more privatised nature reserves, namely Harapan Rainforest and Tambling Wildlife Nature Conservation (see Buergin, 2016; Tambling Wildlife Nature Conservation, 2020 respectively).
2. Last visited on 2nd February 2020. As of June 2020, this website was not functioning.
3. The exact year in which this community-based tourism organisation was established differs in various interviews, ranging from 2012 to 2014.

Acknowledgements

The authors would like to thank the three anonymous reviewers whose valuable contributions allowed us to improve the initial version of this paper. We would also like to thank the Green Life Management and employees, our interpreter Ruddy, host family and interviewees in Batu Katak for sharing their knowledge and experiences with us.

Disclosure statement

No potential conflict of interest was reported by the author(s).

ORCID

Stasja Koot http://orcid.org/0000-0001-8625-7525

References

Agrawal, A. (2005). Environmentality: Community, intimate government, and the making of environmental subjects in Kumaon. *Current Anthropology*, *46*(2), 161–190. https://doi.org/10.1086/427122

Agrawal, A., & Gibson, C. (1999). Enchantment and disenchantment: The role of community in natural resource conservation. *World Development*, *27*(4), 629–649. https://doi.org/10.1016/S0305-750X(98)00161-2

Brockington, D., Duffy, R., & Igoe, J. (2008). *Nature unbound: Conservation, capitalism and the future of protected areas*. London: Earthscan.

Brondo, K. (2013). *Land grab: Green neoliberalism, gender, and garifuna resistance in Honduras*. Tucson: University of Arizona Press.

Buergin, R. (2016). Ecosystem restoration concessions in Indonesia: Conflicts and discourses. *Critical Asian Studies*, *48*(2), 278–301. https://doi.org/10.1080/14672715.2016.1164017

Büscher, B., Sullivan, S., Neves, K., Igoe, J., & Brockington, D. (2012). Towards a synthesized critique of neoliberal biodiversity conservation. *Capitalism Nature Socialism*, *23*(2), 4–30. https://doi.org/10.1080/10455752.2012.674149

Carrier, J., & West, P. (2004). Ecotourism and authenticity: Getting away from it all? *Current Anthropology*, *45*(4), 483–498.

Darier, E. (1999). Foucault and the environment: An introduction. In E. Darier (Ed.), *Discourses of the environment* (pp. 1–33). Wiley-Blackwell.

Deutsch, S. (2020). "Who doesn't like dolphins?!" Neoliberalization, variegated environmentalities, and value alterations in a cross-national comparison of Irrawaddy dolphin conservation. Geoforum, *114*, 159–171. https://doi.org/10.1016/j.geoforum.2020.05.010

Dressler, W., Büscher, B., Schoon, M., Brockington, D. , Hayes, T., Kull, C. A., Mccarthy, J., & Shrestha, K. (2010). From hope to crisis and back again? A critical history of the global CBNRM narrative. *Environmental Conservation*, *37*(1), 5–15. https://doi.org/10.1017/S0376892910000044

Duffy, R. (2006). The politics of ecotourism and the developing world. *Journal of Ecotourism*, *5*(1–2), 1–6. https://doi.org/10.1080/14724040608668443

Duffy, R. (2008). Neoliberalising nature: Global networks and ecotourism development in Madagascar. *Journal of Sustainable Tourism*, *16*(3), 327–344. https://doi.org/10.1080/09669580802154124

Duffy, R. (2013). The international political economy of tourism and the neoliberalisation of nature: Challenges posed by selling close interactions with animals. *Review of International Political Economy*, *20*(3), 605–626. https://doi.org/10.1080/09692290.2012.654443

Duffy, R., Massé, F., Smidt, E., Marijnen, E., Büscher, B., Verweijen, J., Ramutsindela, M., Simlai, T., Joanny, L., & Lunstrum, E. (2019). Why we must question the militarisation of conservation. *Biological Conservation*, *232*, 66–73. https://doi.org/10.1016/j.biocon.2019.01.013

Duffy, R., St John, F. A. V., Büscher, B., & Brockington, D. (2015). The militarization of anti-poaching: Undermining long term goals? *Environmental Conservation*, *42*(4), 345–348. https://doi.org/10.1017/S0376892915000119

Ecotourism Batu Katak. (2015). *Welcome on Batu Katak – ecotourism website*. http://ecotourismbatukatak.com/

Eghenter, C. (2002). Planning for community-based management of conservation areas. In D. Chatty & M. Coclchester (Eds.), *Conservation and mobile indigenous peoples: Displacement, forced settlement, and sustainable development* (pp. 329–346). Oxford: Berghahn.

Fairclough, N. (2012). Critical discourse analysis. *International Advances in Engineering and Technology*, *7*, 452–487.

Fairhead, J., Leach, M., & Scoones, I. (2012). Green grabbing: A new appropriation of nature? *Journal of Peasant Studies*, *39*(2), 237–261. https://doi.org/10.1080/03066150.2012.671770

Farrell, N. (2015). 'Conscience capitalism' and the neoliberalisation of the non-profit sector. *New Political Economy*, *20*(2), 254–272. https://doi.org/10.1080/13563467.2014.923823

Fletcher, R. (2010). Neoliberal environmentality: Towards a poststructuralist political ecology of the conservation debate. *Conservation and Society*, *8*(3), 171–181. https://doi.org/10.4103/0972-4923.73806

Fletcher, R. (2011). Sustaining tourism, sustaining capitalism? The tourism industry's role in global capitalist expansion. *Tourism Geographies*, *13*(3), 443–461. https://doi.org/10.1080/14616688.2011.570372

Fletcher, R., Dressler, W., & Büscher, B. (2014). Introduction - Nature^{TM} Inc.: The new frontiers of environmental conservation. In B. Büscher, W. Dressler, & R. Fletcher (Eds.), *Nature^{TM} Inc.: Environmental conservation in the neoliberal age* (pp. 3–21). Tucson: University of Arizona Press.

Giddens, A. (1984). *The constitution of society: Outline of the theory of structuration*. Cambridge: Polity Press.

Geertz, C. (1973). *The interpretation of cultures*. New York: Basic Books.

Green Life. (2020a). *Green Life Reserve*. http://greenlifeproject.org/green-life-reserve-2/

Green Life. (2020b). *Threats*. http://greenlifeproject.org/threats/

Green Life. (2020c). *Gunung Leuser National Park*. http://greenlifeproject.org/gunung-leuser-np/

Green Life. (2020d). *Save the Rainforest*. http://greenlifeproject.org/save-rainforest/

Green Life. (2020e). *[Map Green Life Reserves] [map]*. http://greenlifeproject.org/wp-content/uploads/2016/04/mapy_anglicky_na_web-1.png

Green Life. (2020f). *Green Life Project*. http://greenlifeproject.org/green-life-project/

Green Life. (2020g). *Green & Blue VISITORS of project*. http://greenlifeproject.org/green-blue-volunteer-program/

Green Life. (2020h). *Tiger Commando*. http://greenlifeproject.org/tiger-commando/

Green Life. (2020i). *Tiger House*. http://greenlifeproject.org/tiger-house-2/

Green Life. (2020j). *Forest for Children*. http://greenlifeproject.org/forest-for-children/

Green Life. (2020k). *Smart Patrol License*. http://greenlifeproject.org/news/smart-patrol-license/

Green Life. (2020l). *Forest for Children as a part of International Ranger Federation (IRF)*. http://greenlifeproject.org/news/forest-children-part-international-ranger-federation-irf/

Harvey, D. (2003). *The new imperialism*. Oxford: Oxford University Press.

Holmes, G. (2007). Protection, politics and protest understanding resistance to conservation. *Conservation and Society*, *5*(2), 184–201.

Holmes, G. (2013). What role do private protected areas have in conserving global biodiversity? *SRI Working Papers*, (46). ISSN 1753-1330. http://eprints.whiterose.ac.uk/76968/

Holmes, G. (2014). What is a land grab? Exploring green grabs, conservation, and private protected areas in Southern Chile. The Journal of Peasant Studies, *41*(4), 547–567. https://doi.org/10.1080/03066150.2014.919266

Hutton, J., Adams, W., & Murombedzi, J. (2005). Back to the barriers? Changing narratives in biodiversity conservation. *Forum for Development Studies*, *32*(2), 341–370. https://doi.org/10.1080/08039410.2005.9666319

Igoe, J., & Brockington, D. (2007). Neoliberal conservation: A brief introduction. *Conservation and Society*, *5*(4), 432–449.

Koot, S., & Van Beek, W. (2017). Ju/'hoansi lodging in a Namibian conservancy: CBNRM, tourism and increasing domination. *Conservation and Society, 15*(2), 136–146. https://doi.org/10.4103/cs.cs_15_30

Li, T. (2008). Contested Commodifications: Struggles over Nature in a National Park. In J. Nevins & N. Peluso (Eds.), *Taking Southeast Asia to market: Commodities, nature and people in the neoliberal age* (pp. 124–139). New York: Cornell University Press.

Luke, T. W. (1995). On environmentality: Geo-power and eco-knowledge in the discourses of contemporary environmentalism. *Cultural Critique, 31*(31), 57–81. https://doi.org/10.2307/1354445

Luke, T. W. (1999). Environmentality as green governmentality. In E. Darier (Ed.), *Discourses of the environment* (pp. 121–151). Oxford: Blackwell Publishers.

Marx, K. (2013). *Capital volumes 1 & 2*. Ware, Hertfordshire: Wordsworth Editions.

Matose, F. (2014). Nature, villagers, and the state: Resistance politics from protected areas in Zimbabwe. In B. Büscher, W. Dressler, & R. Fletcher (Eds.), *Nature™ Inc.: Environmental conservation in the neoliberal age* (pp. 3–21). Tucson: University of Arizona Press.

Matose, F. (2016). Hidden struggles in conservation: People's resistance in Southern Africa. In M. Ramutsindela, G. Miescher & M. Boehi (Eds.), *The politics of nature and science in Southern Africa* (pp. 311–322). Basler: Basler Afrika Bibliographien.

McAfee, K. (1999). Selling nature to save it? Biodiversity and green developmentalism. *Environment and Planning D: Society and Space, 17*(2), 133–154. https://doi.org/10.1068/d170133

McCarthy, J. (2006). *The fourth circle: A political ecology of Sumatra's rainforest frontier*. Palo Alto: Stanford University Press.

Montes, J., Tshering, S., Phuntsho, T., & Fletcher, R. (2020). Cosmological subjectivities: Explorien *'truth'* environmentalities in the Haa Highlands of Bhutan. *Conservation and Society, 0*(0), 0.

Mostafanezhad, M., Norum, R., Shelton, E., & Thompson-Carr, A. (2016). *Political ecology of tourism: Community, power and the environment*. Abingdon, Oxon: Routledge.

Mowforth, M., & Munt, I. (2003). *Tourism and sustainability: Development and new tourism in the third world*. New York: Routledge.

Neumann, R. (1997). Primitive ideas: Protected area buffer zones and the politics of land in Africa. *Development and Change, 28*(3), 559–582. https://doi.org/10.1111/1467-7660.00054

Pearce, F. (2015). To save the rainforest, let the locals take control. In New Scientist, 30. https://www.newscientist.com/article/mg22530100-200-to-save-the-rainforest-let-the-locals-take-control/

Purwanto, E. (2016). *An anti-encroachment strategy for the tropical rainforest heritage of sumatra: Towards new paradigms*. Bogor, Indonesia: UNESCO.

Ramutsindela, M., Spierenburg, M., & Wels, H. (2011). *Sponsoring nature: Environmental philanthropy for conservation*. London: Routledge.

Raw Wildlife Encounters. (2020). *Lembaga Batu Katak*. http://www.rawildlife.com.au/partner/lembaga-batu-katak?A=SearchResult&SearchID=168383236&ObjectID=7255615&ObjectType=35

Rutherford, S. (2007). Green governmentality: Insights and opportunities in the study of nature's rule. *Progress in Human Geography, 31*(3), 291–307. https://doi.org/10.1177/0309132507077080

Sandbrook, C., Luque-Lora, R., & Adams, W. (2018). Human bycatch: Conservation surveillance and the social implications of camera traps. *Conservation and Society, 16*(4), 493–504. https://doi.org/10.4103/cs.cs_17_165

Scoones, I. (2015). *Sustainable livelihoods and rural development*. Rugby, UK: Practical Action Publishing.

Sene-Harper, A., Matarrita-Cascante, D., & Larson, L. R. (2019). Leveraging local livelihood strategies to support conservation and development in West Africa. *Environmental Development, 29*, 16–28. https://doi.org/10.1016/j.envdev.2018.11.002

Serhadli, S. (2020). *Market-based solutions cannot solely fund community-level conservation (commentary)*. https://news.mongabay.com/2020/05/market-based-solutions-cannot-solely-fund-community-level-conservation-commentary/

Spierenburg, M., & Wels, H. (2006). "Securing space": Mapping and fencing in transfrontier conservation in southern africa. *Space and Culture, 9*(3), 294–312. https://doi.org/10.1177/1206331206289018

Stolton, S., Redford, K. H., Dudley, N. (2014). The futures of privately protected areas (Protected Area Technical Report Series No.1). International Union for Conservation of Nature (IUCN). https://portals.iucn.org/library/sites/library/files/documents/PATRS-001.pdf

Sullivan, S. (2002). How sustainable is the communalizing discourse of 'new' conservation? The masking of difference, inequality and aspiration in the fledgling 'conservancies' of Namibia. In D. Chatty & M. Colchester (Eds.), *Conservation and mobile indigenous peoples: Displacement, forced settlement, and sustainable development* (pp. 158–187). Oxford: Berghahn Press.

Sumatra Ecoventures. (2019). *Batu Katak*. https://nl.sumatra-ecoventures.com/batu-katak

Tambling Wildlife Nature Conservation. (2020). *Overview*. https://inilahallam.com/about/

USAID. (2010). *USAID Country Profile. Property Rights and Resource Governance. Indonesia.* https://www.land-links.org/wp-content/uploads/2016/09/USAID_Land_Tenure_Indonesia_Profile_0.pdf

Van Dijk, T. (1993). Principles of critical discourse analysis. *Discourse & Society*, *4*(2), 249–283.

Wadley, R. (2002). The history of displacement and forced settlement in West Kalimantan, Indonesia. In D. Chatty & M. Colchester (Eds.), *Conservation and mobile indigenous peoples: Displacement, forced settlement, and sustainable development* (pp. 313–328). Oxford: Berghahn Press.

West, P. (2016). *Dispossession and the environment: Rhetoric and inequality in Papua New Guinea*. New York: Columbia University Press.

Privately protected areas in increasingly turbulent social contexts: strategic roles, extent, and governance

William Thomas Borrie, Trace Gale and Keith Bosak

ABSTRACT

This paper considers increasing global reliance on privately protected areas (PPAs) and associated nature-based tourism (NBT). A targeted literature review was guided by three research questions, which sought to understand how neoliberal concepts and dynamics manifest in private forms of conservation; how private forms of conservation have manifested in Chile; and, how Chileans' attitudes towards their government's neoliberal policies might affect their expectations of PPAs. The search strategy resulted in 284 resources. Three dimensions of neoliberal approaches to PPAs emerged from the data each highlighting specific vulnerabilities: a loss of the social embeddedness of nature; an imposition of global, capital dynamics; and conflicting discourses and assumptions. Results suggest that, in order to improve long-term support and integration of PPAs and NBT, greater attention needs to be given to social well-being outcomes (including equity and justice concerns), building of social capital, and the preservation of local identities and histories. Additionally, regional and PPA-specific land-use planning needs to incorporate greater public engagement, cross-jurisdictional coordination, and transparent and inclusive decision-making.

Introduction

Conservation experts, in estimating there are tens of thousands of privately protected areas (PPAs) around the world, have observed that these areas have largely evolved on the fringe of regional planning action and international conservation governance (Stolton et al., 2014). Conservation institutions including the IUCN, the World Commission on Protected Areas, and the Convention on Biological Diversity, have all noted the potential and importance of PPAs, calling for governments to more officially recognize and support them. The challenge, then, is the integration of PPAs, and their underlying assumptions, within conservation systems that had been held primarily within the public realm.

Currently, twenty-eight countries include PPAs in the IUCN World Database on Protected Areas (Mitchell et al., 2018). Using the same database, Balmford et al. (2015) demonstrated the importance of protected area visitation and nature-based tourism (NBT) for overall conservation financing. Their study estimated eight-billion annual visits to protected areas worldwide, generating approximately US$600 billion dollars of in-country expenditures. Sustainable tourism is both an important income stream for the ongoing management of PPAs as well as a significant avenue for local employment and support. Over the past two decades, many countries have implemented NBT growth strategies, distinctly increasing the links between conservation, protected areas (public and private), and economic development (Whitelaw et al., 2014). However, reliance upon nature-based tourism for the financing of nature conservation and, in particular, the underwriting of the operation of PPAs, reflects a legitimization and institutionalization of market-based approaches towards conservation. This paper sets out to examine the hidden dynamics and potential controversies with such an reliance on private conservation, which is, essentially, conservation through neoliberalism.

Neoliberal trends in conservation

Büscher et al. (2014) defined neoliberalism as "the widespread trend towards increasing relaxation of state oversight over political-economic affairs and reliance on the 'invisible hand' of the market to efficiently allocate resources across the social landscape" (p. 6). Fletcher (2010) clarified that neoliberalism refers to values, policies, and practices that bring capital market dynamics into social and public realms. It favors the commodification and privatization of land and natural resources (Huiliñir-Curío et al., 2019).

Private conservation initiatives, such as exclusive hunting reserves, have existed for centuries amongst the wealthy and ruling classes. In recent decades, relaxed laws and economic policies have amplified the ability for philanthropists and big non-government organizations (NGOs) to purchase and trade conservation lands. Proponents of PPAs often point to declining government capacity and funding as a constraint for the attainment of global biodiversity conservation goals (Pringle, 2017; Waldron et al., 2013). They purport that PPAs obtain individual and institutional investments that public PAs could not tap and that PPAs can operate in areas that have been inaccessible to government action. Today many of the lands that are high priority for conservation are still not appropriate for government purchase or control such as in resource rich areas and regions where private ownership has predominated (Mitchell et al., 2018). Thus, PPAs, with the capacity to leverage additional financial resources and opportunities to purchase land, can fill gaps in government conservation systems (Figgis et al., 2005).

Extending conservation governance beyond the State to involve many individual landowners, corporations, aboriginal and indigenous communities, and non-government organizations, may open opportunities for greater consideration, legitimacy and voice. However, a breadth of motivations and dynamics is associated with the modern PPA movement (which this paper sets out to highlight) and these may hinder those opportunities for engagement. We suggest that for PPAs to successfully contribute to long-term biodiversity conservation, basic principles of good governance (i.e. transparency, equity, accountability, responsiveness to stakeholders) must be upheld and that these principles do not always align with private ownership models and priorities (Eagles, 2009).

Neoliberal policy, conservation, and nature-based tourism in Chile

Chile is one of the most market-friendly countries in the world and was one of the earliest adopters, under the dictatorial government of Augusto Pinochet (1973–1990), of a neoliberal policy framework (Gale, 2006). The 1980 Chilean national constitution has been described as "a blueprint for the neoliberal world" (Tecklin & Sepúlveda, 2014, p. 213) and set out to open domestic markets to greater international investment, deliberately remove restrictions on trade, and shrink the ability of government to monitor and regulate. Foreign capital investments in key industries (mining, forestry, aquaculture, and some agricultural exports) have grown exponentially (Huiliñir-Curío et al., 2019) and this opened the way for corporations, private interests, and

NGOs to gain substantial control over Chile's natural resources (Taylor, 2003; Tecklin et al., 2011). Since the return to democracy in 1990, Chile's governments have generally continued to implement and support neoliberal-style policies guided by "the discipline of money and markets on both state and society" (Taylor, 2004, p. 5). Combined, over the past four decades, Chile's economic policy has promoted expanded private property rights and extended the commodification of natural resources (Delamaza et al., 2017).

Chile has a long-standing commitment to the framework of the Convention on Biological Diversity which they signed in 1994 shortly after returning to democracy. A 2019 report documents great progress over the last five years towards biodiversity targets (Chilean Ministry of the Environment (MMA), 2019; Villarroel Venturini, 2019). It also highlights the following priorities going forward: the development of the Chilean National Biodiversity Strategy (2017–2030); significant increases in terrestrial and aquatic state protected areas; and the development of legislation to create the Biodiversity Service and the National System of Protected Areas (SBAP). In particular, SBAP aims to integrate all State-owned protected areas under a single public agency and improve the coordination, efficiency, and effectiveness of biological diversity policy with increased involvement and oversight of the country's private conservation initiatives.

Concurrently over the last decade, Chile has enacted a series of laws and policies to encourage and incentivize NBT (Blair et al., 2019; Gale & Ednie, 2019). Tourism Law No. 20.423, and subsequent strategic plans of action, have explicitly linked the role of outside investment, tourism growth, economic development, and protected area viability (Chilean Ministry of Economy Development & Tourism, Chilean Ministry of External Relations, & United Nations Development Programme (UNDP), 2015; Chilean Subsecretary of Tourism, 2017).

Criticism of neoliberalism creates risk for private conservation

However, the past few years have witnessed growing social unrest around the world. France, Spain, Hong Kong, India, Chile, Colombia, Bolivia, Lebanon, Iran, and Iraq, all experienced social uprisings in 2019. Although each movement focused on particular local and regional issues, they shared concerns about the rising inequality and corruption that is frequently attributed to capitalist and neoliberal economic policy (Sehnbruch, 2019; Slobodian, 2019).

And so, perhaps, this social unrest serves as a warning sign for conservation policy and practice. PPAs that are seen as reflecting the values, priorities and political power of private interests, have sparked a backlash against conservation (Louder & Bosak, 2019). Similarly, sustainable tourism development, while espoused as a way to promote conservation and provide local livelihoods, has been criticized as being a driver of global capitalism and a source of conflict with local interests (Fletcher, 2011). Thus, it is beholden upon PPAs and nature-based tourism to navigate these turbulent contexts if they are to be welcomed and successfully contribute to worldwide biodiversity conservation.

Research objectives

This paper explores the consequences of a growing dependence on privately protected areas (PPAs) and nature-based tourism (NBT) to secure long-term biodiversity conservation goals. This paper asks three research questions (RQs):

- RQ1: How do neoliberal concepts and dynamics manifest in PPAs, NBT and other forms of private conservation?
- RQ2: Specifically, how has private conservation manifested in Chile?
- RQ3: How might Chileans' attitudes about their government's neoliberal policy affect their expectations for PPAs and NBT?

OneSearch in Databases:
((ANY FIELD CONTAINS "private protected area" OR ANY FIELD CONTAINS "private protected areas" OR ANY FIELD CONTAINS "integrated systems of protected areas" OR ANY FIELD CONTAINS "conservation financing" OR ANY FIELD CONTAINS "private conservation initiatives" OR ANY FIELD CONTAINS "environmental non-governmental organization" OR ANY FIELD CONTAINS "private conservation initiatives" OR ANY FIELD CONTAINS "conservation landscape" OR ANY FIELD CONTAINS "private protected area incentives" OR ANY FIELD CONTAINS "conservation incentives" OR ANY FIELD CONTAINS "private protected areas" AND "public participation" OR ANY FIELD CONTAINS "private protected areas" AND "tourism" OR ANY FIELD CONTAINS "neoliberalism in Chile" OR ANY FIELD CONTAINS "civil protests" AND "neoliberalism and Chile" OR ANY FIELD CONTAINS "October 18, 2019" AND "Chile" OR ANY FIELD CONTAINS "neoliberalism" AND "inequality" AND "Chile"OR ANY FIELD CONTAINS "reterritorialization" AND " Chile" OR ANY FIELD CONTAINS "private conservation" OR ANY FIELD CONTAINS "capitalist conservation" OR ANY FIELD CONTAINS "Legal rights of conservation" OR ANY FIELD CONTAINS "18/O" OR ANY FIELD CONTAINS "neoliberal and parks")) Limiters - Publication Date: Last 20 years

SciELO:
(BUSCAR "área protegida privada" EN EL CAMPO "todos los indices" OR "áreas protegidas privadas" EN EL CAMPO "todos los indices" OR "AP Privadas" EN EL CAMPO "todos los indices" OR "Sistemas integradas de áreas protegidas" EN EL CAMPO "todos los indices" OR "financiamiento de conservación" EN EL CAMPO "todos los indices" OR "unidades de protección privadas" EN EL CAMPO "todos los indices" OR "iniciativas privadas de conservación" EN EL CAMPO "todos los indices" OR "ONG ambiental" EN EL CAMPO "todos los indices" OR "ONG ambientales" EN EL CAMPO "todos los indices" OR "organización no-gubernamental ambiental" EN EL CAMPO "todos los indices" OR "áreas silvestres protegidas privadas" EN EL CAMPO " todos los indices" OR "iniciativas de conservación privadas" EN EL CAMPO "todos los indices" OR "paisaje de conservación" EN EL CAMPO "todos los indices" OR "incentivos para áreas protegidas privadas" EN EL CAMPO "todos los indices" OR "financiamiento privado para la conservación" EN EL CAMPO "todos los indices" OR "áreas protegidas privadas" AND "participación pública" EN EL CAMPO "todos los indices" OR "áreas protegidas privadas" AND "turísmo" EN EL CAMPO "todos los indices" OR "el neoliberalismo en Chile" EN EL CAMPO "todos los indices" OR "las protestas civiles" AND "el neoliberalismo" AND "Chile" EN EL CAMPO "todos los indices" OR "el 18 de octubre de 2019" AND "Chile" EN EL CAMPO "todos los indices" OR "18/O" "el neoliberalismo" AND "la desigualdad" AND "Chile" EN EL CAMPO "todos los indices" OR "reterritorialización" AND "Chile" EN EL CAMPO "todos los indices" OR "conservación privada" EN EL CAMPO "todos los indices" OR "conservación capitalista" EN EL CAMPO "todos los indices" OR "derecho real de conservación" EN EL CAMPO "todos los indices" OR "neoliberal" AND "parques" EN EL CAMPO "todos los indices")

Figure 1. Search strings for OneSearch and SciELO indexes.

By exploring how neoliberalism underpins the motivations and strategies for PPA and NBT, we hope to expose vulnerabilities in the current PPA model. These, then, provide the basis for recommendations to help private conservation have greater success and support in the coming years.

Methods

This study employed a targeted literature review which aims to provide an in-depth, informative review on a specific topic (Bridges et al., 2015; Parcher et al., 2018; Walton & Rogers, 2017). Targeted literature reviews generally focus on a "knowledgeable selection of current, high-quality articles on the topic of interest" rather than taking a systematic, all-encompassing approach (Huelin et al., 2015, p. 2). The identification of trends and provision of a current state of knowledge then informs future research strategies and agendas. Our methods followed standards generally associated with narrative literature reviews including a balanced overview, a summary of the key outcomes presented, an illumination of supporting assumptions and theory, and identification of future research needs (Eguchi et al., 2020; Newton et al., 2018).

Our targeted review specifically examined research and public policy documents related to neoliberal concepts and dynamics and to the motivations, strategies, and outcomes of private conservation efforts. Special emphasis was given to mechanisms of PPA financing and their ongoing viability. The geographic concentration was both a general world-level review and a specific analysis of Chilean PPA-related literature. We included key articles identified by experts on Chilean protected areas and conservation as well as articles identified from a search using keywords such as: private protected areas, private conservation, conservation financing, conservation incentives, capitalist conservation, neoliberalism, and legal rights of conservation (see full search string in Figure 1). The search strategy was conducted in electronic databases such as Academic Search Complete, JSTOR, Web of Science, EBSCO, BioOne Complete, ProQuest, and SCOPUS (as accessed through the OneSearch platform). A similar strategy was used in the SciELO database for Spanish language texts. The search strategies were also manually applied for identification of government and NGO reports as well as popular press articles. Considering PPA and NBT growth during the past two decades, we focused on articles published between 2000 and 2020.

After duplicates were removed, initial screening was applied to titles and abstracts by three independent reviewers. If two or more of the reviewers agreed that an article did not meet the inclusion criteria (e.g. as being in completely different contexts than conservation, land management or NBT), it was excluded. At least two of the three reviewers then conducted a second analysis of the full text of all articles, summarizing information from articles that supported aspects of the research questions. While the aim of a targeted literature review is to scope a broad range of literatures, it is less a meta-analysis or quantitative effects estimation than it is an identification of trends and current state-of-knowledge. As such, a synthesis of the results aims to provide a summarization and critical appraisal of the resources found (Levac et al., 2010; Snyder, 2019). A particular objective of this study was the development of a dimensionalization of the underlying and sometimes hidden assumptions and dynamics of private conservation, as such a conceptualization was not yet apparent in the conservation literature.

The analysis undertaken was of a qualitative nature. In order to identify and analyze the core ideas and themes, the literature was broken down initially into the history and origins of private conservation; identification of common concerns and criticisms; and examination of relationships between local communities and private conservation actors. As the main ideas and dimensions emerged from the literature, their reliability and authenticity were checked for consistency and the degree to which the literature represents the issue, how it was observed in the Chilean context, and with the authors familiarity and knowledge of the topic. Looking across interdisciplinary sources, the analysis sought commonalities and less-than-obvious connections across multiple studies. The authors sought clustering of ideas rather than a listing or literal reporting of previous data. The groupings in RQ1, specifically, are more scholarly than they are by research component (and not, for example, by empirical setting, research methodology, disciplinary foundation, etc.) Focus, therefore, was on key concepts, definitions and assumptions. It is hoped that the resulting dimensionalization provides guidance for future consideration and implementation of PPAs.

Literature search and study characteristics

The OneSearch database search yielded 940 titles, which was supplemented by 178 titles obtained through expert recommendations (n = 156) and from the Scielo database (22). Of these, 730 titles advanced to the first review and 515 advanced to the full-text screening. Two hundred and eighty four articles were retained for the targeted review, including 154 sources in Spanish and 130 sources in English (See Figure 2).

Academic journal articles (102) and professional reports (110) surfaced within the review, as well as magazine articles (27), book sections (18), and other sources (27). Of these 284 resources, 112 addressed how neoliberal concepts and dynamics (including NBT) manifest in PPA motivations, strategies, and outcomes (RQ1). One hundred and fifty resources addressed how PPA motivations, strategies, and outcomes have manifested in Chile (RQ2). Fifty-six resources addressed how attitudes about Chile's neoliberal economics might manifest, with respect to PPAs and NBT (RQ3). The following subsections address findings with respect to each of these guiding questions.

Results

Manifestations of neoliberal concepts and dynamics within private conservation

Three dimensions help organize the state of knowledge about how neoliberal concepts and dynamics manifest in private conservation. The first dimension refers to the loss of the social

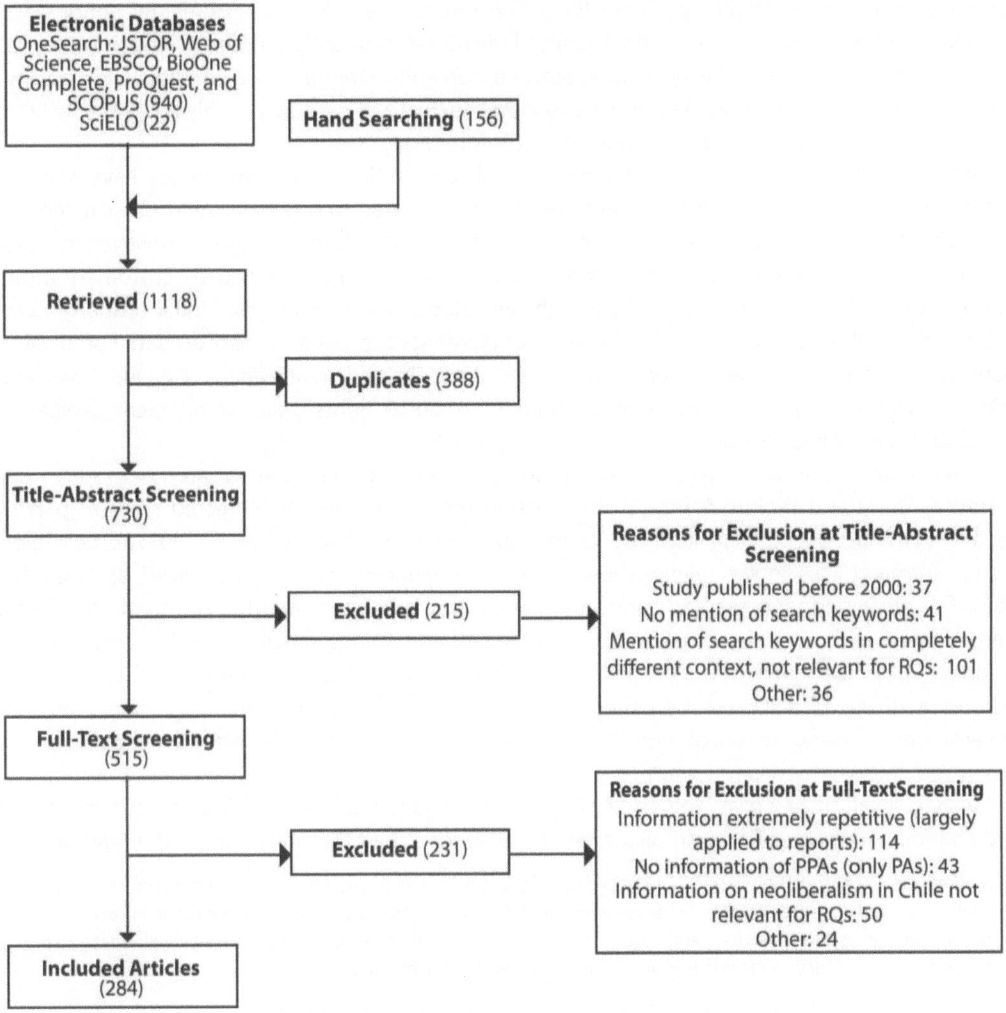

Figure 2. Summary of targeted literature review results.

embeddedness of nature. In contrast to integrated social and cultural contexts of nature, neoliberalism treats resources as distinct, somewhat interchangeable commodities. Secondly, neoliberalism opens nature to the dynamics of capitalism and free trade. Global financial markets are complex and driven by external factors well removed from local contexts. Third, neoliberal assumptions, including an increasing reliance on ecosystem service concepts, may not easily integrate with traditional conservation discourses.

Reduction of the social embeddedness of nature

There is evidence that treating nature as a commodity that can be bought and sold in distant markets reduces the social embeddedness of nature. Personal histories and social entanglements are largely ignored in the face of the external values and priorities that motivate purchases. For example, a remote and underdeveloped place may have long standing cultural traditions associated with natural resource-dependent livelihood activities (i.e. ranching, logging, artisan fishing). By contrast, external purchasers may be attracted to large tracts of land in this place by priorities such as scenic beauty and intrinsic value. Researchers such as Higgins et al. (2014) and Sullivan (2014) have observed that many conservation strategies favored by private investors (such as carbon credits, species credits, payment for ecosystem services (PES), environmental options and futures, mitigation insurance, and biodiversity derivatives) frequently ignore community histories and social entanglements. These market-based conservation strategies are, in effect, commodity-based economic models that buy and trade decontextualized resources. Sullivan (2014, p. 225) termed this marketization "radical geographies of nonlocality."

Berkes (2008) observed that most remote, rural, and indigenous communities have existing institutions and rules that limit short-term self-interests related to overharvesting or extraction in favor of longer term interests that continue the viability of nature and place. However, cultural and natural resources that may have been held communally and governed by customary rights can become formalized and legalized through the establishment of the individual property ownership rights. What was an intertwined system of elements can become disaggregated as distinct resource rights (land, water, mineral, timber, etc.). Tecklin and Sepúlveda (2014) identified the risks of 'asset stripping', a practice of unbundling resource rights and trading them separately based on their individual value.

Other literature noted a reduction of the local oversight of nature associated with PPAs. For example, Anaya and Espírito-Santo (2018) criticized the designation of protected areas as part of an overall reduction of local land-use control and planning. Serenari et al. (2017a), described highly asymmetrical power relationships where PPA landowners were perceived as "wealthy, powerful, privileged, and reaping the returns of private protected area tourism," whereas locals were treated with "a lack of deference, candidness, and transparency" (p. 100). Drescher and Brenner (2018) identified that PPA actors were often less enthusiastic about local control and were increasingly disillusioned by government planning and management. They concluded that governments seeking to encourage PPAs feel pressure to be flexible with their policies and lessen regulatory review and control.

Further, locals may be invited to continue their engagement with PPA lands but in limited and changed capacities. For example, Dressler (2014) identified that in much of southeast Asia:

> market-based conservation is often heralded as the most efficient and effective means to improve rural livelihoods and conserve forests ... State agencies, donors, and NGOs promote the payment of local users to maintain ecosystem services and further their involvement in market ventures to value alternate uses, reducing 'dependency' on forest resources and extensive land uses. (p. 25)

Dressler (2014) also noted that PPAs can also cause significant changes as a result of modernization and technical innovation. Outreach programs designed to teach and instill 'best practices' to surrounding communities can inadvertently hasten the pace and scale of agrarian change (Langewiesche, 1999). Participation for locals may require they shift from hunter-gatherer or agricultural lifestyles to service, repair, or entertainment roles (Spenceley, 2003). In the process, locals may need to abandon livelihoods as independent operators and assume contractual (or daily wage) work as employees. And while locals take on service-oriented roles within PPAs, the project staff for PPAs often come from outside the region as a largely migrant workforce.

Local populations in and around PPAs can experience further dramatic changes in their connections to the land. With respect to NBT, Slocum (2017) suggested that developers, corporate entities, and absentee landlords all crowd out existing local stakeholders and interests. Stronza et al. (2019) described NBT as "a vanguard activity that is likely to create a foothold in culturally and biologically sensitive areas that are later exploited through mass tourism development" (p. 5.6). West and Carrier (2004) identified that NBT serves as a conduit for external economic and social pressure. Spenceley et al. (2019) concurred in the context of southern Africa, finding that once external NBT operators are in place, that they "try to influence protected area management towards more advantageous policies" (p. 76). Spenceley and Meyer (2012) warned that NBT might exacerbate inequalities between local communities and those involved in tourism development.

Local culture and traditions may, themselves, become a form of spectacle or theatre, to be consumed within the experience and gaze of tourists (Gale et al., 2013). Dressler (2014) described these tourist experiences as a commodification of material and immaterial culture as "glossy goods that tourists can readily consume" (p. 4). Further, as Buckley and Mossaz (2018) described, when private NBT operators underwrite PPAs it is their messaging that dominates visitor's attention, often dramatically focused on flagship wildlife species. Similarly, it has been observed that physical and fast-paced visitor activities attract more attention than companionable and leisurely activities (Slattery, 2002). Tourism marketing can also celebrate dated stereotypes such as an agrarianism that has long passed (MacDonald, 2011). Given the transitory nature of most participants in NBT (both visitors and temporary employees), there is less time for understanding and respecting long-standing local relationships with nature.

Bringing nature conservation into the capitalist dynamics of the global marketplace

The second dimension from the literature relates to how the neoliberalization of conservation policy has brought it into the capitalist dynamics of the global marketplace. Accordingly, a number of market-based mechanisms have evolved whereby companies can buy credits to compensate for the impacts associated with their regular activities including payments for ecosystem services, trading of carbon offsets/sequestration and carbon assets (i.e. biochar), biodiversity derivatives securitization, species and wetland banking, and even tradeable *protected area certificates* (Meißner & Grote, 2017). These financial instruments share the common feature of bringing nature into the capitalist dynamics of the global marketplace (Clements & Cumming, 2018; Fletcher, 2010; Fletcher et al., 2014; Higgins et al., 2014). Further, Fletcher (2011) concludes that NBT is one of the "chief means by which capitalism has been able to expand its commodification of natural resources on a global scale" (p. 458).

To attract and maintain international investment, nature must be packaged and presented in the manner that is most understandable and attractive to global purchasers (Matose, 2014). These strategies frequently include a discourse of scarcity and repair – where investment opportunities are marketed as an identification of scarce resources (e.g,. water) and the immediate need for their preservation and restoration (Fairhead et al., 2012; Otsuki et al., 2017; Scoones et al., 2019). PPAs would, for example, be seen as in an urgent state of disrepair thus requiring an influx of economic activity to halt the loss of unique and rare biological resources. Although underlying ecological conditions might warrant such an approach, the concern is that governments and institutions employ these discourses to recover lands from the perceived-to-be inadequate practices of locals. Fairhead et al. (2012) noted that "unsustainable use 'here' can be repaired by sustainable practices 'there', with one nature subordinated to the other" (p. 242).

Meißner and Grote (2017) cautioned how investment vehicles marketed through discourses of scarcity and economies of repair often attract investors with financial incentives such as tax rebates, subsidies, and/or lower regulatory burdens. These can be seen as fair exchange for the lost economic opportunities that the land could have otherwise sustained. However, as Selinske et al. (2017) suggest, "direct payments can potentially increase the cost of conservation over time as landholders come to expect payments, seeing them as an entitlement, or creating dependencies, leading to questions over what happens if or when the incentives cease" (p. 6). Under a neoliberal paradigm, a fundamental search for revenue and profit can dictate the priorities and services of PPAs rather than the long-term stability of ecosystems and species (Arsel & Büscher, 2012; Brockington & Ponte, 2015; Davis & Tilton, 2005).

Conflicting discourses and assumptions

The third major theme that emerged from the literature involved how the discourse of neoliberal conservation undermines more traditional conservation discourses that emphasize aesthetic, cultural, and ethical aspects of nature (Bachmann-Vargas & van Koppen, 2020). While it is easy to assume that PPAs enable both economic growth and ecosystem health, neither can be automatically assured as the economic context and ecological location of a PPA may actually limit any success. And the very notion that conservation can, and should, work alongside economic development should be carefully considered. Given a trajectory of unending growth and cash flow alongside finite natural resources, it leans towards consumption of that resource base. The trade-offs between the two can be irreconcilable, making efforts towards consensus or some sense of win-win somewhat misplaced. Without full accounting of negative social and environmental consequences, conservation can, in effect, be traded-off for economic development. Font and Lynes (2018), for example, in a review of corporate sustainability and responsibility (CSR) in tourism and hospitality, suggest that many companies are making only superficial changes in order to appear more sustainable. Private operators have "defined CSR as secondary to the business growth trajectory" (Font & Lynes, 2018, p. 1030). The challenges surrounding retooling products, services and supply chains towards contemporary sustainability goals, like *carbon neutrality* and *zero negative impact* are significant for the tourism industry (Dinica et al., 2019) and often fall to the wayside if they aren't sufficiently profitable. Regulation, oversight and monitoring are costly for the investors and there is incentive for businesses to circumvent or short-circuit these public processes (Hodge & Greve, 2017; Randle & Hoye, 2016; Spenceley et al., 2019). In the rush to finance and approve private projects, public scrutiny and transparency can be sidelined (Frost & Laing, 2018).

The costs to administer and monitor a system of PPAs may be high and it is unlikely that regional or national management agencies would be allocated additional resources to do so. Rather it is more likely that government agencies would be downsized and have their expertise and funding dispersed. In that process, they lose the ability to influence and oversee conservation and tourist activities and must increasingly delegate that authority to private operators (Hodge & Greve, 2017). Removal and replacement of private contractors can then be costly, impractical and disruptive (Spenceley et al., 2019).

The decisions of who accumulates land, who has access, and how the natural resources get redistributed are fundamentally political ones. Neoliberalism puts markets into a privileged position for the decision of *who* gets ecosystem services (MacDonald & Corson, 2014). International actors may be dictating the accounting of values and benefits, perhaps through projects such as TEEB (The Economics of Ecosystems & Biodiversity, a component of the international Convention on Biological Diversity). And, it can be hard for the poor to see the value of private conservation when they themselves aren't as able to mobilize capital, participate, and gain benefits. Awareness of market mechanisms may be low and collective action may be shunned or disallowed in the market.

Further, PPAs necessarily involve territorialization and partitioning of land. In moving land into a market for buying and selling, PPAs draw attention and access to an area that may have had less visibility. In creating property rights there is an implied degree of control and protection with associated permissions, enforcements, permitted uses, and concession agreements all to follow. Rents can be expected to be extracted from what might have been a commons. The blocking of customary access may be needed to ensure exclusive and managed experiences for ecotourists and/or a high degree of privacy on private properties. Pushing back against imposed control might lead to various forms of resistance such as "false compliance, pilfering, feigned ignorance, slander, arson, sabotage" (Matose, 2014, p. 14). Those who oppose can be seen as obstructionist and their usual activities criminalized (Anaya & Espírito-Santo, 2018).

While PPAs and neoliberalism may use language of freedom, liberty, and choice, the reality for many locals is the opposite. With the retreat of government oversight and capacity, there are fewer rights and avenues of appeal. As Serenari et al. (2017b) suggest "PPAs carry a unique risk of creating a false sense of security and disempowering local people in ways that are less evident in similar efforts promoted by public protected areas" (p. 14). Ramutsindela (2015), using

PPAs in South Africa as an example, points out the attractiveness of training and staff development opportunities for locals but also that the employees may have to live onsite under the gaze, rules, and control of the owners. Migrant labor, more used to a transitory lifestyle and less tied to local patterns of behavior, may depress wages and conditions and further displace locals from their home landscape. Brockington and Ponte (2015) similarly suggests that, "philanthropy allows private owners to structure and control labour, while directly or indirectly affecting the trajectory of land claims" (p. 2201). The process of registration of interests and claims may be complicated and inequitable. Offers of legal representation and technocratic assistance are rightly treated with suspicion. Neoliberalism dominates or tames these settlements, as the price of land increases to a level that excludes or constricts restitution.

Chile: neoliberal dimensions of private conservation in a market-friendly context

The three dimensions discussed in the prior sections represent the results associated with RQ1. In the following sections, we examine how these dimensions have manifested within private conservation efforts in Chile (RQ2). The results section then closes by addressing how attitudes about Chile's neoliberal economics might impact support and integration of PPAs and NBT (RQ3).

Nature conservation in Chile

Chilean neoliberal ideologies have long driven privatization of most aspects of public, social, economic, and environmental policy (Bertrand & Niemann, 2017; Gwartney et al., 2011; Tecklin & Sepúlveda, 2014). Perhaps one of the most telltale signs of Chile's historically neoliberal environmental policy is the lack of formal articulation of environmental protections until well after the return to democracy in 1994. Moreover, there is no one institution overseeing biodiversity conservation priorities nor a single public service dedicated to strengthening protected areas in an integrated manner (BIOFIN Chile, 2017; Chilean Ministry of the Environment (MMA), 2019). Today, Chile's environmental governance system includes at least 14 ministries and 28 public services at a national level, 16 regional governments, and 345 municipalities. While the Ministry of the Environment (MMA) is in charge of environmental policy, plans, and programs; the Environmental Evaluation Service (SEA) administers and interprets environmental impact assessments; and the Superintendency of the Environment (SMA) is responsible for control and compliance.

The exception to Chile's historical environmental ambivalence has been its long-standing range of state run protected areas, dating back to 1907 when its first national reserve was established (Silva, 1996). Currently, about 19.5% of Chile's terrestrial territory and 13.6% of its marine economic exclusive zone (EEZ) are under some level of State conservation protection and are, thus, on track with International Convention on Biological Diversity (CBD) objectives in terms of geographic coverage. However, there are concerns with their ability to meet biodiversity and representative ecosystem goals, as well as their overall management effectiveness (Durán et al., 2013; Fernández et al., 2016; Pauchard & Villarroel, 2002; Petit et al., 2018; Schutz, 2015; Squeo et al., 2012).

In addition to the governance complexities associated with managing this collection of protected areas, funding deficits are a longstanding issue in Chile. In 2010, a joint report of the Argentine Wildlife Foundation, the Food and Agriculture Organisation (FAO) and Spain's national parks agency (Organismo Autónomo Parques Nacionales, OAPN) identified significant funding risks throughout Latin America. It noted that, at the time, Chile invested only around US$0.95 per hectare in protected areas, representing one of the lowest rates in Latin America (Argentine Wildlife Foundation, 2010). During the decade that followed this influential report, Chile's Ministry of the Environment (MMA) has increasingly looked toward privately and community protected areas to help fill gaps. Fernández et al. (2016) noted that "… large private areas can contribute to improving the coverage of ecosystems, species or ecological functions that are absent or underrepresented in the SNASPE, while 'small areas' can form biological corridors or 'stepping stones' that increase connectivity between large areas" (p. 12).

Privately protected areas in Chile

PPAs emerged in Chile in the late 1980s and, despite a lack of formal recognition, regulation, or monitoring on the part of the State, have grown steadily over the past three decades, especially in the south of the country. In 2011, the first attempt to measure the extent of Chilean PPAs identified 377 initiatives, covering nearly 1,500,000 hectares (Chilean Ministry of the Environment (MMA), 2011; Tecklin & Sepúlveda, 2014). However, these results must be taken with caution, as no official definition of PPA parameters existed; thus, counts included all self-declared conservation areas (Root-Bernstein et al., 2013). Chile's web of private conservation projects and areas are owned and managed by a variety of actors including private citizens, foundations, indigenous groups, and environmental NGOs. Most areas are small in size (less than 1,000 hectares), although there are also a handful of large expanses that have been set aside by wealthy Chilean families and businesses and by international foundations and investors (Rivera & Vallejos-Romero, 2015).

Since their emergence, Chilean private conservation efforts have received a mixed response from stakeholders and communities (Jones, 2012; Serenari, 2014). Mendoza et al. (2017) and Núñez et al. (2014) both focused their research and analysis on the ways in which neoliberal policy has contributed to a process of land use change in Chile, identifying how it has contributed to political conflict and struggle. In particular, several of the larger private conservation projects have experienced controversy, conflict, and criticism, coming both from locals and from extractive industries who view them as 'land grabbing' by outsiders and as a taking of valuable resources out of productive use. Further, Serenari et al. (2015) found that local cultural differences clashed with dominant development-based ideologies. They identified inadequate communication, inequitable decision making, and divergent opinions about the purposes, trajectories, and livelihood impacts of PPAs all as factors contributing to the conflict in Chile.

Other researchers have concentrated on understanding local response to the PPA initiatives undertaken by Conservación Patagónica and other organizations established by Doug and Kris Tompkins (i.e. The Conservation Land Trust, the Foundation for Deep Ecology, and Fundación Pumalín). For example, Jones (2012) explored the transformation occurring in the Chacabuco Valley as Conservación Patagónica worked to restore grasslands on the path to eventually establishing the Patagonia National Park. Jones identified local concerns about foreign influence on land-use decisions and also that increased NBT was creating competition for traditional ranching livelihoods. Louder and Bosak (2019) found that the then private park initiative had, "been detrimental to local livelihoods, disrupted systems of production, and elicited emotional responses of pain, sadness, and loss" (p. 161). Gale and Ednie (2019) similarly identified value conflicts with local stakeholder groups, regional and national government entities. A lack of engagement and transparency between locals and the Conservación Patagónica officials contributed to subsequent and continuing bitterness and resentment in the region.

From privatized protection to inclusive conservation-based development? Nature-based tourism in Chile

Many private conservation efforts in Chile include NBT. And, as PPAs become increasingly important and integrated within the national Chilean conservation system, it seems likely that NBT will continue to grow. Formats vary with some private conservation efforts, like the Huilo-Huilo Ecological Reserve in Panguipulli, including upscale tourism resorts. Some limit participation to selected groups and guides through concession-type agreements, while others have employed NBT as a deliberate means to integrate outside actors and enterprises (Serenari et al., 2017b).

New private-public partnerships seem both likely and necessary in order to achieve better connections between private conservation/NBT initiatives and communities. For example, the 2017 agreement between the Tompkins Foundation and Chile included the creation of the *Route of Parks* – a conceptual circuit that connects and packages 17 National Parks within the Magallanes, Aysén, and Lakes Regions (Blair et al., 2019; Otsuki et al., 2017). The Route features coordinated marketing, community development support, and local outreach efforts. The route promotes a collaborative public-private approach to what the Foundation describes as "conservation-based development." As the website describes, "The Route of Parks is a new way of thinking about the immensity of our beloved Patagonia: a vision of economic development based on conservation and ecologically-minded tourism" (www.rutadelosparques.org/en/). Still, the Route's success will ultimately depend on the willingness of local communities to incorporate its logic and strategies within their own planning, development, and priorities. For example, in early 2020, a publicly funded Territorial Integration Program (PTI), was created to focus on the integration of local interests with the Route of Parks Concept. This initiative has received funding for three years through the support of the government economic development agency (CORFO), the Lakes Regional Government, the National Tourism Agency (Sernatur), the National Forest Corporation (CONAF), and the Ministry of the Environment (MMA), as well as private participation from universities, six associated communities, and tourism operators (ptirutadelosparques.org). If initiatives like PTI prove successful, and are validated by local communities, they are likely to be replicated within other parts of the increasingly integrated public-private Chilean system.

Moving forward: implications for Chile and beyond

Many of the issues seen with PPAs in Chile, and with neoliberalism in general, are concerns that can be raised equally across the full range of conservation areas. Both state-sponsored and privately-funded protected areas need to acknowledge the political and economic dynamics that occur when identifying and designating new areas. Relaxed oversight, lack of strategic coordination, insecure funding mechanisms, deteriorating basic infrastructure, importation of foreign priorities and expertise, and the like are all common challenges regardless of ownership.

Good principles of governance (eg. transparency & public accountability, equitable distribution of costs and benefits, engagement and responsiveness with stakeholders) must be genuinely and steadfastly pursued. However, our review has highlighted how PPAs and NBT, in particular, have embedded priorities, language and dynamics that reflect neoliberal policy, often exacerbating and amplifying already existing deficiencies. We conclude similarly to Ramutsindela (2015) who observed PPAs as a mechanism for the introduction of neoliberalism and not just a product of it. Neoliberalism, however, hides and obfuscates the very system and consequences that it embraces. For example, many PPAs have been created and structured in parallel to the economic development of associated natural resources (i.e. sustainable nature-based tourism, agriculture related initiatives, commercialization of ecosystem services, and/or acceptance or trading of conservation related incentives). Dressler (2014) noted that these types of market-based conservation are "often heralded as the most efficient and effective means to improve rural livelihoods and conserve forests" (p. 25).

The outcome of these capitalist processes (as seen in the literature and the example in Chile) is that there is a loss in the social embeddedness of nature as the local becomes supplanted with global markets and conservation agendas. This is evidenced in the external pressures and priorities imposed on communities, often by external owners with differing values. When decisions about local land-use are made elsewhere and without transparency and local accountability, the resulting asymmetrical power relations can create conflict. When NBT is layered into this process, sometimes portrayed as an 'opening-up' of local economies and 'unleashing' of entrepreneurial activity, it can further crowd out local social and economic connections with nature. Local people go from producers and providers to employees. Furthermore, local culture and traditions can become commodified as tourists seek to engage with the 'authentic' culture of the area.

Simultaneously, the partitioning and territorializing of land asserts the primacy of private property rights. Wealthy landowners are able to document their ownership when locals are not; leading to conflicts over land claims and or restitution for loss of land and/or access to what were once seen as common lands. When NBT is introduced into this dynamic, the expectation is towards the generation of rents on the property. The enclosure of land for conservation and then opening of that same area for tourists excludes locals while providing tourists with a private enclave to gaze at nature imagined, not from a local perspective but from that of the global elite. Meanwhile, locals bear the costs of tourism (i.e. loss of access, identity, connection to the landscape, economic opportunities) without reaping the benefits. The exclusion of locals in the process of privatizing and enclosing land and then opening it for tourism restructures the local economy and erodes local connections to the land. This devalues local knowledge, capacity, leadership and social capital. While these dynamics look similar to what occurs with PA development in general, there are some distinct differences. When state-owned PAs are developed, local people can assert themselves as citizens with certain rights and the state can be held accountable. In the case of PPAs there can be no such guarantee of effective governance.

Although community and local stakeholder participation has been identified as being of high importance for PPA and NBT sustainability, successful implementation of these practices is still not the norm within neoliberal approaches. And, yet, if NBT is "the most prominent form of neoliberal conservation globally at present" (Fletcher, 2014, p. 89) then the disparities created through neoliberalism must be addressed. Growing frustrations suggest a need for sustainable NBT to explicitly link market-based approaches, improvements in biodiversity protection, and improved quality of life. The capacity and viability of different PPA and NBT governance systems, and their attendant funding streams, is rarely assessed and there is often no accountability for their differing effectiveness (Eagles, 2014; Varghese, 2008). Social well-being outcomes of PPAs and NBT initiatives could be assessed, for instance, on the basis of their "protection or respect of local cultures, heightened visitor awareness, local participation and ownership of business ventures, enhanced local pride and sense of empowerment" (Serenari et al., 2017b, p. 1793). NBT, as a potential tool for poverty reduction, has a rather mixed track record (Dinica, 2017; Mbaiwa, 2008; Spenceley & Meyer, 2012). As a result, we urge a focus towards development of social capital within surrounding communities such as capacity building and entrepreneurial skills training, preparation and participation in planning and decision-making, and a heightened sense of engagement and stewardship towards the natural environment.

We believe that one of the distinct roles for PPAs may be in bringing these concerns to the fore and explicitly working towards integrated governance. As Mitchell et al. (2018) argue, long-term conservation and permanence of PPAs is, "best secured through legal instruments empowering governance … linked to an institutionalized conservation agent or legal status through an appropriate government agency" (p. 33). Jones (2014) similarly called for reduced leakage of economic benefits outside of the surrounding communities and for promotion of a diversity of livelihoods to be built into PA governance frameworks (such as through transparency, fair and legal adjudication, establishment of collaborative platforms, cross-jurisdictional coordination and peer enforcement). Having an integration of conservation approaches (public, private, NGO, local, and international), along with their different ecological and social imperatives, confers greater long-term resilience and adaptability. Overlapping and inter-related dependencies, as well as reinforcing interactions, strengthens the governance framework in the face of future disturbances and threats (Jones, 2013). The future of private protected areas seems bright but extra efforts will need to be taken to ensure their success. The insights gained from a consideration of the contentious, underlying neoliberal assumptions help show the way.

Disclosure statement

No potential conflicts of interest are reported by the authors.

Funding

This research was supported by Chile's National Research and Development Agency (ANID) under ANID's Regional Program R17A10002.

ORCID

William Thomas Borrie http://orcid.org/0000-0003-3788-5220
Trace Gale http://orcid.org/0000-0001-9392-3653

References

Anaya, F. C., & Espírito-Santo, M. M. (2018). Protected areas and territorial exclusion of traditional communities: Analyzing the social impacts of environmental compensation strategies in Brazil. *Ecology and Society*, *23*(1):8. https://doi.org/10.5751/ES-09850-230108

Argentine Wildlife Foundation. (2010). *Manual de capacitación sobre sostenibilidad financiera para áreas protegidas en Latinoamérica [Training manual on financial sustainability for protected areas in Latin America]*. Food and Agriculture Organization of the United Nations (FAO).

Arsel, M., & Büscher, B. (2012). Nature™ Inc.: Changes and continuities in neoliberal conservation and market-based environmental policy. *Development and Change*, *43*(1), 53–78. https://doi.org/10.1111/j.1467-7660.2012.01752.x

Bachmann-Vargas, P., & van Koppen, C. S. A. (2020). Disentangling environmental and development discourses in a peripheral spatial context: The case of the Aysén region, Patagonia, Chile. *Journal of Environment and Development*, *29*(3), 366–390. https://doi.org/10.1177/1070496520937041

Balmford, A., Green, J. M. H., Anderson, M., Beresford, J., Huang, C., Naidoo, R., Walpole, M., & Manica, A. (2015). Walk on the wild side: Estimating the global magnitude of visits to protected areas. *PLoS Biology*, *13*(2), e1002074. https://doi.org/10.1371/journal.pbio.1002074

Berkes, F. (2008). *Sacred ecology: Traditional ecological knowledge and resource management* (2nd ed.). Taylor and Francis.

Bertrand, J., & Niemann, T. (2017). Aproximación a la naturaleza jurídica del derecho real de conservación en Chile. *Foro Jurídico (Lima)*, *16*, 163–178. http://revistas.pucp.edu.pe/index.php/forojuridico/article/view/19869

BIOFIN Chile. (2017). *Policy brief: Biodiversidad en Chile. Propuestas para financiar su conservación y uso sostenible [Policy brief: Biodiversity in Chile. Proposals for financing conservation and sustainable use]*. https://www.cl.undp.org/content/chile/es/home/library/environment_energy/biodiversidad-en-chile–propuestas-para-financiar-su-conservacio.html

Blair, H., Bosak, K., & Gale, T. (2019). Protected areas, tourism, and rural transition in Aysén, Chile. *Sustainability*, *11*(24), 7087. https://doi.org/10.3390/su11247087

Bridges, J., Berger, Z., Austin, M., Nassery, N., Sharma, R., Chelladurai, Y., Karmarkar, T. D., & Segal, J. B. (2015). Public reporting of cost measures in health: An environmental scan of current practices and assessment of consumer centeredness. Agency for Healthcare Research and Quality (US). https://pubmed.ncbi.nlm.nih.gov/25763451/

Brockington, D., & Ponte, S. (2015). The green economy in the global south: Experiences, redistributions and resistance. *Third World Quarterly*, *36*(12), 2197–2206. https://doi.org/10.1080/01436597.2015.1086639

Buckley, R. C., & Mossaz, A. (2018). Private conservation funding from wildlife tourism enterprises in sub-Saharan Africa: Conservation marketing beliefs and practices. *Biological Conservation*, *218*, 57–63. https://doi.org/10.1016/j.biocon.2017.12.001

Büscher, B., Dressler, W., & Fletcher, R. (Eds.). (2014). *Nature™ Inc.: Environmental conservation in the neoliberal age*. University of Arizona Press.

Chilean Ministry of Economy Development & Tourism, Chilean Ministry of External Relations, & United Nations Development Programme (UNDP). (2015). *Apoyo a la implementación del plan de acción: Turismo sustentable en áreas protegidas del Estado [Support for the implementation of the action plan: Sustainable tourism in protected areas of the state]*. Government of Chile.

Chilean Ministry of the Environment (MMA). (2011). *Sistematización y recopilación de información directa de dos bases de datos pertenecientes a estudios sobre las iniciativas de conservación privadas [Systematization and direct data collection from two databases for studies on private conservation initiatives]*. Government of Chile.

Chilean Ministry of the Environment (MMA). (2019). *Sexto Informe Nacional de Biodiversidad de Chile ante el Convenio sobre la Diversidad Biológica (CDB)* [Sixth national biodiversity report of Chile to the *convention on biological diversity (CBD)]*. Government of Chile.

Chilean Subsecretary of Tourism. (2017). *Turismo sustentable en Áreas Protegidas del Estado: Una apuesta de presente y futuro [Sustainable tourism in state protected areas: A commitment to the present and future]*. Government of Chile. http://www.subturismo.gob.cl/wp-content/uploads/2018/03/TURISMO-SUSTENTABLE-EN-AP.pdf.

Clements, H. S., & Cumming, G. S. (2018). Traps and transformations influencing the financial viability of tourism on private-land conservation areas. *Conservation Biology: The Journal of the Society for Conservation Biology, 32*(2), 424–436. https://doi.org/10.1111/cobi.12999

Davis, G. A., & Tilton, J. E. (2005). The resource curse. *Natural Resources Forum, 29*(3), 233–242. https://doi.org/10.1111/j.1477-8947.2005.00133.x

Delamaza, G., Maillet, A., & Martínez Neira, C. (2017). Socio-territorial conflicts in Chile configuration and politicization (2005-2014). *European Review of Latin American and Caribbean Studies*, (104), 23–46.

Dinica, V. (2017). Tourism concessions in national parks: Neo-liberal governance experiments for a conservation economy in New Zealand. *Journal of Sustainable Tourism, 25*(12), 1811–1829. https://doi.org/10.1080/09669582.2015.1115512

Dinica, V., Lund-Durlacher, D., & Reiser, D. (2019). Challenges for tourism—Transitioning to corporate sustainability and responsibility. In D. Lund-Durlacher, V. Dinica, D. Reiser & M. S. Fifka (Eds.), *Corporate sustainability and responsibility in tourism* (pp. 3–27). Springer.

Drescher, M., & Brenner, J. C. (2018). The practice and promise of private land conservation. *Ecology and Society, 23*(2). https://doi.org/10.5751/ES-10020-230203

Dressler, W. (2014). Capitalizing conservation on Palawan Island, the Philippines. In B. Büscher, W. Dressler, & R. Fletcher (Eds.), *Nature^{TM} Inc.: Environmental conservation in the neoliberal age* (pp. 25–43). University of Arizona Press.

Durán, A. P., Casalegno, S., Marquet, P. A., & Gaston, K. J. (2013). Representation of ecosystem services by terrestrial protected areas: Chile as a case study. *PLOS One, 8*(12), e82643. https://doi.org/10.1371/journal.pone.0082643

Eagles, P. F. (2009). Governance of recreation and tourism partnerships in parks and protected areas. *Journal of Sustainable Tourism, 17*(2), 231–248. https://doi.org/10.1080/09669580802495725

Eagles, P. F. J. (2014). Research priorities in park tourism. *Journal of Sustainable Tourism, 22*(4), 528–549. https://doi.org/10.1080/09669582.2013.785554

Eguchi, Y., Wong, G., Akhtar, O., & Sumida, Y. (2020). Non-invasive diagnosis of non-alcoholic steatohepatitis and advanced fibrosis in Japan: A targeted literature review. *Hepatology Research, 50*(6), 645–655. https://doi.org/10.1111/hepr.13502

Fairhead, J., Leach, M., & Scoones, I. (2012). Green grabbing: A new appropriation of nature? *Journal of Peasant Studies, 39*(2), 237–261. https://doi.org/10.1080/03066150.2012.671770

Fernández, P., Moya, D., Palma, M., Nuñez-Avila, M., Corcuera, E., Farias, A., Pliscoff, P., Palma, J., Barrientos, M., Sepúlveda, C., Canessa, R., Pinto, A., Rivas, F., Rosenman, E., Donoso, C., Caiplillán, C., Rojas, R., Antezana, R., Olivares, L. & Vliegenthart, A. M. (2016). *Diagnóstico y caracterización de las iniciativas de conservación privada en Chile* [Diagnosis and characterization of private conservation initiatives in Chile]. Ministerio del Medio Ambiente.

Figgis, P., Humann, D., & Looker, M. (2005). Conservation on private land in Australia. *Parks, 15*(2), 19–29.

Fletcher, R. (2010). Neoliberal environmentality: Towards a poststructuralist political ecology of the conservation debate. *Conservation and Society, 8*(3), 171–181. https://www.jstor.org/stable/26393009

Fletcher, R. (2011). Sustaining tourism, sustaining capitalism? The tourism industry's role in global capitalist expansion. *Tourism Geographies, 13*(3), 443–461. https://doi.org/10.1080/14616688.2011.570372

Fletcher, R. (2014). Taking the chocolate laxative: Why neoliberal conservation 'fails forward. In B. Büscher, W. Dressler, & R. Fletcher (Eds.), *NatureTM Inc.: Environmental conservation in the neoliberal age* (pp. 87–107). The University of Arizona Press.

Fletcher, R., Dressler, W., & Büscher, B. (2014). Introduction. In B. Büscher, W. Dressler, & R. Fletcher (Eds.), *Nature^{TM} Inc.: Environmental conservation in the neoliberal age* (pp. 3–22). University of Arizona Press.

Font, X., & Lynes, J. (2018). Corporate social responsibility in tourism and hospitality. *Journal of Sustainable Tourism, 26*(7), 1027–1042. https://doi.org/10.1080/09669582.2018.1488856

Frost, W., & Laing, J. (2018). Public–private partnerships for nature-based tourist attractions: The failure of seal rocks. *Journal of Sustainable Tourism, 26*(6), 942–956. https://doi.org/10.1080/09669582.2017.1423319

Gale, T. (2006). *Finding meaning in sustainability and a livelihood based on tourism: An ethnographic case study of rural citizens in the Aysén Region of Chile (Publication No. 3276498).* [Doctoral dissertation, West Virginia University]. ProQuest Dissertations and Theses Global.

Gale, T., Bosak, K., & Caplins, L. (2013). Moving beyond tourists' concepts of authenticity: Place-based tourism differentiation within rural zones of Chilean Patagonia. *Journal of Tourism and Cultural Change, 11*(4), 264–286. https://doi.org/10.1080/14766825.2013.851201

Gale, T., & Ednie, A. (2019). Can intrinsic, instrumental, and relational value assignments inform more integrative methods of protected area conflict resolution? Exploratory findings from Aysén, Chile. *Journal of Tourism and Cultural Change*, 1–21. https://doi.org/10.1080/14766825.2019.1633336

Gwartney, J., Lawson, R., & Hall, J. (2011). *Economic freedom dataset. Economic Freedom of the World: 2011 Annual Report.* Fraser Institute.

Higgins, V., Dibden, J., Potter, C., Moon, K., & Cocklin, C. (2014). Payments for ecosystem services, neoliberalisation, and the hybrid governance of land management in Australia. *Journal of Rural Studies, 36*, 463–474. https://doi.org/10.1016/j.jrurstud.2014.10.003

Hodge, G. A., & Greve, C. (2017). On public–private partnership performance: A contemporary review. *Public Works Management & Policy, 22*(1), 55–78. https://doi.org/10.1177/1087724x16657830

Huelin, R., Iheanacho, I., Payne, K., & Sandman, K. (2015). What's in a name? Systematic and non-systematic literature reviews, and why the distinction matters. *The Evidence Forum*, 34–37.

Huiliñir-Curío, V., Zunino, H. M., & De Matheus e Silva, L. F. (2019). Exclusión y desigualdad en localidades próximas a la Reserva Ecológica Privada Huilo-Huilo en el sur de Chile [Exclusion and inequality in localities near the Huilo-Huilo Private Ecological Reserve in southern Chile]. *ACME: An International Journal for Critical Geographies, 18*(2), 335–363.

Jones, C. (2012). Ecophilanthropy, neoliberal conservation, and the transformation of Chilean Patagonia's Chacabuco Valley. Oceania, *82*(3), 250–263. https://doi.org/10.1002/j.1834-4461.2012.tb00132.x

Jones, P. J. (2014). *Governing marine protected areas: Resilience through diversity*. Routledge. https://doi.org/10.4324/9780203126295

Jones, P. J. S. (2013). A governance analysis of the Galápagos Marine Reserve. *Marine Policy, 41*, 65–71. https://doi.org/10.1016/j.marpol.2012.12.019

Langewiesche, W. (1999). Eden: A gated community. *Atlantic Monthly, 283*(6), 84–103. https://www.theatlantic.com/magazine/archive/1999/06/eden-a-gated-community/304919/

Levac, D., Colquhoun, H., & O'Brien, K. K. (2010). Scoping studies: Advancing the methodology. *Implementation Science: IS, 5*, 69–69. https://doi.org/10.1186/1748-5908-5-69

Louder, E., & Bosak, K. (2019). What the gringos brought: Local perspectives on a private protected area in Chilean Patagonia. *Conservation and Society, 17*(2), 161–172. https://doi.org/10.4103/cs.cs_17_169

MacDonald, E. (2011). A landscape ... with figures: Tourism and environment on Prince Edward Island. *Acadiensis, 40*(1):70–85. https://journals.lib.unb.ca/index.php/Acadiensis/article/view/18562

MacDonald, K. I., & Corson, C. (2014). Orchestrating nature. InB. Büscher, W. Dressler, & R. Fletcher (Eds.), *Nature™ Inc.: Environmental conservation in the neoliberal age* (pp. 44–65). University of Arizona Press.

Matose, F. (2014). Nature, villagers, and the state: Resistance politics from protected areas in Zimbabwe. In B. Büscher, W. Dressler, & R. Fletcher (Eds.), *Nature™ Inc.: Environmental conservation in the neoliberal age* (pp. 66–86). University of Arizona Press.

Mbaiwa, J. E. (2008). The realities of ecotourism development in Botswana. In A. Spenceley (Ed.), *Responsible tourism: Critical issues for conservation and development* (pp. 205–224): Routledge.

Meißner, N., & Grote, U. (2017). Motives, opportunities, and risks for private sector investment in protected areas with international importance: Evidence from German companies. *Environment, Development and Sustainability, 19*(1), 199–219. https://doi.org/10.1007/s10668-015-9722-2

Mendoza, M., Fletcher, R., Holmes, G., Ogden, L. A., & Schaeffer, C. (2017). The Patagonian imaginary: Natural resources and global capitalism at the far end of the world. *Journal of Latin American Geography, 16*(2), 93–116. https://doi.org/10.1353/lag.2017.0023

Mitchell, B. A., Stolton, S., Bezaury-Creel, J., Bingham, H. C., Cumming, T. L., Dudley, N., Fitzsimons, J. A., Malleret-King, D., Redford, K. H., & Solano, P. (2018). *Guidelines for privately protected areas* (Best Practice Protected Area Guidelines Series No. 29). IUCN.

Newton, R., Rouleau, A., Nylander, A. G., Loze, J. Y., Resemann, H. K., Steeves, S., & Crespo-Facorro, B. (2018). Diverse definitions of the early course of schizophrenia-a targeted literature review. *NPJ Schizophrenia, 4*(1), 21. https://doi.org/10.1038/s41537-018-0063-7

Núñez, A., Aliste Almuna, E., & Bello, Á. (2014). El discurso del desarrollo en Patagonia-Aysén: La conservación y la protección de la naturaleza como dispositivos de una renovada colonización. Chile, siglos XX-XXI [The discourse of development in Patagonia-Aysen: Nature conservation and protection]. *Scripta Nova: Revista Electrónica de Geografía y Ciencias Sociales, 18*, 46. https://revistes.ub.edu/index.php/ScriptaNova/article/download/15035/18388

Otsuki, K., Schoneveld, G., & Zoomers, A. (2017). From land grabs to inclusive development? Geoforum, *83*, 115–118. https://doi.org/10.1016/j.geoforum.2017.05.001

Parcher, B., Gabriel, S., Ruiz, K., Ariely, R., & Mosnaim, G. (2018). Digital technology interventions and their outcomes in moderate-to-severe asthmatic patients: A targeted literature review. *Journal of Allergy and Clinical Immunology, 141*(2), AB115. https://doi.org/10.1016/j.jaci.2017.12.366

Pauchard, A., & Villarroel, P. (2002). Protected areas in Chile: History, current status, and challenges. *Natural Areas Journal, 22*(4), 318–330.

Petit, I. J., Campoy, A. N., Hevia, M. J., Gaymer, C. F., & Squeo, F. A. (2018). Protected areas in Chile: Are we managing them? *Revista Chilena de Historia Natural, 91*(1), 8. https://doi.org/10.1186/s40693-018-0071-z

Pringle, R. M. (2017). Upgrading protected areas to conserve wild biodiversity. Nature, *546*(7656), 91–99. https://doi.org/10.1038/nature22902

Ramutsindela, M. (2015). Extractive philanthropy: Securing labour and land claim settlements in private nature reserves. *Third World Quarterly, 36*(12), 2259–2272. https://doi.org/10.1080/01436597.2015.1068112

Randle, E. J., & Hoye, R. (2016). Stakeholder perception of regulating commercial tourism in Victorian National Parks, Australia. *Tourism Management, 54*, 138–149. https://doi.org/10.1016/j.tourman.2015.11.002

Rivera, C., & Vallejos-Romero, A. (2015). La privatización de la conservación en Chile: repensando la gobernanza ambiental [The privatization of conservation in Chile: Rethinking environmental governance. *Bosque (Valdivia), 36*(1), 15–25. https://doi.org/10.4067/S0717-92002015000100003

Root-Bernstein, M., Montecinos Carvajal, Y., Ladle, R., Jepson, P., & Jaksic, F. (2013). Conservation easements and mining: The case of Chile. *Earth's Future, 1*(1), 33–38.

Schutz, J. (2015). *A GIS assessment of ecoregion representation in Chile's existing and proposed integrated network of protected areas* [Master's thesis]. The University of Montana. https://scholarworks.umt.edu/etd/4532/

Scoones, I., Smalley, R., Hall, R., & Tsikata, D. (2019). Narratives of scarcity: Framing the global land rush. Geoforum, *101*, 231–241. https://doi.org/10.1016/j.geoforum.2018.06.006

Sehnbruch, K. (2019, October 30). How Pinochet's economic model led to the current crisis engulfing Chile. *The Guardian*. https://www.theguardian.com/world/2019/oct/30/pinochet-economic-model-current-crisis-chile

Selinske, M. J., Cooke, B., Torabi, N., Hardy, M. J., Knight, A. T., & Bekessy, S. A. (2017). Locating financial incentives among diverse motivations for long-term private land conservation. *Ecology and Society, 22*(2). https://doi.org/10.5751/ES-09148-220207

Serenari, C. (2014). *Conservation and development in Chilean Patagonia: Assessing value orientations, cultural truths, and rural change in the context of public and private protected areas* [Doctoral dissertation]. North Carolina State University. http://repository.lib.ncsu.edu/ir/bitstream/1840.16/9496/1/etd.pdf

Serenari, C., Peterson, M. N., Leung, Y. F., Stowhas, P., Wallace, T., & Sills, E. O. (2015). Private development-based forest conservation in Patagonia: Comparing mental models and revealing cultural truths. *Ecology and Society, 20*(3). https://doi.org/10.5751/ES-07696-200304

Serenari, C., Peterson, M. N., Wallace, T., & Stowhas, P. (2017a). Indigenous perspectives on private protected areas in Chile. *Natural Areas Journal, 37*(1), 98–107. https://doi.org/10.3375/043.037.0112

Serenari, C., Peterson, M. N., Wallace, T., & Stowhas, P. (2017b). Private protected areas, ecotourism development and impacts on local people's well-being: A review from case studies in Southern Chile. *Journal of Sustainable Tourism, 25*(12), 1792–1810. https://doi.org/10.1080/09669582.2016.1178755

Silva, E. (1996). Democracy, market economics, and environmental policy in Chile. *Journal of Interamerican Studies and World Affairs, 38*(4), 1–33.

Slattery, D. (2002). Resistance to development at Wilsons Promontory National Park (Victoria, Australia). *Society & Natural Resources, 15*(7), 563–580. https://doi.org/10.1080/08941920290069191

Slobodian, Q. (2019, November 11). Democracy doesn't matter to the defenders of economic freedom. *The Guardian*. https://www.theguardian.com/commentisfree/2019/nov/11/democracy-defenders-economic-freedom-neoliberalism

Slocum, S. L. (2017). Operationalising both sustainability and neo-liberalism in protected areas: Implications from the USA's National Park Service's evolving experiences and challenges. *Journal of Sustainable Tourism, 25*(12), 1848–1817. https://doi.org/10.1080/09669582.2016.1260574

Snyder, H. (2019). Literature review as a research methodology: An overview and guidelines. *Journal of Business Research, 104*, 333–339.

Spenceley, A. (2003). *Tourism, local livelihoods and the private sector in South Africa: Case studies on the growing role of the private sector in natural resources management* (Sustainable Livelihoods in Southern Africa Research Paper 8). Institute of Development Studies.

Spenceley, A., & Meyer, D. (2012). Tourism and poverty reduction: Theory and practice in less economically developed countries. *Journal of Sustainable Tourism, 20*(3), 297–317. https://doi.org/10.1080/09669582.2012.668909

Spenceley, A., Snyman, S., & Eagles, P. F. J. (2019). A decision framework on the choice of management models for park and protected area tourism services. *Journal of Outdoor Recreation and Tourism, 26*, 72–80. https://doi.org/10.1016/j.jort.2019.03.004

Squeo, F. A., Estévez, R. A., Stoll, A., Gaymer, C. F., Letelier, L., & Sierralta, L. (2012). Towards the creation of an integrated system of protected areas in Chile: Achievements and challenges. *Plant Ecology & Diversity, 5*(2), 233–243. https://doi.org/10.1080/17550874.2012.679012

Stolton, S., Redford, K. H., & Dudley, N. (2014). *The futures of privately protected areas: Developing capacity for a protected planet*. Protected Area Technical Report Series. IUCN.

Stronza, A. L., Hunt, C. A., & Fitzgerald, L. A. (2019). Ecotourism for conservation? *Annual Review of Environment and Resources*, *44*(1), 229–253. https://doi.org/10.1146/annurev-environ-101718-033046

Sullivan, S. (2014). Nature on the Move III: (Re)countenancing an animate nature. In B. Büscher, W. Dressler, & R. Fletcher (Eds.), *Nature™ Inc.: Environmental conservation in the neoliberal age* (pp. 222–245). University of Arizona Press.

Taylor, M. (2003). The reformulation of social policy in Chile, 1973-2001. *Global Social Policy*, *3*(1), 21–44. https://doi.org/10.1177/1468018103003001009

Taylor, M. (2004). Labor reform and the contradictions of "growth with equity" in postdictatorship Chile. *Latin American Perspectives*, *31*(4), 76–93. https://doi.org/10.1177/0094582X04266258

Tecklin, D. R., Bauer, C., & Prieto, M. (2011). Making environmental law for the market: The emergence, character, and implications of Chile's environmental regime. *Environmental Politics*, *20*(6), 879–898. https://doi.org/10.1080/09644016.2011.617172

Tecklin, D. R., & Sepúlveda, C. (2014). The diverse properties of private land conservation in Chile: Growth and barriers to private protected areas in a market-friendly context. *Conservation and Society*, *12*(2), 203–217. https://www.jstor.org/stable/26393155

Varghese, G. (2008). Public–private partnerships in South African national parks: The rationale, benefits and lessons learned. In A. Spenceley (Ed.), *Responsible tourism: Critical issues for conservation and development* (pp. 97–112). Routledge.

Villarroel Venturini, P. (2019). *Áreas protegidas privadas en Chile, ¿una 'tragedia hardiniana'? [Private protected areas in Chile, a "Hardinian tragedy"?]*. http://www.cienciaenchile.cl/areas-protegidas-privadas-en-chile-una-tragedia-hardiniana/

Waldron, A., Mooers, A. O., Miller, D. C., Nibbelink, N., Redding, D., Kuhn, T. S., Roberts, J. T., & Gittleman, J. L. (2013). Targeting global conservation funding to limit immediate biodiversity declines. *Proceedings of the National Academy of Sciences of the United States of America*, *110*(29), 12144–12148. https://doi.org/10.1073/pnas.1221370110

Walton, A. L., & Rogers, B. (2017). Workplace hazards faced by nursing assistants in the United States: A focused literature review. *International Journal of Environmental Research and Public Health*, *14*(5), 544. https://doi.org/10.3390/ijerph14050544

West, P., & Carrier, J. (2004). Ecotourism and authenticity: Getting away from it all? *Current Anthropology*, *45*(4), 483–498. https://doi.org/10.1086/422082

Whitelaw, P. A., King, B. E. M., & Tolkach, D. (2014). Protected areas, conservation and tourism - Financing the sustainable dream. *Journal of Sustainable Tourism*, *22*(4), 584–603. https://doi.org/10.1080/09669582.2013.873445

The use of tourism as a social intervention in indigenous communities to support the conservation of natural protected areas in Mexico

Gerda Warnholtz, Neil Ormerod and Chris Cooper

ABSTRACT
Tourism has been considered a tool for sustainable development (SD) of rural/indigenous communities living within natural protected areas (NPAs) since the 1980s. This article examines the use of tourism as a SD intervention in the management of NPAs. Based on the theory of social change and on the realist methodological approach to social research, the study focuses on the case of Mexico. It seeks to establish causal chains to identify mechanisms that trigger (un)expected outcomes within a specific sociocultural context. The research distinguishes between development *of* tourism, and SD *through* tourism. Preliminary findings from on-going research in the indigenous community of Sta. Cruz Tepetotutla in Oaxaca, Mexico point to the necessity for an alternative approach to the design, application and assessment of tourism interventions. It is argued that specific contextual features trigger mechanisms which lead to (un)expected outcomes. Considering traditional social organization is fundamental to achieve improvement in collective well-being, biodiversity preservation and cultural heritage. The article seeks to help decision-makers improve their outcomes from interventions, and to generate further discussion on the use of tourism for the management of NPAs, and on the role tourism can play in the improvement of the living conditions of the communities.

Introduction

Tourism, and specifically community-based tourism (CBT), has been considered a tool for social improvement and environmental protection since the 1980s (Murphy, 1983). However, research has shown that these projects have had poor outcomes, both for the benefit of the host communities, and for conservation (Spenceley, 2008). So far, mainstream tourism studies in this field draw particularly upon the development of tourism, based on pro-poor tourism (Bennett et al., 1999) and CBT trends – respectively rooted in pro-poor growth theory (Perry et al., 2006) and community-based and community-driven growth theories (Mansuri & Rao, 2004). As such, these theories focus on the development *of* the tourism business and an eventual/relative benefit for the local poor. Consequently, social, cultural, environmental and political wellbeing is neglected. This is especially critical when tourism is part of a social intervention strategy both, for social change and for conservation. Thus, these interventions seek environmental conservation and social improvement *through* tourism projects.

This approach calls for a revision of the key concepts of sustainable development, tourism, and the tourism phenomenon in relation to natural protected areas (NPAs) and their management within a closed and finite ecosystem, which is at the same time complex, diverse and dynamic (Daly & Farley, 2010, pp. 61–76). Drawing on this framework, the discussion about the possible contradiction between the ideas of sustainable development *of* tourism (as an economic activity), and sustainable development *through* tourism generates further discussion and future research agendas that will facilitate an understanding of the real role tourism can play – or not – both in the improvement of the living conditions of the communities, and the conservation of biodiversity.

Distinguishing between development *of* tourism, and sustainable development *through* tourism, this study is based on the theory of change as the foundation of the interventions (Aromatario et al., 2019), and on the realist approach to qualitative research, which seeks to establish causal chains to evidence the mechanisms that can trigger given outcomes within specific sociocultural contexts (Maxwell, 2012). The study analyses the case of Mexico and how it has attempted to use tourism as an intervention to offer indigenous communities sustainable development opportunities as part of the management of NPAs (CONANP, 2011, 2018b). The final part of this paper presents preliminary results of a case study using a mixed method, qualitative approach to a particular tourism-based sustainable development intervention (TBSDI) in Santa Cruz Tepetotutla, Oaxaca, Mexico, a Chinantec indigenous community which has participated in such initiatives for over 12 years.

In conclusion, we propose an alternative approach to tourism-based interventions focused on local sustainable development *through* tourism. This approach considers the local sociocultural and socioeconomic context, as well as the host community's needs and expectations into their theory of social change when planning, designing, implementing and evaluating interventions. For this purpose, the traditional social organization and philosophy (*comunalidad*), as well as local participation, are fundamental to achieve the objectives of improving the quality of life while strengthening local institutions and conserving the physical infrastructure (Barkin, 2009; Martínez Luna, 2011). This approach will improve the outcomes of the interventions, including the conservation and rehabilitation of their ecosystems, the deep symbiotic relationship that indigenous communities have with their traditional territories and their cultural heritage.

Conceptual framework

Natural protected areas and indigenous communities

In the early 1970s, concern for a constantly deteriorating natural environment gave way to a report commissioned by the Club of Rome. The report was developed by the MIT and in 1972, the report was published as *The Limits to Growth* (Meadows et al., 1972). The UN-UNESCO *Man and the Biosphere* (MAB) project emerged as a multidisciplinary programme combining natural and social sciences to improve human livelihoods and protect natural ecosystems through socially and culturally appropriate strategies (UNESCO/IBEROMaB, 2010). Consequently, NPAs and biosphere reserves, and their corresponding management programmes proliferated at a national and international level. The purpose was, and still is, to preserve, protect and restore the world's ecosystems through specific strategies, often implying the control of land use and natural resources.

In many cases, conservation strategies are centrally designed and determined, without considering the effects on the communities living within or near the NPAs (Méndez-López et al., 2019). The dominant conservation model often advocates the immobilization and isolation of the territory, which causes divergence between conservation purposes and local communities' interests. The traditional relationship between humans and nature, especially in indigenous communities has been significantly altered (D. B. Bray et al., 2007; Chape et al., 2008). In extreme cases, conservation policies have even forced the displacement of the peoples living within or adjacent to the NPAs (Desmet, 2014; Melubo & Carr, 2019). This has affected the communities' socio-cultural, economic and political structures and interactions, causing conflict, within and between the communities, and with the NPAs' management teams. It has also affected the reproduction of their culture, causing the loss of their intangible cultural heritage (Monterroso, 2008).

Estimated at around 370 million, indigenous populations represent a small minority of the world's population. However, and disproportionately, it represents approximately one third of the world's poor (Hall & Patrinos, 2010). Their traditional territories cover up to 24% of the Earth's surface, and about 80% of the planet's healthy ecosystems overlap with traditional indigenous territories (Carino et al., 2009; Watanabe, 2008). A large proportion of the world's healthy ecosystems is located in Asia, Africa and Latin America, which is home to the majority of indigenous communities (UNEP-WCMC, 2018, p. 8). In this sense, historically, indigenous communities have depended on local environments and their resources to survive and have accumulated knowledge about the management of their territories as complex ecological systems, and at the same time, their entire socio-cultural system has been shaped and determined by their surrounding environment. Consequently, their symbiotic relationship with the natural environment is the basis of their economic, political and social relations. Their strong and deeply rooted social, cultural, economic and internal political organizations have shaped their (our) identity and their particular cosmogonies, which are the essence of their (our) intangible cultural heritage (Arizpe, 2006; Stavenhagen, 2010). Hence, these social groups are the guardians of the majority of the world's traditional knowledge and intangible cultural heritage, adding relevance and complexity to the conservation of their territories.

Thus, the combination of poverty, cultural heritage and natural resources at risk necessarily calls for careful planning, design, application and assessment of policies. Therefore, environmental, sustainable development and poverty reduction agencies, governments and NGOs have established a strong agenda of social interventions based on the UN's sustainable development paradigm (Brundtland, 1987), which was reviewed as the foundation for the UN's Millennium Development Goals (UN., 2000) and more recently for the UN's Sustainable Development Goals (United Nations, 2015).

Seeking better solutions to this challenge, scholars from different disciplines suggest considering traditional natural management knowledge, transmitted and refined from generation to generation, as complementary to that of environmentalists, and advised its incorporation into their conservation strategies (Esfehani & Albrecht, 2018). The inclusion of the communities' rights into a number of strategies should improve conservation and avoid conflict. As a result, in recent years, local populations have been involved with the conservation of their own territories and popular empirical knowledge has been gradually incorporated, in combination with academic knowledge, into the design of conservation strategies (Toledo, 2005).

Tourism, indigenous communities and NPAs

Ecotourism, and its variations have been incorporated into NPA management strategies since the 1970s (de Kadt, 1979). It is generally assumed that tourism can be used as a tool for local development, while preserving the natural and cultural heritage of the host community. Diversification of economic activities, the generation of additional income for the community and the appreciation of local cultural and natural heritage are considered as mechanisms to improve local livelihoods (Stone & Nyaupane, 2018). The study of this issue has been mainly based on two main concepts: Pro-poor tourism and community-based tourism.

By the end of the Twentieth Century, the idea of combining the expansion of tourism with the economic growth of host communities within the pro-poor growth (PPG) model, left the benefit of the poor as a relative and proportional issue (Kakwani & Pernia, 2000). Derived from this general economic growth model, the notion of pro-poor tourism (PPT) emerged from a study commissioned in 1999 by the Department of International Development of the United Kingdom (Bennett et al., 1999). The results of the study led to the inclusion of tourism as an effective tool for poverty alleviation and sustainable development within the UN-Millennium Goals Agenda (UN., 2000). This approach is a guideline for the growth and expansion of tourism (Chok et al., 2007; Saayman & Giampiccoli, 2016). This necessarily implies that the PPT approach may be applied to the tourism sector and its development in general, but it is difficult to consistently relate it to the benefit of the host communities, or to interventions seeking social change.

An additional concept which is deeply related with tourism expansion and environmental protection is that of community-based tourism (CBT), which emerged in the early 1980s, as a response to mass tourism (Murphy, 1983) and is related to the community-based development trend (de Kadt, 1979; Zapata et al., 2011). Even though there is no agreement about its definition, CBT is considered as a for-profit economic activity, and associated with community-based enterprises. In CBT projects, community members' participation, self-determination, empowerment and inclusion are considered fundamental for their success (Dangi & Jamal, 2016; Strasdas, 2005). However, these concepts are rooted in the relative justice definition (Sachs, 2010) typically involving individuals and not the social group as a whole, privileging economic growth, and its derived access to services and given assets, rather than increasing rights and freedoms, or generating communal well-being and/or benefits.

The triple-bottom line approach (TBL) has also been used to plan, implement and analyse ecotourism and its use for conservation. It intends to go beyond the issues of profit, return on investment and stakeholder benefit by considering local social and environmental dimensions to business management (Elkington, 1997; Wood, 2004). It seeks the balance of social, environmental and economic dimensions for the sustainable development of the enterprise through social responsibility measures (Slaper, 2011). When applied to tourism, it keeps the focus on the business and the sustainable development *of* tourism.

Hence, CBT and PPT and the TBL approaches have been regarded as effective elements to develop more responsible tourism. They share the concepts of *poverty alleviation* and *sustainable development*, as well as the idea of tourism as a for-profit economic activity. However, they have been used as fundamental conceptual premises, not only to plan, design and apply tourism development policies at a macro-economic level, but also for externally-funded TBSDIs within NPAs at a micro-economic level (Butler & Hinch, 2013; Honey, 2008), centring the attention on the for-profit roots of tourism and the *trickling down* of its benefits, concentrating on the development *of* tourism, rather than on the community's sustainable development *through* tourism.

In this sense, despite the ever-increasing resources invested in them, while conservation initiatives have been consistently studied, the outcomes of these interventions remain insufficiently studied, and evidence points at a high failure rate (Dodds et al., 2018; Wardle et al., 2018). The majority of academic studies have been carried out by universities and research centres specialised in tourism development and management in developed countries (Harbor & Hunt, 2021; Job et al., 2017; Spenceley & Meyer, 2012) and most study cases in developing countries, where a great number of TBSDIs have been applied. Despite the difference between the objectives of interventions based on tourism – sustainable local development, poverty alleviation and the protection of their natural and cultural heritage – and those of tourism as a commercial activity (Dodds et al., 2018), the literature demonstrates that research has mainly focused on the business success/failure perspective (Mendoza-Ramos & Prideaux, 2018; Palomino-Villavicencio et al., 2016; Pereiro, 2016), rather than on interventions for social change. In summary, most academic studies assume that the main objective is the development of tourism and consider the well-being of the communities a relative and eventual consequence.

Tourism as a form of social intervention

Social interventions are organised efforts to generate social change (Abram, 2010). The purpose is generally to improve human welfare through the use of specific strategies and intervention techniques. These initiatives are externally financed and/or supported programmes to generate pre-established social modifications in favour of the social group they target, and are usually related to development and public policy (Pawson, 2013). These projects depend on public or private donor funds for their planning, design and implementation. As resources are scarce, the monitoring and evaluation of these interventions has become crucial to measure efficiency and efficacy in the achievement of their goals, as well as to generate knowledge to improve their performance (Wholey et al., 2010).

Social interventions seek to directly or indirectly modify the socio-political and cultural elements within the social group, which necessarily affects their interactions and balances (Strengers & Cecily, 2015). The application of an intervention within a given sociocultural context may activate (in)visible mechanisms which lead to (un)expected outcomes. This is particularly sensitive in geographically isolated and marginalised indigenous communities. Consequently, the assessment of complex interventions, such as TBSDIs in indigenous communities, poses methodological and conceptual challenges, which are further increased by the combination of indigenous socio-cultural contexts and environmental protection objectives.

Designing, applying and evaluating tourism-based sustainable development interventions (TBSDIs) in indigenous communities living within and/or near NPAs

Tourism and local sustainable development are essentially different issues and pursue objectives that may be incompatible; while one seeks the accumulation of capital and economic benefit through the sale of services, the other focuses on the integral development – social, cultural, economic, political and environmental – of the communities that host it. Furthermore, tourism is inherently complex and multidisciplinary with activities carried out in plural contexts, which can lead to a variety of positive and/or negative effects in various dimensions – socio-cultural, economic, political and environmental (Scheyvens, 2014). Therefore, tourism development does not necessarily coincide with the sustainable development of the host community.

Designing, implementing and assessing an intervention for social change in complex and diverse contexts is particularly difficult. It is argued that this issue, which has been practically overseen by academics and consultants, is fundamental for the achievement of a TBSDI and its objectives. It is assumed then, that the theory of social change in a TBSDI applied in an indigenous community in, or adjacent to an NPA, is to improve local livelihoods, while at the same time, preserving their culture and ecosystem, *through* the tourism activity. Therefore, the main objective of the study was to contrast the objectives of the intervention – the theory of change underlying the project (Aromatario et al., 2019; Weiss, 1995) – with the (un)expected outcomes and explain how, why and under which circumstances these outcomes were obtained (Pawson, 2013; Pawson & Tilley, 2010). Tourism-based sustainable development interventions are applied in highly diverse contexts and are influenced by the different stakeholders – development agencies, intermediaries, environmentalists, the tourism sector, and beneficiary communities – as well as by the power struggles between these stakeholders, at every stage of its design and implementation (Julian, 2016; Nepal & Saarinen, 2016).

Each of these stakeholders have different interpretations about the relevance and purpose of the intervention and about the processes implemented to achieve the desired effects. Stakeholders may have different understandings of the intervention, according to (i) their personal and institutional background and interests, (ii) their concept of sustainable development and environmental protection, (iii) their understanding of tourism as an income-generating economic activity, and (iv) what the welfare of the social group involved implies. These issues necessarily generate a continuous formulation and reformulation of the nature of the intervention and its outcomes (Gillespie & Cornish, 2010). Moreover, in the case of Mexico, documentary research evidenced, on one hand a lack of explicit objectives, and on the other, little information about how the intended change ought to be achieved (CDI, 2013; CONANP, 2009, 2011, 2018b; INPI, 2019; SEMARNAT/CONANP, 2007).

For decades, at the international level, decision makers of social interventions – and of TBSDIs – have favoured the quantification and standardization of the intervention's processes (what and how much), neglecting not only the perspectives of the target social groups, but also the causal chains leading to given outcomes (why and how); favouring an approach to economic growth and relative justice, which gives more relevance to individual-related notions of self-determination, equity and empowerment (Saayman & Giampiccoli, 2016; W. Sachs, 2010). By contrast, indigenous communities privilege absolute justice, which is directly related to autonomy, solidarity, self-sufficiency, productive diversification, participative decision-making, common labour and sustainable management (Esteva, 2010; Helliwell et al., 2013). Consequently, TBSDIs should base their theory of social change on the improvement of collective wellbeing, rather than that of the individual.

The study of the relationship between poverty and tourism has usually been limited to economic, and to a lesser extent, gender and communal issues. It has used a methodological approach based on processes and results, focusing on the analysis of the way in which interventions are applied, and in the quantitative measurement of results in financial and business management terms, without explaining why and how an intervention is successful or not, in terms of the expected social, environmental, cultural and/or economic change (Frenzel, 2013; Pawson & Tilley, 2010; Stame, 2004). These assessment approaches aim on one hand, to ensure that the protocols and rules of operation are followed, and on the other, to measure the economic impacts of tourism in the communities, while neglecting the analysis of the (un)expected change caused by the intervention (Banerjee & Cicowiez, 2015).

Another frequently overlooked issue, which increases the complexity of TBSDIs, is that interventions are embedded in dynamic contexts (Pawson & Tilley, 2004); TBSDIs are influenced by a large number of internal and external variables – such as other social interventions operating simultaneously, sudden changes at local, regional or national level or natural disasters – which may affect the social and natural context and in consequence, their outcomes. Therefore, attributing specific results to a single intervention becomes extremely difficult (Byrne, 2013; Magalón de Salazar, 2011; Pawson, 2013). Hence, regardless of the way in which the initial intervention model was designed, a TBSDI is not a static entity, but a dynamic system that adapts and transforms not only from one socio-cultural and socio-environmental context, but also as it passes from one to another actor of the design-implementation-evaluation process (Mark & Shotland, 1985; Pawson, 2013). This often leads to gaps between the initial conceptualization of the intervention and the reality of a TBSDI, its expected impacts, the actual outcomes, and the mechanisms that cause them, which are often concealed in the so-called black box (Pawson, 2013; Zapata et al., 2011).

The impacts of TBSDIs can be measured at different levels (individual, household, community or region) and in multiple dimensions – economic, environmental, social, cultural, political. Thus, designing, applying and assessing an intervention requires a methodological approach that helps to unveil the mechanisms triggered within specific contexts, leading to certain (un)expected outcomes. This allows the interactions between the TBSDI, the dynamic context and the (un)expected change in the beneficiary social group to be established (Chok et al., 2007; Gascón, 2015).

The case of Mexico

Methodology and methods

This case study is based on preliminary findings of a larger, on-going study analysing the use of tourism as interventions for social change in NPAs. The research sought to understand a tourism-based intervention as part of a set of environmental, conservation and local sustainable development strategies. It used an innovative qualitative research approach to unveil the mechanisms activated through intervention processes within the sociocultural context of the benefited social group, and to explain the processes of change generated in the benefitted community (Trau, 2012; Zapata et al., 2011).

Data were collected through different methods. Firstly, secondary data were used for the literature review, which was carried out between October 2019 and July 2020. The documents came from official tourism-based programmes developed by the Mexican Government, government policies, peer-reviewed journal articles, books, annual reports and online sources. Secondly, primary data were collected though different research tools: semi-structured interviews, focus groups, participant observation and a field diary to understand how and why the tourism-based intervention yielded (un)expected outcomes. The fieldwork was carried out between December 2019 and May 2020. This involved a variety of stakeholders (decision-makers, practitioners, consultants, community leaders and members directly involved in the projects) and their particular social contexts.

Participants' ages ranged between 18 and 68 years, of these, 12 were women and 49, male. Respondents were categorised as:

- External: Decision-makers (designers and planners of TBSDIs in Mexico City [4 + 2], in the State of Morelos [2] and Oaxaca [4]);
- Liaisons: intermediaries [2] and consultants [3]; and
- Internal: Community leaders (State of Morelos [1], Oaxaca [2], Santa Cruz Tepetotutla [4] and community members directly involved in the project [4]. Focus group participants [2 × 12 = 24].

Data analysis consisted of transcribing the field diary and the interviews and audio recordings of focus groups. After several readings, data were translated from Spanish into English and systematically organised, coded, and categorised with the aid of NVivo, a data organization software. Data were then analysed based on the hypotheses in the form of causal chains, to further refine the foundational theories of tourism interventions. This analysis was based on critical discourse analysis as the analysis of discourse (written and/or spoken) considering the way social power, dominance, and inequality are practiced and reproduced (van Dijk, 2008).

Mexican indigenous communities, land tenure system, and natural and cultural heritage

Mexico is one of the five most biologically diverse countries worldwide. It is also host to at least 68 different linguistic groups and around 364 variations of them (Berger et al., 2020). Its indigenous population represents around 10% of the country's total, of which around 10 million live in extreme poverty; many of them marginalised and geographically isolated (Berger et al., 2020). However, about 70% of the national territory is socially owned by rural/indigenous communities. This means that private property in their territories is practically non-existent and that these social groups own circa 80% of the healthy ecosystems and are the custodians of most of the intangible cultural heritage of the nation (Arizpe & Nalda, 2003; Battlori Guerrero, 2005; Reyes et al., 2012).

Despite their diversity, the integration of their internal social, cultural, economic and political structures and institutions can be defined as *comunalidad*. This social system is made up of two dimensions: the first is the notion of *community* as a complex system composed by evident traits: a social group that, besides sharing a common history, also shares a present and a future. The community is understood through its symbolic relationship with the natural environment, and through the interactions within the social group. Collective interactions are considered more relevant than individual relationships; the rules of coexistence are collectively established and interpreted based on nature itself and on knowledge passed from one generation to another (Díaz Gómez, 1994; Martínez Luna, 2013).

The second dimension is *communality* as the foundation of the *community*. This is determined by abstract notions of the traditional territory. Decisions are made through democratic consensus reached within the communal assembly. Work has a social value, and common work – *tequio* – is voluntary, unpaid work in public service. Rites and ceremonies are the expression of the community's cultural heritage and cohesion. Consequently, to understand the apparent elements of the *community*, it is necessary to understand the intangible notion of the *communal* as collective and collaborative. The interactions between these two dimensions, and among the different elements and their natural territory, represent the community's complex operation system. Moreover, the communal way of life is not always explicit, but is tacitly present, and usually becomes evident when an individualistic attitude emerges (Martínez Luna, 2011; Rendón Monzón, 2003). The collective essence defining indigenous communities has been amply documented by different anthropological studies (see Barthas, 1997; Bonfil Batalla, 1995; Rendón Monzón, 2003, among others).

Thus, indigenous life is only possible in a specific natural territory, which is symbolically owned, understood and appropriated; a territory inhabited by people, nature and often by supernatural forces which interact within it. Social relationships are ritually determined and explained through myths and other narratives. The territory is commonly owned and is home to the community, which is composed of families, intertwined by kinship, power and ritual links. Community members build their community life on reciprocity and participation as a fundamental rule. Reciprocity and participation are based upon common, voluntary labour, on the exercise of power, and on rituals and festivities, which serve to generate cohesion, achieve collective goals and solve common problems (Barabas, 2004). Community members express their will of being part of their community through voluntary collaboration, reciprocity and participation. Common labour is considered not only a moral duty, but also a trait of identity, engagement and rootedness. For this reason, those who refuse common labour or mutual help – *tequio, faena*, or *fagina* – reject the charges or duties they are elected for, or cease to participate in the festivities, tacitly implying they do not want to be part of the community, and are willing to lose their rights (Millán & Valle, 2003).

This kind of social organization is not exclusive of indigenous social groups. Many non-indigenous rural communities have the same traditions of reciprocity, participation and common labour, and make decisions through their assembly, due to the historically common ownership of the land and their symbiotic relationship with it. What distinguishes these communities from indigenous communities is their ethno-cultural elements – the local organization based on the common language, their particular world views (usually of Pre-Hispanic origin), their corresponding rituals and festivities, the use of traditional clothing, the local cuisine, among many other aspects. It could be stated that indigenous/rural communities base their social life upon the fundamental elements of common life, and that the difference is defined by the degree of conservation or loss of the ethnic elements (Bonfil Batalla, 2012; Warman, 2003; Zizumbo Villarreal, 2013).

These elements, combined with geographic isolation, generate an economic organization which is weakly linked to the mainstream market, in which social value and barter are more relevant than money and price (fundamental for capitalist economic relations). Therefore, money is not their main exchange factor (Carlos & Pardo, 2019; Cassano et al., 2003). Subsistence economy, common land tenure and the practical absence of paid labour gives their work a social value. Furthermore, leading subsistence economies and their geographical isolation hinders the notions of accumulation or financial planning. Consequently, their relationship with the mainstream market is marginal, as is their understanding of service-based economic activity. For these social groups, money is only useful to relate with the *outside* and acquire goods or services that are not locally available. Hence, the insertion of tourism in these social structures may pose a serious challenge for the activity to succeed, as well as a huge risk of disruption for the local socio-cultural and socio-economic traditional structures (Barkin, 2009; Carlos & Pardo, 2019).

Mexican Strategic Framework for Sustainable Tourism in Protected Natural Areas

Natural resources protection in Mexico dates to the 1920s (CONANP, 2018a). Since then, the conservation of biodiversity has consistently advanced. By 2018, Mexico had a total of 182 federal NPAs covering 90,839,521.55 ha (CONANP, 2020). Regulated by the General Law of Ecological Balance and Environmental Protection, they form the National System of Natural Protected Areas, and since 2000, they are managed by the National Commission of National Protected

Areas (CONANP), dependent on the Ministry of Environment and Natural Resources (SEMARNAT) (CONANP, 2018b).

The federal NPAs are distributed in six categories with different management strategies, according to their biophysical and social characteristics, therefore, each has different regulations for tourism activities: Areas of Flora and Fauna Protection (40), Natural Resources Protection Areas (8), Natural Monuments (5), National Parks (67), Biosphere Reserves (44) and Sanctuaries (18). An additional kind of NPA, which is legally acknowledged, is the areas voluntarily destined for conservation (AVDC). Until 2017, CONANP had certified a total of 388 AVDCs covering a total of 417,562.27 hectares (CONANP, 2018b). Finally, there is a high number of parks and reserves at a regional level, which are heterogeneous in legal-administrative terms and vary considerably depending on each local legislation (CONANP, 2018b).

Despite the existence of a legal framework, environmental protection policy in Mexico faces a series of challenges: the increasing pressure on natural resources due to the colonization of ecologically fragile areas, the advance of agrarian borders and international and national extractive interests. In addition, the complex land tenure system, composed by ejidos and indigenous communal land, means that the land boundaries are often uncertain and therefore in dispute (Chapela y Mendoza & Barkin, 1995; Ortega-Rubio, 2018). This situation not only facilitates the entry of foreign actors interested in the irresponsible use of natural resources (Moreno-Barajas et al., 2019), but also gives way to corruption (Morales Garcia & Morales Garcia, 2017). These issues constitute a major management problem for the NPAs, as well as for the practice of sustainable and economically viable tourism (Ortega-Rubio, 2018). Hence, the Mexican government recognized nature conservation as a State policy and as part of the National Development Plan (Poder Ejecutivo, 2013). The main approach of the 2019-2024 environmental policy regarding protected areas is to achieve sustainable development and conservation, and CBT has been consistently considered as part of this approach (CONANP, 2018b).

The case of Santa Cruz Tepetotutla, Oaxaca, Mexico

The Chinantla region in the State of Oaxaca, Mexico, is recognized as the third best preserved cloud forest in the country. Since 2004, the indigenous Chinantec community of Santa Cruz Tepetotutla voluntarily destined 9,670 ha of its 12,000 ha territory for conservation as an AVDC (CONANP, 2020). According to the community leaders, they started to receive payment for environmental services in 2014.

Interviews with the elders demonstrated that the community has a deeply rooted sense of conservation of their natural environment:

> *We may not have resources, but we have what is elemental. In the forest, no… you might not believe me, but in the forest, there is everything; in the forest, there is a great amount of edible plants for the communities… plants… many things, then… in the forest […]*
>
> *[…] The forest is water catchment, isn't it? And this is fundamental. Without water, there is no life. Hence, that is what, hmm, we have procured in our communities […]*
>
> *[…] Because we have had more than enough reasons. This is ours, and belongs to the community, and we do not want to destroy it; we want to inherit this to our future children.*

The community leads a subsistence economy. They produce corn, chillies and beans for their own consumption and grow coffee and vanilla. They also produce honey and some preserves. However, they have not yet managed to commercialise their products and access to the area is extremely difficult.

In 2008, *Fundación Grupo Modelo* financed a TBSDI in the community seeking to support the conservation of the Papaloapan river basin and its pristine cloud forests, while providing the community with opportunities to improve their livelihoods. Based on its outstanding natural and cultural heritage, the project managers (PMs) stated that they intended to attract researchers and ecotourists willing to pay fairly for comfortable lodging facilities and good food (the area has one of the most interesting and rich cuisines in the country).

The project consisted of three interrelated initiatives: a preserves workshop, a set of backyard greenhouses and an ecolodge to host visitors. According to the PM, the financing agency agreed with the Assembly, which is the highest authority in the community, that three groups of single mothers would be in charge of the initiatives. Additionally, five male guides were democratically appointed to manage three eco-trails. All participants were provided with professional training and equipment – the researcher was involved in this process for nine months. The community contributed with unpaid labour (*tequio*) to the construction of the infrastructure and provided food and lodging for the training staff.

Over five years later, the researcher returned to the community to find that the outcomes were far from the social change it intended to generate. Since then, research has been carried out to understand why and how these outcomes were generated.

According to the PMs, although the community participated in it, the intervention was designed considering the tourists' potential demands, rather than the community's circumstances, needs and expectations, following the CBT (development *of* tourism) conceptual framework. During fieldwork, the researcher observed internal conflict and poor service quality.

In the ecolodge, western-style beds with sheets and blankets, warm running water and toilets, as well as the gas range were practically unknown to the appointed staff. Almost no home in the community had these items and facilities: they sleep on *petates* (straw mats) on the floor, there is no warm water, they have latrines outside their homes and most of them cook on an open wood fire.

Community leaders stated that after the funding agency left in 2008, the Assembly decided not to comply with the agreement of maintaining the group of women as managers of the ecolodge and continue with their mores and customs. They voted for the creation of the Tourism Committee, appointing a group of young men for the three-year job. The appointed staff (2020) stated that they hadn't been trained to provide the service and argued they were unwilling to work, because they consider *cleaning and cooking as chores for women.*

By contrast, the women are almost incapable of thinking of selling the food they cook for money. In this sense, a young mother stated:

> *We have it from birth. My mother taught me since I was little. First you do errands and from there we learn to wash the nixtamal, grind it on the metate*[1]*, to make tortillas, and to put it on the comal. I let my girls play with the dough since they were very little. As you can see, the youngest* (3 years old) *can make gorditas*[2]*!*

Labour has a social value and is determined by gender and tradition. It is strongly related to the population's identity. When asked about paid work, respondents reported that in the beginning (2008), they did not think of economic income; their main concern was that their young men and women were forced to migrate, they wanted them to stay.

> *[…] It is with great grief we see how our children are migrating, but we cannot stop them, because there is no option of… of how to and say: you know? stay! They are forced to leave, for there is no alternative to retain them […]*

However, this has changed (2020), the young men engaged in the project stated they do have interest in earning money, but it became clear that their linkage with the mainstream market is still limited. The community has not been able to commercialise their products, and in general, for them it is unclear how work should be priced, how products are sold, or what they should charge for the services they provide. These issues may seem simple when assessed from a western-urban perspective but are extremely complex for them to understand and culturally are deeply disrupting for the community. Furthermore, individual paid work, revenue distribution, economic benefit, surplus and financial planning are totally incompatible with a subsistence economy and contradict the value system of common labour, social benefit and solidarity.

Discussion and conclusion: an alternative approach to tourism used as part of NPA management

The research has so far demonstrated that decision-makers assumed tourists would be willing to pay fairly for high quality lodging and food services. This should improve the women's livelihoods through the increase of their economic income. However, accessibility, economic and cultural issues have demonstrated that the geographical, socioeconomic and cultural contexts play a fundamental role in the outcomes of the intervention.

The community expressed that they related tourism to what they had seen on TV: *like in Acapulco* and they expected mass tourists to arrive. Furthermore, they stated that in the beginning, they did not consider income as fundamental; based on their concept of common labour, they expected their young members would have attractive activities, so they would not be forced to migrate to the cities or abroad. The community expressed their concern for their families, the future of the community and the conservation of their rootedness and cultural identity. In this sense, PMs, consultants and the community did not fully share their views about what they expected from the intervention.

The research also evidenced that this perspective has evolved over the years. Community members who migrated have returned and their experience has influenced the social group, generating relevant cultural modifications at different levels. Examples are that young men have become interested in getting their job paid, that women are starting to participate in the charges and that Internet connectivity has arrived to facilitate communication with possible tourists; although computers are very scarce, by February 2020, almost every household had a cell phone.

This case shows that the business management analysis approach alone is insufficient to understand how, under which circumstances and for whom a TBSDI with specific social change theories and objectives, works. Hence, the study of these interventions calls for a totally different analytical perspective; one that allows the understanding of the causal chains that lead to given outcomes, within specific social, environmental, economic, political and cultural contexts. It has been established that understanding tourism as a social intervention seeking social betterment while protecting the natural and cultural heritage poses important conceptual and methodological challenges and that the context in which TBSDIs operate is highly complex. The task calls for an adequate conceptual framework, and a methodological approach which allows the mechanisms contained in the so-called black box of the intervention to be unpacked (Aromatario et al., 2019). This approach also needs to encompass indigenous communities' sociocultural and socio-environmental contexts – *comunalidad* (Martínez Luna, 2016). These elements constitute the foundation, values, premises and expectations of the actors involved in the TBSDIs.

In this sense, tourism used as an intervention for social change also necessarily shifts the attention from the tourism business, to the modifications sought by the intervention. The scope moves from the business success/failure factors, to consider highly complex social, economic, cultural, political and environmental interactions, influenced by interests and power struggles. Therefore, a deep knowledge of the sociocultural, socioenvironmental and socioeconomic context is necessary for the planning, design, implementation and evaluation of the intervention, to develop a solid theory of the change that is expected, and the strategies to achieve it. Consequently, the emphasis should lie upon the community's needs and expectations and the tourism project should be regarded as a means. In other words, the planning and design of a TBSDI for an indigenous community living within or adjacent to a NPA needs to be carried out in terms of expected/unexpected outcomes of the intervention and of the social change to be achieved. The success/failure of the business should be considered as part of this strategy.

Furthermore, TBSDIs as part of NPA management require very careful planning with full and active participation of the community, prioritising real informed consent and involvement of the social group. This implies real and deep-reaching communication between the stakeholders. In this sense, the degree of intersubjectivity (Gillespie & Cornish, 2010) on what is being discussed, expected and done is fundamental. Hence, it is essential that programme planners, designers, decision-makers and practitioners consider the particular communal organization of the social group, their conception of collective well-being over that of the individual, the relevance of solidarity, collaboration, communal decision-making and social value of labour. This will ensure that the delicate balances which underpin the cultural resources that give the community identity and rootedness are not disrupted, so protecting the local cultural structures and institutions. The importance of common land tenure, social relations, rituals and traditions, and the community's symbolic and symbiotic relationship with its natural environment must also be recognized. In the end, their cultural heritage constitutes their identity and uniqueness. Their cultural identity is their main tourism asset and simultaneously their opportunity to gradually incorporate what they need from Western culture and technology to survive as unique, distinct social groups.

The research also indicates that TBSDIs as part of NPA management strategies should be guided by alternative economic theories, such as ecological economics (Barkin & Lemus, 2014; Daly & Farley, 2010), which provide alternatives for these social groups to establish links with the mainstream market, thereby retaining their own values and knowledge through *comunalidad* as their traditional sociocultural, political and economic organization. Their collaboration and solidarity features are important assets to make common interest prevail over that of individual benefit through the implementation of cooperative structures (Barkin & Lemus, 2016, pp. 257–286; Esteva, 2009; V. M. Toledo, 2004).

The analysis proposes the use of realist theory-driven methods to plan, design, implement and assess TBSDIs as part of natural conservation projects, which advocate the consideration of the context in which the interventions take place as well as the possibilities for triggering given mechanisms to yield the expected outcomes. The study also demonstrates the need for further research on this matter, as the iterative nature of the methodological approach is based on the accumulation of knowledge, the detection of patterns, and the refinement of the theories of how social change can be achieved, which should lead to constant improvement of the design, implementation and assessment of the processes and outcomes of these interventions.

Notes

1. Grinding stone
2. Thick, small tortilla-like flatbreads

Disclosure statement

No potential conflict of interest was reported by the author(s).

References

Abram, S. (2010). Anthropology, Tourism and Intervention? In J. Scott & T. Selwyn (Eds.), *Thinking Through Tourism* (1st ed., pp. 231–254). Berg Publishers.

Arizpe, L., Nalda, E. (2003). Cultura, patrimonio y turismo. http://132.248.35.1/cultura/2003/modulo 2/Lectura4.html

Arizpe, L. (2006). Los debates internacionales en torno al patrimonio cultural inmaterial. *Cuicuilco, 13*(38), 13–27. http://www.brasilplural.ufsc.br/35103802Arizpe.pdf

Aromatario, O., Van Hoye, A., Vuillemin, A., Foucaut, A. M., Pommier, J., & Cambon, L. (2019). Using theory of change to develop an intervention theory for designing and evaluating behavior change SDApps for healthy eating and physical exercise: The OCAPREV theory. *BMC Public Health, 19*(1), 1–12. https://doi.org/10.1186/s12889-019-7828-4

Banerjee, O., & Cicowiez, M. (2015). *A quantitative framework for assessing public investment in tourism e An application to Haiti * (No. 180)*. Argentina. https://doi.org/10.1016/j.tourman.2015.05.015

Barabas, A. M. (2004). La territorialidad simbólica y los derechos territoriales indígenas: reflexiones para el Estado pluriétnico. *Alteridades, 14*(27), 105–119.

Barkin, D. (2009). Principles for constructing alternative socio-economic organizations: Lessons learned from working outside institutional structures. *Review of Radical Political Economics, 41*(3), 372–379. https://doi.org/10.1177/0486613409336352

Barkin, D., & Lemus, B. (2014). Rethinking the social and solidarity society in light of community practice. *Sustainability (Sustainability,), 6*(9), 6432–6445. https://doi.org/10.3390/su6096432

Barkin, D., & Lemus, B. (2016). Local Solutions for Environmental Justice. In F. de Castro, B. Baud, & H. Michiel (Eds.), *Environmental Governance in Latin America: Conflicts, Projects and Possibilities* (pp. 257–286). Palgrave Macmillan.

Barthas, B. (1997). La Comunidad Indígena como Organización: El caso de la Huasteca. In *Encuentro de la Asociación de Estudios Latinoamericanos (LASA)* (pp. 1–14). CLACSO, Consejo Latinoamericano de Ciencias Sociales. http://biblioteca.clacso.edu.ar/ar/libros/lasa97/barthas.pdf

Battlori Guerrero, A. (2005). Perspectiva Educativa Ambiental del Turismo Cultural y Alternativo en México. *Boletín de Los Sistemas Nacionales Estadísticos y de …, 1*(1), 52–65. http://scholar.google.com/scholar?hl=en&btnG=Search&q=intitle:Perspectiva+Educativa+Ambiental+del+Turismo+Cultural+en+M?xico#0

Bennett, O., Roe, D., & Ashley, C. (1999). Sustainable Tourism and Poverty Elimination Study A report to the Department for International Development. UK.

Berger, D. N., Bulanin, N., García-Alix, L., Jensen, M. W., Leth, S., Alvarado, E., … Wessendorf, K. (2020). *The Indigenous World 2020* (D. Mamo, Ed.) (34th ed.). The International Work Group for Indigenous Affairs (IWGIA).

Bonfil Batalla, G. (1995). El Etnodesarrollo: sus premisas jurídicas, políticas y de organización. In *Obras Escogidas de Guillermo Bonfil Batalla. Tomo.* (Vol. *2*, pp. 464–480). INAH/INI.

Bonfil Batalla, G. (2012). *México Profundo: Una civilización negada (1a ed., 7a)*. Random House Mondadori.

Bray, D. B., Merino, L., Barry, D., Boyer, C. R., Segura Warnholtz, G., Gerez-Fernández, P., … Ambinakudige, S. (2007). *Los Bosques Comunitarios de México.* Manejo Sustentable de Paisajes Forestales (D. Bray, L. Merino, & D. Barry, Eds.). Instituto Nacional de Ecología/SEMARNAT.

Brundtland, G. H. (1987). Report of the World Commission on Environment and Development : Our Common Future. Norway.

Butler, R., & Hinch, T. (2013). *Tourism and Indigenous Peoples: Issues and Implications.* (R. Butler & T. Hinch, Eds.). Routledge.

Byrne, D. (2013). Evaluating Complex Social Interventions in a. *Evaluation, 19*(3), 217–228. https://doi.org/10.1177/1356389013495617

Carino, J., Champagne, D., Collings, N., Cunningham, M., Dorough, D. S., Kipuri, N., & Trask, M. (2009). *State of the World's Indigenous Peoples.* United Nations/Department of Economic and Social Affairs.

Carlos, J., & Pardo, R. (2019). El papel del trueque en la transición a otros mundos posibles en el sector Zoogocho y la ciudad de Oaxaca, México. *Otra Economía, 12*(22), 66–88.

Cassano, D., Coraggio, J. L., Cortesi, J., González Bombal, I., Hintze, S., Ilari, D., & Silva, J. (2003). *Trueque y Economía Solidaria.* (S. Hintze, Ed.). Editorial Prometeo Libros.

CDI. (2013). Reglas de Operación del Programa de Turismo en Zonas Indígenas (PTAZI). *Diario Oficial de La FederaciónDiario Oficial de La Federación*, 83.

Chape, S., Spalding, M., & Jenkins, M. (2008). *The World's Protected Areas. Status, values and prospects in the 21st Century.* UNEP World Conservation Monitoring Centre. University of California Press.

Chapela y Mendoza, G., & Barkin, D. (1995). *Monarcas y Campesinos: Estrategia de desarrollo sustentable en el oriente de Michoacán.* Centro de Ecología y Desarrollo.

Chok, S., Macbeth, J., & Warren, C. (2007). Tourism and Sustainable Development: Exploring the Theoretical Divide. *Current Issues in Tourism, 10*(2-3), 144–165. https://doi.org/10.2167/cit303

CONANP (2009). *Marco Estratégico de Turismo Sustentable en Áreas Naturales Protegidas de México.* DF.

CONANP (2011). Estrategia Nacional para un desarrollo sustentable del turismo y la recreación en las áreas protegidas de México. SEMARNAT/CONANP. Mexico, DF: Secretaría de Medio Ambiente y Recursos Naturales (SEMARNAT)-Comisión Nacional de Áreas Naturales Protegidas (CONANP). http://www.conanp.gob.mx/pdf_publicaciones/TurismoEstrategia.pdf

CONANP (2018a). *100 años de conservación en México: Áreas Naturales Protegidas*. SEMARNAT-CONANP.

CONANP (2018b). Marco Estratégico de Turismo Sustentable en Áreas Naturales Protegidas de México. Mexico City, Mexico: Secretaría de Medio Ambiente y Recursos Naturales (Semarnat). https://www.conanp.gob.mx/acciones/advc/MarcoEstrategico.pdf

CONANP (2020). *Áreas Destinadas Voluntariamente a la Conservación de junio de 2002 a agosto de 2019*. https://advc.conanp.gob.mx/wp-content/uploads/2020/06/Listado-ADVC-junio-2020.pdf

Daly, H. E., & Farley, J. (2010). *Ecological economics: principles and applications. Ecological Economics*. Island Press. http://indomarine.webs.com/documents/Ecological_Economics_Principles_And_Applications.pdf

Dangi, T. B., & Jamal, T. (2016). An integrated approach to "sustainable community-based tourism. *Sustainability (Sustainability ,), 8*(5), 475. https://doi.org/10.3390/su8050475

de Kadt, E. (1979). *Tourism*: Passport to Development? Perspectives on the Social and Cultural Effects of Tourism in Developing Countries (E. de Kadt Ed.). Oxford University Press.

Desmet, E. (2014). Conservación y Pueblos Indígenas: Un análisis socio-jurídico (Cuadernos Deusto de Derechos Humanos No. 75). Spain.

Díaz Gómez, F. (1994). Comunidad y comunalidad. In *Antología sobre Cultura Popular e Indígena. Lecturas del Seminario Diálogos en la Acción, Segunda Etapa* (pp. 365–373). Consejo NAcional para la Cultura y las Artes (CONACULTA)/Culturas Populares e Indígenas.

Dodds, R., Ali, A., & Galaski, K. (2018). Mobilizing knowledge: determining key elements for success and pitfalls in developing community-based tourism. *Current Issues in Tourism, 21*(13), 1547–1568. https://doi.org/10.1080/13683500.2016.1150257

Ejecutivo, P. (2013). *Plan Nacional de Desarrollo 2013-2018.*

Elkington, J. (1997). *Cannibals with forks: The triple bottom line of 21st century business*. Capstone Publishing Limited.

Esfehani, M. H., & Albrecht, J. N. (2018). Roles of intangible cultural heritage in tourism in natural protected areas. *Journal of Heritage Tourism, 13*(1), 15–29. https://doi.org/10.1080/1743873X.2016.1245735

Esteva, G. (2009). Más allá del desarrollo: la buena vida. *Revista América Latina En Movimiento*, (445).

Esteva, G. (2010). Development. In Wolfgang Sachs (Ed.), *Development Dictionary*. (2nd ed., pp. 1–23). Zed Books.

Frenzel, F. (2013). Slum tourism in the context of the tourism and poverty (relief) debate. *Erde*, https://doi.org/10.12854/erde-144-9

Gascón, J. (2015). Pro-Poor Tourism as a Strategy to Fight Rural Poverty: A Critique. *Journal of Agrarian Change, 15*(4), 499–518. https://doi.org/10.1111/joac.12087

Gillespie, A., & Cornish, F. (2010). Intersubjectivity: Towards a dialogical analysis. *Journal for the Theory of Social Behaviour, 40*(1), 19–46. https://doi.org/10.1111/j.1468-5914.2009.00419.x

Hall, G., & Patrinos, H. (2010). *Indigenous Peoples, Poverty and Development*. USA.

Harbor, L. C., & Hunt, C. A. (2021). Indigenous tourism and cultural justice in a Tz'utujil Maya community. *Journal of Sustainable Tourism, 29*(2-3), 214–220. https://doi.org/10.1080/09669582.2020.1770771

Helliwell, J., Layard, R., & Sachs, J. (2013). *World Happiness Report 2013*. (J. Helliwell, R. Layard, & J. Sachs, Eds.). UN Sustainable Development Solutions Network. www.unsdsn.org

Honey, M. (2008). *Ecotourism and Sustainable Development: Who owns Paradise?*. Island Press.

INPI (2019). Reglas de Operación del Programa de Infraestructura Indígena: Instituto Nacional de los Pueblos Indígenas. Diario Oficial de La Federación, pp. 1–123. https://www.dof.gob.mx/nota_detalle.php?codigo=5583031&fecha=28/12/2019

Job, H., Becken, S., & Lane, B. (2017). Protected Areas in a neoliberal world and the role of tourism in supporting conservation and sustainable development: an assessment of strategic planning, zoning, impact monitoring, and tourism management at natural World Heritage Sites. *Journal of Sustainable Tourism, 25*(12), 1697–1718. https://doi.org/10.1080/09669582.2017.1377432

Julian, R. (2016). Is it for donors or locals? The relationship between stakeholder interests and demonstrating results in international development. *International Journal of Managing Projects in Business, 1*(1), 125–130. https://doi.org/10.1108/17538370810883819

Kakwani, N., & Pernia, E. (2000). What is pro-poor growth? *Asian Development Review, 18*(1), 1–16. https://www.researchgate.net/profile/Nanak_Kakwani/publication/253876792_What_is_Pro-poor_Growth/links/540ea4360cf2d8daaacd54a5.pdf

Magalón de Salazar, L. (2011). *Reflexiones y posiciones alrededor de la evaluación de intervenciones complejas*. Santiago de Cali, Colombia: Programa Editorial Universidad del Valle. http://www.fundacionfundesalud.org/pdffiles/reflexiones-y-posiciones-alrededor.pdf

Mansuri, G., & Rao, V. (2004). Evaluating community-based and community-driven development: a critical review of the evidence. *The World Bank Research Observer, 19*(1), 1–39. https://doi.org/10.1093/wbro/lkh012

Mark, M. M., & Shotland, R. L. (1985). Stakeholder-based evaluation and value judgements. *Evaluation Review, 9*(5), 605–626. https://doi.org/10.1177/0193841X8500900504

Martínez Luna, J. (2011). Eso que llaman comunalidad, 2011.

Martínez Luna, J. (2013). *Textos Sobre el Camino Andado. Tomo I.* Oaxaca, Mexico: CSEIIO, IEEPO, Centro de Apoyo Técnico al Movimiento Popular Oaxaqueño, Coalición de Maestros y Promotores Indígenas de Oaxaca (CMPIO)-Plan Piloto, Coordinación Estatal de Escuelas de Eduación Secundaria Comunitaria Indígena (CEEESCI), Congreso Nacional d.

Martínez Luna, J. (2016). *Textos Sobre el Camino Andado. Tomo II.* Coalición de Maestros y Promotores Indígenas de Oaxaca (CMPIO)-Plan Piloto, Coordinación Estatal de Escuelas de Eduación Secundaria Comunitaria Indñigena (CEEESCI), COngreso Nacional de Educación Indígena Intercultural (CNEII).

Maxwell, J. A. (2012). *A realist approach for qualitative research.* Sage Publications, Inc.

Meadows, D. H., Meadows, D. L., & Behrens, W. E. III, (1972). *The limits to growth.* Universe Books/Potomac Assocoates.

Melubo, K., & Carr, A. (2019). Developing indigenous tourism in the bomas: critiquing issues from within the Maasai community in Tanzania. *Journal of Heritage Tourism, 14*(3), 219–232. https://doi.org/10.1080/1743873X.2018.1533557

Méndez-López, M. E., García-Frapolli, E., Ruíz-Mallén, I., Porter-Bolland, L., Sánchez-González, M. C., & Reyes-García, V. (2019). Who participates in conservation initiatives? Case studies in six rural communities of Mexico. *Journal of Environmental Planning and Management, 62*(6), 1045–1064. https://doi.org/10.1080/09640568.2018.1462152

Mendoza-Ramos, A., & Prideaux, B. (2018). Assessing ecotourism in an Indigenous community: using, testing and proving the wheel of empowerment framework as a measurement tool. *Journal of Sustainable Tourism, 26*(2), 277–291. https://doi.org/10.1080/09669582.2017.1347176

Millán, S., & Valle, J. (2003). *La Comunidad sin Límites.* Estructura social y organización comunitaria en las regionaes indígenas de México. (S. Millán & J. Valle, Eds.). INstituto Nacional de Antropología e Historia.

Monterroso, I. (2008). Comunidades Locales en Áreas Protegidas: Reflexiones sobre las políticas de conservación en la Reserva de Biosfera Maya. In G. Merino Alvarado, G. C. Ramos Delgado, D. Domínguez, C. Campello do Amaral Mello, I. Monterroso, & G. Wilde (Eds.), *Gestión Ambiental y Conflicto Social en América Latina.* (pp. 227–263). CLACSO, Consejo Latinoamericano de Ciencias Sociales.

Morales Garcia, A. D., & Morales Garcia, J. J. (2017). Combate efectivo de los delitos contra la biodiversidad en méxico como una herramienta de conservación de la biodiversidad. *Nómadas. Revista Crítica de Ciencias Sociales y Jurídicas, 51*(2), 407–428. https://doi.org/10.5209/NOMA.55473

Moreno-Barajas, R., Talavera-Garduño, K., Rivera-Morales, S., & Hernández-Ramírez, N. (2019). Evaluación de la situación actual de las Áreas Naturales Protegidas del Estado de México. *Quivera Revista de Estudios Territoriales, 21*(2), 113–129. https://doi.org/10.36677/qret.v21i2.11974

Murphy, P. E. (1983). Tourism as a community industry-an ecological model of tourism development. *Tourism Management, 4*(3), 180–193. https://doi.org/10.1016/0261-5177(83)90062-6

Nepal, S., & Saarinen, J. (2016). *Political Ecology and Tourism.* (S. Nepal & J. Saarinen, Eds.). Routledge.

Ortega-Rubio, A. (2018). *Mexican Natural Resources Management and Biodiversity Conservation.* In A. Ortega-Rubio (Ed.). *Recent case studies.* Springer International Publishing AG.

Palomino-Villavicencio, B., Gasca-Zamora, J., & López-Pardo, G. (2016). El Turismo Comunitario en la Sierra Norte de Oaxaca: Perspectiva desde las Instituciones y la gobernanza en territorios indígenas. *El Periplo Sustentable, 30*(Enero/Junio), 6–37.

Pawson, R. (2013). *The science of evaluation. a realist manifesto (1st.).* Sage Publications.

Pawson, R., & Tilley, N. (2010). *Realistic evaluation (1st/13th).* SAGE.

Pawson, R., Tilley, N. (2004). Realist Evaluation (pp. 1–36). http://www.communitymatters.com.au/RE_chapter.pdf

Pereiro, X. (2016). A review of Indigenous tourism in Latin America: reflections on an anthropological study of Guna tourism (Panama). *Journal of Sustainable Tourism, 24*(8-9), 1121–1138. https://doi.org/10.1080/09669582.2016.1189924

Perry, G. E., Arias, O. S., López, J. H., Maloney, W. F., & Servén, L., (2006). *Poverty Rduction and Growth: Virtuous and vicious circles.* The World Bank.

Rendón Monzón, J. J. (2003). La Comunalidad: Modo de vida en los pueblos indios. Tomo I. Mexico. Consejo Nacional para la Cultura y las Artes/Dirección General de Culturas Populares e Indígenas.

Reyes, J. A., Gómez, J. P., Muis, R. O., Zavala, R., Ríos, G. A., & Villalobos, O. (2012). *Atlas de propiedad social y servicios ambientales en México.* IICA. http://www.iica.int.

Saayman, M., & Giampiccoli, A. (2016). Community-based and pro-poor tourism: Initial assessment of their relation to community development. *European Journal of Tourism Research, 12,* 145–190.

Sachs, W. (2010). *The Development Dictionary.* A guide to Knowledge as Power. (Wolfgang Sachs, Ed.) (2nd ed.). Zed Books.

Scheyvens, R. (2014). Tourism and Poverty Alleviation. In R. Sharpley & D. J. Telfer (Eds.), *Tourism and Development: Concepts and Issues.* (2nd ed., pp. 118–139). Channel view Publications.

SEMARNAT/CONANP (2007). *Programa de Turismo en Áreas protegidas 2006 - 2012.* (M. J. Sánchez Lira, Ed.). SEMARNAT/CONANP.

Slaper, T. F. (2011). The Triple Bottom Line: What Is It and How Does It Work? The Triple Bottom Line Defined. *Indiana Business Review, 86*(1), 4–8. http://www.ibrc.indiana.edu/ibr/2011/spring/article2.html

Spenceley, A. (2008). *Responsible Tourism.* Critical Issues for Conservation and Dvelopment. (A. Spenceley, Ed.). Earthscan.

Spenceley, A., & Meyer, D. (2012). Tourism and poverty reduction: Theory and practice in less economically developed countries. *Journal of Sustainable Tourism, 20*(3), 297–317. http://www.tandfonline.com/doi/abs/10.1080/09669582.2012.668909 https://doi.org/10.1080/09669582.2012.668909

Stame, N. (2004). Theory-Based Evaluation and Types of Complexity. *Evaluation, 10*(1), 58–76. https://doi.org/10.1177/1356389004043135

Stavenhagen, R. (2010). *Los Pueblos Originarios: El debate necesario. Instituto de Estudios y Formación de la CTA.*, Ediciones CTA, Consejo LAtinoamericano de Ciencias Sociales CLACSO.

Stone, M. T., & Nyaupane, G. P. (2018). Protected areas, wildlife-based community tourism and community livelihoods dynamics: spiraling up and down of community capitals. *Journal of Sustainable Tourism, 26*(2), 307–324. https://doi.org/10.1080/09669582.2017.1349774

Strasdas, W. (2005). Community-based Tourism: Between self-determination and market realities. In *Tourism Forum International at the Reiseforum.* The International Ecotourism Society (TIES, Germany).

Strengers, Y., & Cecily, M. (2015). Introduction. Social practices, interventon and sustainability: beyond behaviour change. In Y. Strengers & M. Cecily (Eds.), *Social Practices, Interventon and Sustainability: Beyond behaviour change.* (pp. 1–12). Routledge.

Toledo, V. (2005). Repensar la conservación : ¿áreas naturales protegidas o estrategia bioregional? *Gaceta Ecológica, 77*, 67–83.

Toledo, V. M. (2004). Principios Etnoecológicos para el desarrollo sustentable de comunidades campesinas e indígenas. Retrieved April 26, 2017, from http://ecologiasocial.com/2004/04/principios-etnoecologicos-para-el-desarrollo-sustentable-de-comunidades-campesinas-e-indigenas/

Trau, A. M. (2012). Beyond Pro-Poor Tourism: (Re)Interpreting Tourism-Based Approaches to Poverty Alleviation in Vanuatu. *Tourism Planning & Development, 9*(2), 149–164. (November 2014), https://doi.org/10.1080/21568316.2011.630750

UN. (2000). General Assembly. In *UN General Assembly, 18 September 2000* (p. 9.

UNEP-WCMC (2018). *2018 United Nations List of Protected Areas. Supplement on protected area management effectiveness.* UNEP-WCMC.

UNESCO/IBEROMaB (2010). *Plan de Acción para la Red de Comités Nacionales MaB y Reservas de Biósfera de Iberoamérica y el Caribe 2010-2020.*

United Nations (2015). *Transforming our world: the 2030 Agenda for Sustainable Development. General Assembley 70 session.* (Vol. 16301). https://doi.org/10.1007/s13398-014-0173-7.2

van Dijk, T. A. (2008). *Discourse and Context. A sociocognitive approach.* Cambridge University Press.

Wardle, C., Buckley, R., Shakeela, A., & Castley, J. G. (2018). Ecotourism's contributions to conservation: analysing patterns in published studies. *Journal of Ecotourism, 0*(0), 1–31. https://doi.org/10.1080/14724049.2018.1424173

Warman, A. (2003). *Los Indios de México en el Umbral del Milenio (1a).* Fondo de Cultura Económica.

Watanabe, Y. (2008). *Comunidades indígenas y biodiversidad.* Washington, D. C., USA.

Weiss, C. H. (1995). Nothing as Practical as Good Theory: Exploring Theory-based Evaluation for Comprehensive Community Initiatives for Children and Families. In J. I. Connell, A. C. Kubisch, L. B. Schorr, & C. H. Weiss (Eds.), *New Approaches to Evaluating Community Initiatives: Concepts, Methods, and Contexts.* (pp. 65–92). The Aspen Institute.

Wholey, J. S., Hatry, H. P., & Newcomer, K. E. (2010). *Handbook of Practical Program Evaluation (3rd.).* Jossey-Bass.

Wood, M. E. (2004). A Triple Bottom Line Framework for Sustainable Tourism Development for International Donors. *Epler-Wood International, Burlington, VT.* http://scholar.google.com/scholar?hl=en&btnG=Search&q=intitle:A+Triple+Bottom+Line+Framework+for+Sustainable+Tourism+Development+for+International+Donors+Defining+Indicators+for+Conservation,+Community+and+Local+Enterprise+Development#0

Zapata, M. J., Hall, C. M., Lindo, P., & Vanderschaeghe, M. (2011). Can community-based tourism contribute to development and poverty alleviation? Lessons from Nicaragua. *Current Issues in Tourism, 14*(8), 725–749. https://doi.org/10.1080/13683500.2011.559200

Zizumbo Villarreal, L. (2013). *Las Paradojas del Desarrollo Local y del Turismo.* (1st ed.). Universidad Autónoma del Estado de México, Miguel Ángel Porrúa.

Social and cultural capitals in tourism resource governance: the essential lenses for community focussed co-management

Muhammad Shoeb-Ur-Rahman, David Simmons, Michael C. Shone and Nazmun N. Ratna

ABSTRACT

This paper investigates the relative importance of social and cultural capitals in (re)shaping a tourism resource governance approach for supporting sustainable livelihood outcomes. We apply a 'Capitals Co-management for Sustainable Livelihood Framework (CCSLF)' in a remote case study setting of Chittagong Hill Tracts (CHT), Bangladesh. Within this setting, culturally distant communities living within an environment of historical political distrust provide a highly nuanced context to our investigation. Based on data collected through multiple qualitative methods, we found that power structures and politics play the central role in resource allocation, and thereby hindering community involvement in the decision-making processes. As an engagement mechanism, bridging social capital can play a significant role in facilitating the establishment of co-management structures, through which the community can exercise greater say over the deployment of common property or community resources in tourism development. For a tourism destination comprised of diverse social groups and associated cultural capitals, a co-management structure is seen as having its best application at micro- or local community-levels. This investigation challenges destination area planners to explore the various dimensions of social and cultural capitals to support (community) tourism development.

Introduction

The significance of community capitals for sustainable livelihoods, quality of life (QoL), and community wellbeing is gaining attention in the sustainable tourism literature (Çakir et al., 2018; Moscardo, 2014; Rahman, 2019; Shen et al., 2008; Tao & Wall, 2009). Community resources contributing to sustainable livelihoods have been coined by numerous terms in the literature and include 'livelihood assets' (Ashley, 2000; DFID., 1999), 'community capitals' (Flora et al., 2004; Stone & Nyaupane, 2018), and 'rural capitals' (Bosworth & Turner, 2018; Castle, 1998). Tourism projects can generate positive results towards the enhancement of stock and flow of these resources, which are required to improve community livelihoods (Moscardo, 2014; Stone & Nyaupane, 2016; 2018). The literature, however, largely fails to identify a mediating strategic tool

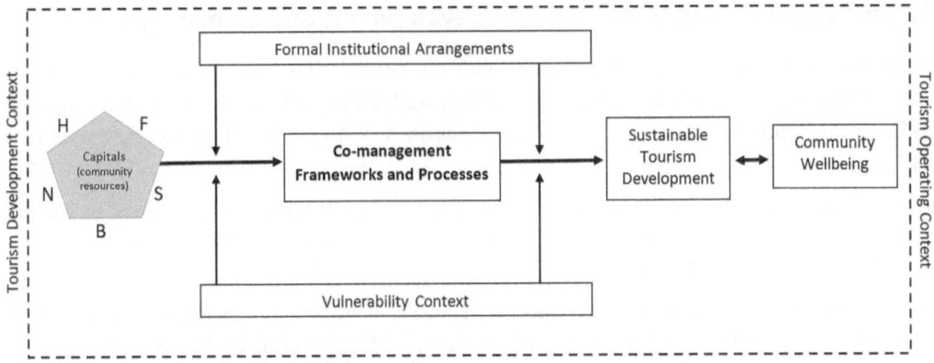

Figure 1. Capitals co-management for sustainable livelihood framework. Source: Adapted from Rahman et al., 2020, p. 232.

to operationalise tourism resource governance and broader stakeholder participation in tourism resource decision-making. This gap in the literature has been identified by Moscardo (2011), for example, who notes that destination community residents are often excluded from planning processes and tourism resource governance in favour of external agents who are commonly government representatives.

To address this particular gap, Rahman et al. (2020) proposed a 'Capitals Co-management for Sustainable Livelihood Framework' (CCSLF) which positioned 'co-management' of community resources (capitals) as a decentralised and devolved platform to harness community resources (capitals) and promote, via sustainable tourism development, community wellbeing outcomes. The CCSLF itself has six elements: livelihood capitals, formal institutional arrangements, co-management frameworks and processes, vulnerability context, and livelihood outcomes (see Figure 1). Importantly, all of these elements must be addressed within the local tourism and destination context (Plummer et al., 2006).

The elements presented in Figure 1 depict community resources as 'livelihood capital inputs' to be transformed and/or mediated within a (tourism) co-management structure, which in turn is framed by formal institutional arrangements to reduce community vulnerability. At the output level, mutually inclusive goals of community wellbeing and sustainable tourism development are identified as core outcomes.

While the literature identifies various forms of capital, CCSLF adapts a set of five livelihood capitals (natural, built, financial, human, and social) within the existing frames of livelihood assets, community capitals, and rural capitals (Bosworth & Turner, 2018; Castle, 1998; DFID., 1999; Flora et al., 2004). To elaborate, CCSLF expands social capital to include consideration of cultural and political capitals to redefine the set of livelihood capitals from Flora et al. (2004) original seven community capitals. In this vein, authors initially classified cultural capital as associated with 'social capital' (Castle, 1998), while the notion of political capital is captured by separating formal and informal institutional aspects (Shen et al., 2008).

This paper addresses the co-management research gap identified within the literature via the lenses of social and cultural capitals. It achieves this via testing Rahman et al. (2020) CCSLF model within a developing country case study setting of the Chittagong Hill Tracts (CHT)[1] of Bangladesh. Here culturally distant Bengali[2] and Indigenous[3] communities provide a highly contested and politically nuanced destination context. Within this CHT setting, a large-scale, government-backed, migration of Bengalis initiated in the late 1970s provides an interesting and contested set of political, governance, and resource dimensions for the research context. Moreover, it is claimed that within this destination setting Indigenous groups are not well represented in different decision-making fora and their voices remain underheard (Uddin et al., 2019). This paper, therefore, extends the CCSLF model to interrogate specifically the role of social and cultural capitals in promoting a tourism resource governance approach.

Social and cultural capitals for tourism development and co-management

Although the overlaying nature of social and cultural capitals has been identified in the literature (Castle, 1998; Flora et al., 2004), exploring a conceptual divide would be meaningful to shed light on collaborative tourism resource governance approach. Putnam (2000, p. 19) defines social capital as follows:

> *Whereas physical capital refers to physical objects and human capital refers to the properties of individuals, social capital refers to connections among individuals – social networks and the norms of reciprocity and trustworthiness that arise from them.*

The Organization for Economic Co-operation and Development (OECD, 2001, p. 41) offers a practitioner lens (albeit based on Putnam) to define social capital as "networks together with shared norms, values and understandings that facilitate cooperation within or among groups". Unlike social capital that emphasises the structural and cognitive aspects to be built (Castiglione et al., 2008), cultural capital highlights symbolic (e.g., cultural backgrounds, languages, rituals) and material (e.g., dresses, paintings, instruments, monuments) features to be transferred from generation to generation (Bourdieu, 1986; Macbeth et al., 2004). However, Bourdieu (1986) claimed that the institutionalised state of cultural capital in the form of educational qualifications is not transmissible. Another view is provided by Dalziel et al. (2009, p. 14), who define cultural capital as "the set of values, norms, traditions and behaviour which individual group members and groups as a whole can develop into assets or resources to leverage economic, political and social gains". Apart from their conceptual difference, both social and cultural capitals are directly linked to facilitate community livelihoods by actively engaging destination communities by means of securing economic benefits such as supporting more entrepreneurial initiatives (Karlsson, 2005).

In this paper, we focus on three basic dimensions of social capital: its *scope*, *forms*, and *channels* (Castiglione et al., 2008; Musavengane & Kloppers, 2020). Scope focuses on the unit of observation, which at times can be focussed on various layers of observation including the micro, meso, and macro. Analysis of the forms of social capitals includes structural and cognitive aspects. While structural elements are objective constructs such as social networks, cognitive features are more subjective and intangible (e.g., values, norms of reciprocity, and trust which enables cooperative activity). Two types of descriptors linking these elements are identified by Putnam (2000, pp. 22–24) as "bonding" and "bridging". Bonding social capital refers to the productive benefits associated with the relationships within a single group or structure, such as relations among members of an ethnic group. Bridging social capital originates from examining the relationships between people who are different in terms of socio-economic status, generation, or entities. Woolcock (2000) identifies the third descriptor – 'linking social capital' – which defines the payoffs or benefits associated with the relationships people have with those in power. Without sufficient bonding and bridging social capital, linking capital may lead to ineffective community associations and/or ineffective democratic governance (Titeca & Vervisch, 2008). On this note, the third dimension of social capital indicates the 'channels' through which development is affected and links with elements (e.g., information sharing, collective action, and decision-making) to bring about mutual benefits (Castiglione et al., 2008). Figure 2 illustrates the conceptual framework on social capital and (tourism) co-management that underpins the current research focus.

This research investigates how different forms of social capital in a diverse cultural setting inform a tourism co-management structure and process. Accordingly, cognitive and structural components have been considered for evaluation via the lens of bonding, bridging, and linking social capitals (Musavengane & Kloppers, 2020; Ooi et al., 2015). Bonding and bridging functions are the result of horizontal networks and relationships, whereas linking social capital is a result of vertical networks (i.e., with those in 'power' and/or decision-making roles) (Titeca & Vervisch,

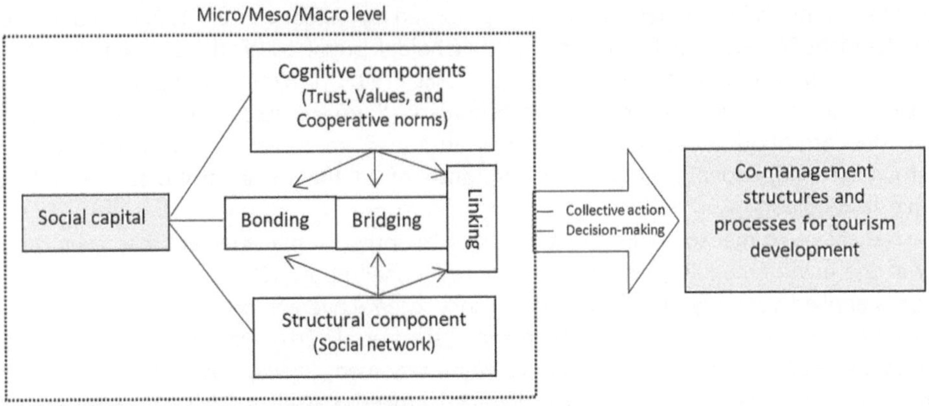

Figure 2. Conceptual framework on social capital and tourism co-management.

2008). Eventually, the nature and extent of social capital is directed to tourism resource governance structures or, more specifically, tourism co-management structures and processes (Rahman et al., 2020). In particular, social and process dimension of tourism resource management, including key tourism stakeholder participation in decision-making, has been sought out to test at a community (meso) level within a 'contested' research setting (i.e., the presence of conflicting social groups). Building social capital at the community level enhances resilience to withstand governance shocks and facilitates collaborative resource management for community-based tourism development (Musavengane & Kloppers, 2020).

Within the existing sustainable livelihood frameworks, social capital is argued to foster connections and coordination across stakeholder groups that eventually help people to pursue their desired livelihood strategies (Scoones, 1998). In the context of securing livelihoods through tourism development, the impacts of 'social' capital are more salient where social capital forms the basis of collaboration and community participation (Nunkoo, 2017; Ooi et al., 2015; Stone & Nyaupane, 2017). It is argued that cultural capital can influence the (re)production of social capital (Bourdieu, 1986). In that respect, cultural capital complements social capital and lays a stable foundation from which to generate tourism products, and also supports the extent of effectiveness for community engagement (Dalziel et al., 2009; Macbeth et al., 2004). However, cultural capital within an ethnically diversified setting may challenge the social inclusion requirements for ensuring sustainable development through participatory governance approach (Sachs, 2012).

Although the governance of tourism resources is particularly affected by a local or destination context (Wang & Bramwell, 2012), the most common practice is top-down planning at destination level (Andriotis et al., 2019). Here, the emphasis is on the exercise of power, which is believed to be held mostly by public institutions. Such approaches fail to adopt truly participative processes. That is, they do not involve multiple actors in decision-making to ensure 'inclusive governance' and therefore share power inequitably (Ooi et al., 2015). Rahman et al. (2020) advocate co-management as an alternative to the traditional top-down processes of tourism resource management to better involve key stakeholders in resource decision-making and implementation processes. Plummer and FitzGibbon (2004) observe that co-management is usually accompanied with a common pool of resources comprising human, economic, cultural, and natural resources.

Co-management responds to the community's desire for involvement in decision-making and fosters cooperation among the involved stakeholders by means of reciprocity or shared benefits. 'Stakeholder' identification is therefore a core consideration of co-management, especially in a setting which incorporates diverse cultural identities and multiple governance structures. In such a situation, ethnic diversity may increase 'interest group polarisation', which in turn provides benefits mostly to those who are in power at the cost of society as a whole (Barnes-Mauthe et al., 2013; Mavridis, 2015). To ensure equitable representation in a co-management structure, a stakeholder identification model such as that proposed by Mitchell et al. (1997) remains helpful. The model identifies seven different types of stakeholder premised on three stakeholders' attributes: power, urgency, and legitimacy (Mitchell et al., 1997, p. 874). Stakeholders who possess only one attribute are known as *latent* stakeholders (dormant, discretionary, and demanding) and gain less attention; those who possess two attributes are *expectant* stakeholders (dominant, dangerous, and dependent), whereas the presence of all the three attributes are defined as *definitive* stakeholders.

As a variant of co-management, the concept of 'adaptive co-management' has been reported widely in the environmental governance literature (Fennell et al., 2008; Laplaza et al., 2017). The concept is applied variously in tourism with a more recent coverage on protected areas management to include tourism destination governance (Islam et al., 2018; Nyaupane et al., 2020). The principles and features identified for adaptive co-management overlap those identified under the general umbrella of 'co-management'. An iterative process based on social learning and resilience building is at the core of all co-management frameworks and processes (Sessin-Dilascio et al., 2015).

A precondition of co-management is the need to facilitate 'empowerment', which provides additional requirements for the inclusion of Indigenous cultures (Carr et al., 2016; Castro & Nielsen, 2001). In this vein, 'social' and 'political' empowerment is highlighted to balance government and community participation in a co-management structure (Joo et al., 2020; Pomeroy & Berkes, 1997; Scheyvens, 1999). There are necessarily trade-offs between full participation and efficiency of process, and also effectiveness of contribution (Simmons, 1994). The ideal level of participation is not necessarily located at the 'full participation' end of the scale. As such Wall (1999, p. 281) claimed that "... true co-management may be difficult to find and various forms of co-optation may be more common." For instance, the CCSLF (Rahman et al., 2020) upholds Carlsson and Berkes (2005, p. 68) second typology of co-management (out of four typologies) as a theoretical base, which framed co-management as an intercepting joint organisation of state and community. Such an approach, in principle, adheres to the 'polycentric' governance systems involving multiple stakeholders rather than a 'monocentric' governance by a particular stakeholder group (Nyaupane et al., 2020).

The effectiveness of a co-management structure and process is largely determined by the depth of, and linkages within, dimensions of social capital (Bodin & Prell, 2011; Plummer & FitzGibbon, 2006). By the same token cultural capital may either promote or inhibit social capital's functioning for community-inclusive resource governance approaches including tourism (Karlsson, 2005; Macbeth et al., 2004; Ooi et al., 2015). For example, cultural capital can be realised through psychological empowerment and/or disempowerment, such as forming a sense of cultural pride or affected negatively and demotivated by tourism-related projects (Schmidt & Uriely, 2018). This can filter and reshape social capital and the extent of effective community participation.

Methods

Study area

In order to test the CCSLF model, we employed a single-embedded case study strategy (Yin, 2014). Specifically, the Chittagong Hill Tracts (CHT) of Bangladesh was selected as the focal area for our study. It is a hilly area located in the south-eastern part of Bangladesh sitting under the broader Chittagong Division. The CHT region has a population of 1.6 million and comprises a total land area of 13,294 square kilometres (UNICEF Bangladesh, 2019). It holds significant geopolitical values in that it represents 10 percent of the total land area of Bangladesh and shares land borders with two neighbouring countries: India and Myanmar (Royhan, 2016). Historically, the CHT remains the ancestral home of Indigenous populations who are simultaneously

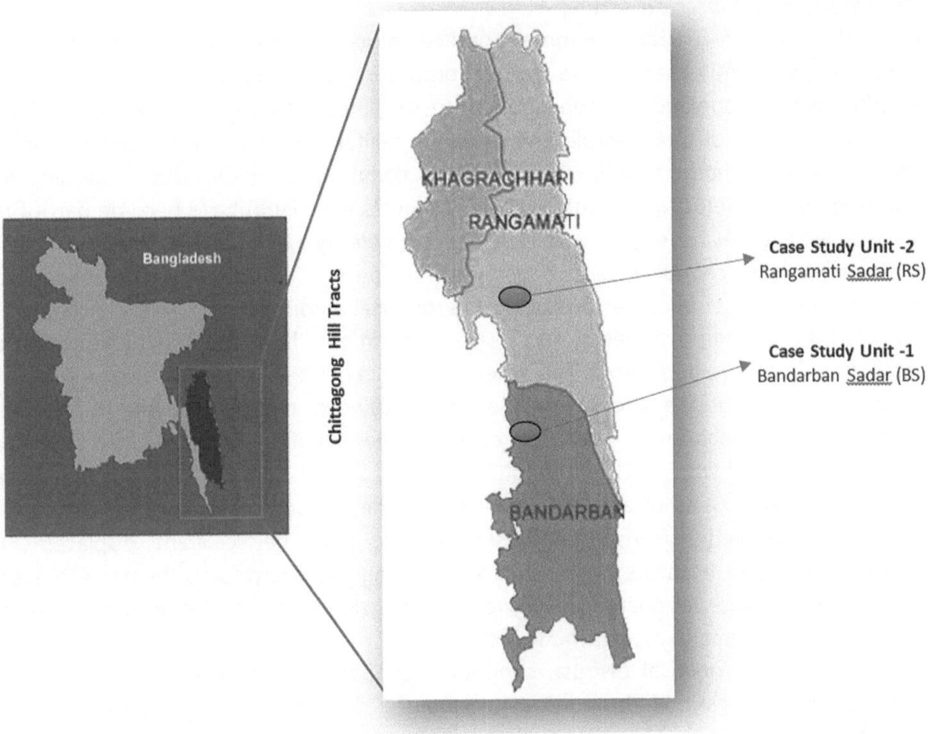

Figure 3. Location of study units within Eastern Bangladesh.

identified by several other terms, such as *Adivasi*, tribal, *Pahari* (people who live in hills), and small ethnic communities (as used by the government of Bangladesh). Eleven ethnic groups, *Chakma, Marma, Tripura, Tanchangya, Chak, Pankhu, Mro, Bawm, Lushai, Kyang,* and *Khumi,* reside in the CHT (Ministry of Chittagong Hill Tracts Affairs, 2020; UNICEF Bangladesh, 2019). Figure 3 provides a location map of the CHT study setting.

Two '*Upazilas*'[4] namely Bandarban *Sadar*[5] (BS) and Rangamati *Sadar* (RS) from the CHT were the focus of our investigation. The choice of two parallel cases within a common national and regional context provides a unique opportunity to examine Indigenous communities' capabilities and capacities. This was particularly useful since the concentration of Indigenous communities vary considerably within each district and sub-district (Bangladesh Bureau of Statistics, 2013a; 2013b; 2013c). The two case study areas are the central points of their respective districts and serve as the common gateway to tourist destinations and/or sites of the region. There are no official statistics available to reference overall tourism development, however increasing numbers of hotels and other accommodation facilities over the last decade indicate the tourism potential of these areas. Tourism in this region is predominantly centred on nature and experiencing ethnic lifestyles (Shamsuddoha et al., 2011).

Although local tourism's governance is ostensibly delegated to the respective Hill District Councils (HDCs), the Office of the Deputy Commissioner within each hill district under the general administrative structure still has significant influence in tourism development decisions. The military administration has also become directly involved in tourism projects. Chakma and Chakma (2016) have argued these projects as a mechanism to displace Indigenous communities and to conceal their much-debated land rights issue. Tourism development opportunities are further compounded by a lack of market accessibility as the central government imposes entry restrictions to the region for foreign visitors.

Conflict, unrest, and contested setting

An extended period of central government-backed large-scale migration commencing in the late-1970s has significantly changed the regional proportion of Bengali residents from 13 percent in 1971 (the year of Bangladesh Independence) to 41 percent by 1981 (Adnan, 2004; Dhamai, 2014). The deployment of a large-scale and enduring military presence in the CHT has acted to reinforce the political marginalisation of Indigenous communities within this study area (Gray, 1994). Consequently, interest-group conflict over resources (e.g., land) have become part of regular discourse leading to insurgency and violence. Chowdhury (2008) argues the root cause of insurgency stems from central government's failure to formally acknowledge the various ethnic groups as 'Indigenous' communities and acknowledge local autonomy.

In response to growing insurgency, successive governments from the mid-1970s to the mid-1990s transformed the region into a militarised zone (Gray, 1994). Finally, a 'Peace Treaty' (the 'Chittagong Hill Tracts Peace Accord') was enacted between the government of Bangladesh (GoB) and Indigenous factions in 1997. The overall implementation status of the Peace Accord remains highly disputed within CHT, although the Accord nonetheless has considerable influence over the operational aspects of any endeavour inside the region. Importantly for this research, tourism projects backed by institutional sponsorship, and the resultant displacement of Indigenous communities from their 'contested lands', bring into sharp focus the need to examine tourism's role within the designated Peace Accord alongside the need to develop and explore the effectiveness of (co)management frameworks.

Taken together, the political distrust, communal conflicts, overlapping administrative structures, and restricted accessibility have led us to label the destination as 'contested' and to some extent 'protected'.

Data collection

This study has been undertaken using an integrated programme of qualitative methods. Data were collected from the study areas during December 2015 to July 2016. The 'stakeholder identification theory' and the 'CCSLF framework' (Mitchell et al., 1997; Rahman et al., 2020) led to the identification of two categories of research participants namely community residents and institutional representatives. The 'community' participant category, which includes both Indigenous and migrated (Bengali), has been identified as expectant stakeholders. In juxtaposition, institutional representatives include key institutional personnel (based on power and position) representing formal institutions, both political (public) and economic (private) in nature, and at local, regional, and central levels. Correspondingly, the 'institutional' category of participants includes a wide range of latent, expectant, and definitive stakeholders. In the case of definitive stakeholders, respective institutional representatives could also be local residents.

To ensure the richness of information, as well as to facilitate methodological triangulation and research validity, we applied multiple methods of data collection. Firstly, wide representation from diverse communities and institutions was sought (see Table 1). In total, 52 research participants were interviewed with an average duration of 75 minutes each. We utilised stratified purposeful sampling which was later complemented by snowball sampling (Patton, 2002) to address this goal. The presence of Indigenous communities and their customary decision-making processes reinforced our choice in using a snowball technique. Although a snowball technique is recommended to conduct research in a conflict-prone environment, it may lead to selection bias and lack of generalisability due to limitations within social networks (Cohen & Arieli, 2011). To overcome such limitations, we spent considerable time in the field to balance the insider–outsider perspectives and observe the relative strength of each sub-group and their internal dynamics (Jennings, 2010).

Table 1. Profiling research participants.

Method	Participants' category	Sample coding information	Number (n)
Interview	Community Resident Indigenous	CRIBB, CRIRM	17
	Community Resident Bengali	CRBBB, CRBRM	7
	Institutional Representative Central Government	IRCG, IRBPC, IRBTB, IRMOCAT, IRMOCHTA	7
	Institutional Representative Regional Government	IRRG	2
	Institutional Representative Local Government	IRLG	6
	Institutional Representative Economic and Others	IRTRO, IRTRP, IRWCC, IRDFO, IRNGO, IRARMY, IRICI, IRTA	13
	Total number of participants (n)		52
Focus Group Discussion (FGD)	Interviewees' sub-set	FGD_BB	6
		FGD_RM	8
	Field-driven students' set	YFG_BB	8
		YFG_RM	8

Secondly, in order to triangulate data collected from individual interviews noted above, a total of four Focus Group Discussions (FGDs) were undertaken with two groups in each study location. The first set of FDGs comprised a subset of interviewees selected on the basis of indicated willingness and availability to participate. All of these informants were over the age of 30 years. A second set of FGDs comprising student informants referred by the Principals of tertiary-level educational institutions from the two case study areas. The inclusion criteria for the second set of FGDs considered age limit (from 18 to 29 years), reference by the Principals, willingness, and availability to participate.

Thirdly, participant observation was undertaken during discrete stages of the primary data collection, and which included participation in a range of formal administrative meetings and workshops in the respective case study areas. In the course of interviews and focus groups, observations were made as a complete observer (Jennings, 2010). However, invited attendances at the meetings and workshop based on 'peripheral membership' enabled the lead author to observe closely how (tourism) decisions are made within the settings (Jennings, 2010). Lastly, supplementary data were obtained from secondary sources, including secondary data sourced from a range of documents, including local books describing the case sites, government reports, policy and planning documents, promotional brochures, meeting minutes, and a number of unsorted documents collected from the offices of institutional informants.

Given the cross-language issues, the interviews and FGDs were transcribed in Bengali and translated in English and communicated to the participants to ensure their viewpoints were captured accurately in the research (Esfehani & Walters, 2018; Johnson, 1997). Accordingly, all participants were offered the opportunity to review the transcriptions of their interviews, 16 interview participants (nine institutional representatives and seven community residents) took up this opportunity as did one focus group participant.

Data analysis

For formal data analysis of interviews and FGD data, open codes were developed to link ideas in line with the underlying meanings in the transcripts (Saldaña, 2016). Following open coding, second cycle codes were developed to form themes, which were subsequently grouped to align with themes based on the literature (base themes) such as social capital, cultural capital, and co-management (Saldaña, 2016). Data reduction was then undertaken to focus on the most relevant points from the transcripts (Miles & Huberman, 1994). In the next phase, transcribed data (ensuing themes and codes) were assembled to inform conclusions, which were later verified in view of the literature and/or available empirical evidence from other sources. Thereafter, the triangulated findings have been reported, following data-derived and base themes. Observational data were analysed using description and constructionist approaches (Marvasti, 2014). While descriptive analysis was targeted at explaining the setting, the constructionist approach involved understanding the social context from the participants' viewpoint. Finally, secondary documents were analysed through a separate content analysis (Sarantakos, 2012).

Findings

The findings highlight different dimensions of social capital, the contribution of cultural capital, within the desired CCSLF co-management structure (Figure 2). Key dimensions of social capital are reported in respect of bonding, bridging, and linking social capital while emphasising cognitive and structural components.

Bonding social capital

We found that the *bonding social capital*, or more particularly the presence of cognitive and structural components within an individual community, is reported as strong in both sub-districts. All community participants (representing different Indigenous groups and migrant Bengali) as well as locally resident institutional representatives, noted that each local group has its own distinct set of values and norms. They each maintain their associations for networking purposes and unifying local group members to protect and uphold their rights. For instance, one Indigenous community participant (22_CRIRM_17) claimed that:

> *We have Tripura Welfare Foundation. This Foundation helps members of our community when they need.*

During the interviews, we also observed that members of any particular community know each other by name. However, our observations reveal that such (structural) bonding is the strongest within 'minority' (whose members are not that large in numbers) communities.

Bridging social capital

Bridging social capital reflects cognitive and structural components across different communities. Our investigation identified and reported two patterns of relationship between Indigenous and Bengali communities, and across Indigenous communities. Cognitive values and cooperative norms across Indigenous communities were reported to be variable and evolving. For example, a number of participants (e.g., 09_CRIBB_06; 24_CRIRM_01_NOTES; 04_IRTRP_01; 10_IRICI_01; 24_IRTA_01) reported that inter-ethnic marriage which is now increasing, was a previously uncommon practice. Such a shift, in turn, is broadening and strengthening internal networks within Indigenous communities. Alternatively, some participants claimed that the shared values and norms of reciprocity between Indigenous and Bengali communities were reported as minimal. Again, these were reportedly associated with the political ideologies of people and relative position of communities in general. In this light the claim of a community participant (17_CRIRM_12) in RS is noteworthy:

> *The minor* (smaller) *communities are not involved in politics … that is why their relationship is quite good internally as well as with Bengalis. You just saw people surrounding my place are Bengali Muslims and they are giving me the Iftar* (Food served during Ramadan to break fasting).

Notwithstanding the above, our overall assessment is that the relationship between Indigenous and Bengali communities (i.e., bridging capital) was poor, with the significant demographic influx from the late 1970s referenced consistently by participants as being a significant causal factor. For example, in an interview, one Indigenous participant (20_CRIRM_14-15) commented that *"No communities will accept forced migration. If you live in Dhaka district and we forcefully transfer 500,000 people from Dinajpur district, how would you feel? We don't deny the natural migration processes"*. Almost all community participants referred to the historical background of Bengali–Indigenous community conflict while often noting that Indigenous communities generally have a good relationship with 'Adi' (older) Bengalis who have been residing in the region for years[6].

Bridging across Indigenous communities has been found to be stronger, although challenged by language diversities, at a broader scale. However, inequitable representation of Indigenous communities in the decision-making body (e.g., Hill District Council's executive committee), coupled with unbalanced distribution of benefits, raised a notion of there being 'minority' and 'larger' ethnic groups. Such an idea is, in turn, reinforced by the relative size of each Indigenous community's network as well as control over (tourism) resources. Power centralisation accruing to the larger Indigenous communities such as 'Chakma' are reported as alienating the minority indigenous communities. In such a situation, the smaller communities (e.g., Lushai, Khumi, and Chak) are reportedly 'losing' their democratic access to decision-making.

> *If you observe closely structural representation at the state-imposed administration- those communities which are minor* (smaller) *such as my one 'Khumi' has no representation at the HDC. I don't believe other communities can represent and communicate our pain and concern.* (06_CRIBB_03)

Similarly, another small Indigenous community participant (15_CRIRM_11) added that "there are differentiations among the Indigenous communities too. I cannot reach my deserved place, even though I have that ability". These divisions are also attributed to wider political outcomes, such as the Peace Treaty:

> *The peace treaty is what? It basically came out of a political game. Santu Larma* (an Indigenous leader who is from the Chakma community and represents the PCJSS) *signed the peace treaty but he does not necessarily represent all the Indigenous communities … Through this process, they gain individually and for their community mostly.* (08_CRIBB_05)

However, it was noted by many 'Indigenous' participants holding roles from across the community and institutions (e.g., 03_CRIBB_02, 10_IRICI_01, 19_IRRG_02, 21_CRIRM_16, 27_IRMOCHTA_01_NOTES) that 'educational qualification(s)' are the critical input required to promote and support meaningful engagement from members of larger groups. On a similar note, Shelly (1992) found that during the Pakistan period (1947–1971) government activities to intensify education were highly appreciated by the larger Indigenous communities, especially 'Chakma'. Accordingly, their (Chakmas) literacy rate reached over 50 percent by 1970, which was significant in comparison with other communities in the CHT (Shelly, 1992). Thus, the 'institutionalised state' of cultural capital supports the building and strengthening of social capital (Bourdieu, 1986).

We found that the strength of bridging social capital varies across the two geographical case areas within CHT. A comparative evaluation reveals that Bandarban has more favourable bridging capital than in Rangamati. For example, one youth focus group participant (YFG_BB) in BS commented on the comparative strength of network and trust issues in general:

> *Here in Bandarban, we are residing in harmony. This is happening as we are having a compromising mentality and not politically involved that much. If you observe in Rangamati, then you will notice frequent strikes and communal conflict-events. Let me give you an example. In our village, we've 52 families. Of these, Tanchangya are 13, Hindus* (Tripura and Bengali) *are 15 and the rest are Bengali Muslims.*

From interviews, a few participants (e.g., 02_CRIBB_01, 15_CRIRM_11, 16_CRBRM_06, 18_IRLG_04) however reported that although an apparent good relationship can be observed at the personal level, that relationship could be of limited use when communal issues arise. Thus, (regional) politics might be seen to outweigh or supersede personal relationships in general. In this connection, one participant from the youth focus group session in RS (YFG_RM) discussed an event that exemplifies how a social issue could be turned into a communal (or, in broader sense, political) conflict. The participant said that:

> *During the last water festival, we were observing that a few Bengali boys splashing water over a group of Indigenous girls with some bad gestures. After few minutes, one of the girls slapped a boy. We were also dissatisfied, but if we slapped instead it could easily be politicised as a conflict between Pahari and Bengali. Here* (in RS), *it is quite easy to give any conflict a communal flavour.*

Throughout the field-based research, the poor trust and relationship issues between the Bengali and Indigenous communities were commonly acknowledged. The central government representatives interpreted the context from a community dominance perspective, wherein they observed that the 'Marmas' (an Indigenous community) work more in collaboration than the 'Chakmas' (another Indigenous community) with the Bengalis. This situation was in-line with the reporting of fewer problems in Bandarban (where there is a Marma majority within the Indigenous communities) than in the other two hill districts (where there is a Chakma majority). Such a view was endorsed by a hotel owner (22_IRTRO_03) who stated that *"... If you want me to employ from Indigenous communities, I'm eager to employ from minor communities and even from Marma community rather than the Chakmas"*. Thus, cognitive and structural components of bridging social capital are critically linked to the (tourism) resource decision-making.

Linking social capital

Our findings indicate mixed outcomes from our assessment of linking social capital. First, a leadership-oriented societal culture is more prevalent in the Indigenous communities than in the Bengali community. For example, one Indigenous community participant (02_CRIBB_01) reported that *"in case of any individual and/or social conflicts, we mitigate by involving Karbaris and Headmen. When they fail to bring any result, we go to the Raja (Circle Chief)"*. Such a view was reinforced by a reported experience of a local government representative (13_IRLG_01):

> *I was the UNO* (Upazila Nirbahi/Executive Officer) *in Alikadam for eight months. During that period, I had to witness complain[ts] everyday regarding various social issues. However, in these eight months I only found a single complain from an Indigenous community member while the rest were from Bengalis. What does it indicate? Don't they have any problem among themselves- surely, they do, but they mitigate at lower-level or within their traditional system.*

This observation also highlights the strength of bonding social capital within Indigenous communities, in general, in the exercise of traditional governance. In contrast, Indigenous and Bengali participants provided a negative evaluation of the role of elected/selected members who represent the decision-making bodies in public organisations for resource allocations such as respective HDC's executive committee. They reported that the leaders are disconnected from the community and from individual concerns. Reflecting on their roles, one community participant (07_CRIBB_04) claimed that "they are [too] busy think[ing] about how much benefit they will get in-person by building a bridge or culvert in this area. They don't have time to think about general people". Participant observation data complement this finding by noting that peoples' attitude towards those who are charged with resource allocation decisions is negative. This is partly due to the fact of the high level of 'power centralisation' within the (formal) public institutions, observations such as this reinforce the requirement for a decentralised and shared system of decision-making, where a true representation of communities can be sought/secured and enacted.

Within the local tourism development context, formal public institutions usually control the facilitation and governance of tourism capitals indicating a top–down approach with government-based management. In the field, we noted the lack of coordination among different stakeholders and the influence of ideological differences between regional political parties and government administration on tourism resource allocation. However, responsible authorities from respective HDCs confirmed that they are in a process of developing tourism plans for respective districts within the CHT to ensure better coordination.

The contribution of cultural capital

The presence of diverse cultural groups has been identified a fundamental resource in CHT to attract tourists. Participants from all stakeholder categories highlighted Indigenous cultures, traditions, and rituals (e.g., water festival and Rajpunnah) as important points of attraction. In general, traditional lifestyles reflecting different exotic Indigenous cultures, languages, attires, dances, values, and traditions emphasise cultural capital and its role in these tourism destinations. Two issues have been broadly reported by the research participants to highlight 'cultural capital': cultural identity, and cultural integrity.

A majority of participants from all stakeholder groups claimed that diverse cultural identities (with specific reference to Indigenous communities) is an asset for tourism development in CHT. A sense of cultural pride has been strongly reported among relatively small Indigenous communities. For example, one such participant (13_CRIBB_10) said that *"I feel disheartened seeing sometimes Chakmas and Marmas are performing our dances (Kheyang) in different programs while introducing themselves as Kheyang. Audiences do not understand this because dresses are different, and they rarely differentiate"*. Participants from tourism-related organisations observed that exhibition of unique (Indigenous) culture could have broadened livelihood opportunities both for community people and businesspeople. However, such possibilities are largely constrained by political sensitivities

> *I used to arrange occasionally Indigenous cultural songs and dances for my resort's guests. Through these, poor people adjacent to my resort could earn some money. Later on, I was notified by the district administration not to do so.* (03_IRTRO_01)

Local government representatives recognised inequitable benefit-sharing to some specific cultural groups might create discontent at a broader community level. Nonetheless, most community interviewees desired a community-level management approach to exhibit their culture. As such, one participant from the Bengali community (01_CRBBB_01) emphasised the necessity of cultural displays which respect local culture and reinforce local community involvement on a 'cultural resource ownership' basis. In this context the role of local government is highlighted as one of facilitator. One participant representing an Indigenous cultural institution commented (10_IRICI_01) that *"Government should support involved communities to exhibit local (Indigenous) cultures and traditions as a tourism product"*.

The linkage to decision-making is identified through need to protect cultural integrity, and accordingly is brought into the discussion with the need of protection and preservation of various cultural elements (e.g., language). One Indigenous community participant (20_CRIRM_14-15) argued that:

> *The cultural aggression is taking a charge over our language. The local places' names are replaced with Bengali names. We feel our identity is in threat. This can't promote wellbeing for community. If the community people get the opportunity to participate in (a) decision-making forum and define their geographical locations by themselves and if they can develop policies to protect their own culture, which is acknowledged by the government bodies, only then community will feel good and cooperate with tourism development.*

The implication of the above statement is clearly linked to institution-sponsored tourism projects that not only dislocated Indigenous communities from their usual place of residence but also affected negatively their emotional attachments to individual culture (Chakma & Chakma, 2016). The accumulated impacts lead to a negative attitude (and, potentially, negative behaviours) towards tourism development. However, the impacts might be seen in a more positive light if the communities were provided with an opportunity to participate meaningfully in tourism decision-making and businesses. By way of example one participant representing a tribal administration (24_IRTA_01) advocated a micro-level management structure with a 'protection-budget' from the government.

Desired co-management structure

Taken together the findings of this research indicate that community empowerment and providing opportunities for community members to make destination-based resource allocations is essential for sustainable livelihood outcomes. Accordingly, a community-based management structure is preferred over conventional top–down government-based management. For example, one participant representing a traditional administration entity (24_IRTA_01) observed that:

> *I personally think that it* (sustainable tourism) *only can be possible through community-led tourism, which is owned by the communities, fundamentally family-based, cooperative-based, village-based, or local association-based.*

Almost all participants acknowledge the cultural and political sensitivities of the CHT and admit the necessity for an alternative source of resource governance to develop tourism sustainably and to ensure community wellbeing. In this regard, participants from focus groups (FGD_BB) formed a consensus to report that "a shared structure for managing tourism resources can be imagined. To make it happen, hill district councils need to take strong standpoint. Investors will invest in joint ownership projects. The government will develop roads and human resources". A critical observation reveals that the participants not only find the necessity of a shared resource governance structure but also talk about the possible roles the involved parties should play. On a similar note, interview participants (e.g., 04_CRBBB_02, 11_CRIBB_08, 21_CRIRM_16, 05_IRWCC_01, 24_IRCCC_01, 25_IRBPC_02) highlight the importance of community-focussed and government-supported structure. This point is well summarised by one community participant (11_CRIBB_08):

> *…a co-management structure can be sought out that will engage community people, associations, and government institutions. Traditional administrative structures as well as formal administrative structures must be considered… this kind of structure must render benefits to the community. In essence, it will be successful if community and local government work together in which government will cooperate not control (the) overall process.*

There is no identifiable co-management structure for tourism development currently functioning within the research setting. However, composition of members in the 'District Law & Order Committee', a general discussion and decision-making forum Chaired by the respective Deputy Commissioner, accommodates both the community residents and formal institutional representatives. We observed more representation by the institutional participants and more importantly lack of engagement of the community residents in this broad forum. Accordingly, the research findings indicate several features towards achieving a functional co-management structure and process for tourism resource decisions, despite the various issues relating to social and cultural capitals across the research settings. Some of these issues included equitable representation of communities (particularly smaller Indigenous communities), increased linkages among stakeholders, cooperation (and competition), and sharing benefits.

In practice, a community-owned approach must guide development, in which the local community acts as a centre of control for destination-based tourism resources, while vertical linkage with the governments is inevitable for the empowerment of local people (to increase their capacities, to provide expertise, to ensure proper monitoring etc.). Both 'political' and 'social' empowerment should facilitate active community participation in tourism resource decisions (Joo et al., 2020). Political empowerment supports actions by community residents who feel empowered as and when they perceive their opinions are considered to shape tourism development (Scheyvens, 1999; Timothy, 2007). Therefore, political empowerment is clearly linked to linking social capital. The two CHT cases examined for this study indicate a low level of political empowerment within the resource decision-making structures. For example, one institutional representative (05_IRWCC_01) mentioned that *"we can observe a broader structure of co-management in our Hill District Council, but this is not functional. We are called only during the meeting; we are unsure whether the administration addresses our concern or not in many respects. We are also unsure about the role of Army administration or CHT Development Board ... "*.

In contrast with political empowerment, social empowerment facilitates shared norms while bringing community residents into a closer network for the sake of tourism development (Scheyvens, 1999). To ensure social empowerment and enhance trust and reciprocity, a co-management structure as suggested by the research participants is assumed to facilitate policy inputs and equitable outcome benefits for community residents. One regional government representative (19_IRRG_02), claimed that *"the problem is army managing the tourism in CHT ... a combined structure can only be successful when it provides benefits to the community members. If the community members get benefits in respect to their relative involvement, the society as a whole will ensure security, which [is] lack[ing] here"*. However, one central government representative (28_IRBTB_02) observed there is considerable difficulty in ensuring equitable benefits from tourism in an ethnically diversified region such as the CHT.

Discussion

The original CCSLF assumed cultural capitals under a broad expression of social capital (Rahman et al., 2020). However, this study has challenged us to consider that the separation of cultural capital from social capital allows social capital to provide a more precise focus in analysing its impact on (tourism) co-management structures and processes. Moreover, such a separation facilitates the scoping of cultural capital influence on social capital to extend the proposed CCSLF. This is particularly meaningful when the relative strength and uniqueness associated with different cultural elements (e.g., self-identity, language, attire, dances) are being brought forward as tourism products.

The findings highlight the role of both cultural and social capitals in guiding co-management approaches. The research findings identify the desirability of community-based co-management, which encourages community self-governance and recognises vertical and horizontal linkages with government and non-government institutions (Pomeroy & Berkes, 1997).

Ethnic diversity, which is inherently a cultural capital element, has a direct impact on the expression of social capital and has implications for co-management structures and processes. The findings confirm that interest group polarisation within an ethnically diverse setting impacts negatively on trust and networks (Mavridis, 2015). Consequently, it challenges structures and process functions across collaborative management platforms. Zurba et al. (2012, p. 1130) found it difficult to develop and implement an effective co-management structure in situations where "... there is a mix of resource users, appropriateness of management mechanisms is perceived differently by mainstream and Indigenous cultures, and there exist multiple centres of authority". These complex factors are evident within the current research setting. To overcome this challenge, encouraging small or micro governance foci that "... roughly fit the ecological boundaries of the problems they are designed to address" (Ostrom, 2005, p. 258). In a situation that is in part challenged by cognitive and structural bridging issues based on differing ethnic orientations, a local level focus is similarly recommended. Such a 'micro-focus' necessarily indicates concentration on a particular community and/or destination. This signifies a 'monocentric' governance approach to avoid complexities in broader-level resource decisions, and to identify coordination units at local level (Nyaupane et al., 2020). At this scale, the promotion and development of bridging capital across multiple communities will result, arguably, in tourism development outcomes with stronger community alignment and connection.

The effectiveness of co-management is discernible with a fewer number of committee members that enable collaborative and participative functioning while facilitating high level of trust and reciprocity among members (Huda, 2008). Given the cultural diversity and strength of bonding (or alternatively weak bridging) social capital as evidenced in the CHT, activity at this level reinforces the focus of co-management at a micro level. Micro level management is also believed to ensure clearer accountability through equitable participation of different community representatives, which is absent from the existing top–down approach to decision-making. Acknowledging the realities of cultural and social capitals in the current research context, a two-tier project-based structure of co-management has been identified appropriate for generating sustainable livelihood outcomes through the application of CCSLF.

At the local (site/community) destination level, a project-based co-management process following a community-based structure has been identified as desirable, while at broader local government level, an extended structure for coordination and integration of co-management policies and activities is recommended. The project level is targeted for a single community and/or destination or few communities having strong structural bridging capital (e.g., two minor Indigenous communities such as Bawm and Lushai). At this scale, a community-based co-management committee (CBCC) structure serves the operational aspects of a tourism co-management process. Broadly, the contextual elements inform an extended co-management structure. This is because this administrative body needs to associate closely with policy formulation to guide general operational issues and shape effective functioning of project level co-management. For the extended structure, the importance of cognitive and structural bridging has been reported in the findings of this research. Thus, building trust among the various actors at the very beginning, and then maintaining that trust, is essential for the effective performance of extended co-management council/committee (ECC).

The implication of operational levels of co-management is one of delegated authority in which central government delegates authority to local government, which in turn delegates authority to project level CBCCs. In order to build social capital at the community level, local government would need to sacrifice some autonomy, support establishing and maintaining both horizontal and vertical linkages, and require reciprocity and responses to ensure optimum 'return' on community involvement. Balanced autonomy and strong horizontal linkages reinforce generalised reciprocity and democratic governance (Warner, 2001). On the one hand, an ECC should develop relevant policies based on the inputs from a CBCC, thus ensuring bottom-up engagement and empowerment. On the other hand, ECC must inform and enforce the policies of the central government via regional policy-making bodies. Additionally, central government must have linkages to the extended co-management. Local government may ask for assistance from the central government directly or via the regional government. It should be noted that an ECC must also ensure balanced community representation and benefit-sharing from the community-based co-management projects. A lack of representativeness hinders the 'social inclusion' requirement of sustainable development and limits the accountability goals of a co-management process (Berkes, 2010; Sachs, 2012). Considering the key features that have emerged from this study, we develop a process function framework (see Figure 4) for co-management of tourism resources (in CHT). Importantly, it draws attention to the necessary social processes that need to be developed and supported.

In Figure 4, 'context' is associated with the nature of resources to be managed whereas 'conditions' refer to the surrounding circumstances (e.g., trust, dynamics of social networks, past record of conflict) that impact upon the operation of co-management. An important finding is that both the context and conditions make it difficult to establish or operationalise the existing co-management within CHT. In practice, current co-management procedures largely fail to consider the prevailing unique context and complex operating conditions in the CHT. Thus, they remain contestable in their initiation and implementation.

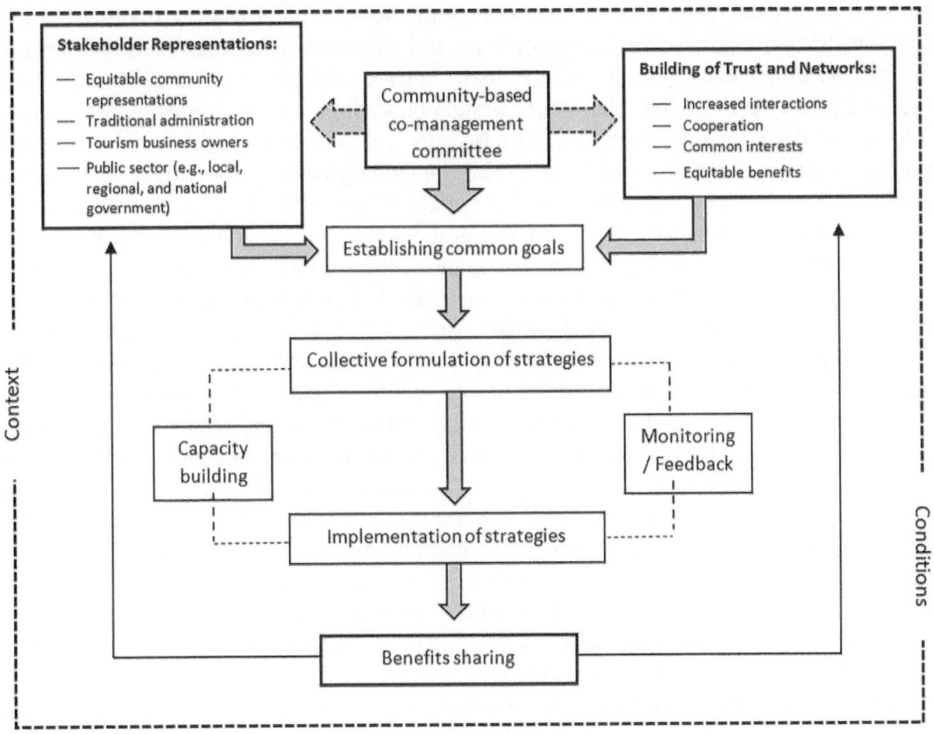

Figure 4. Process functions in support of co-management.

A set of common goals should ideally direct the formulation of appropriate strategies as well as the execution of those strategies. Strategies necessarily consider if, and how, the co-management model can accommodate discord and agreements among actors. Strategy formulation and implementation stages are linked with an iterative monitoring (feedback loop) to enable continued investment in trust and relationship-building (Sessin-Dilascio et al., 2015). A feedback loop allows the co-management process to respond continuously to problems while offering flexibly in process and capacity. Such an iterative approach aligns with social learning and resilience building models. Finally, the benefits derived through this process must be shared among stakeholders on a mutually agreed basis, which has been argued to increase trust and build operational linkages.

To summarise, a destination-oriented and community-focussed governance structure must balance broader stakeholder involvement while enabling and giving community voice, collaborative policy and planning, and sharing tourism benefits equitably among stakeholders. This is particularly important to progress community livelihood outcomes such as community wellbeing through sustainable tourism development (Rahman, 2019).

Conclusion and implications

This study concludes that social and cultural capitals have a considerable impact upon the functioning of co-management of tourism resources. To reach such a conclusion, the study adopted and tested the CCSLF (Rahman et al., 2020) within a 'contested' destination setting of the Chittagong Hill Tracts (CHT) of Bangladesh. The CHT exhibits diverse cultural identities (e.g., distinct Indigenous groups and broader Bengali community) and historical political distrust which have combined to create communal conflict. This has resulted, ultimately, in the militarisation of the region. Set against this, regions of high tourism potential commonly attract high levels of central interest, and at times control. Within this context, bonding social capitals (i.e., within a particular group) have been reported stronger and more favourably than bridging (i.e., across groups) and linking capitals. Weak or poor trust and networks have been identified and underpinned by two dimensions of relationships: firstly, between Bengali and Indigenous communities; and secondly, between smaller and larger Indigenous communities. These in turn identify the need for a two-tier co-management: one at a broader local government level involving local government agencies and other relevant stakeholders including community representatives from the lower-order structures, paralleled by a second structure that must rest at a particular community and/or destination level with a focus on implementation and outcomes. The community-level focus will utilise the strength of bonding capital or favourable structural bridging capital, as reported across minority Indigenous communities. However, the extended local government level requires building on trust and networks across communities, as well as linking capital, to negotiate issues with the regional and central governments. Social and cultural capitals influence the process functions of co-management, and potentially inform structures as noted above.

This study extends the theoretical underpinning of the CCSLF and aligns with the fourth typology of Carlsson and Berkes (2005, p. 68) that positions co-management as a 'community-nested system'. Nonetheless, considering the contextual realities of an ethnically diverse region where land ownership remains contested and politically fraught coupled with prevailing rigid top–down resource governance structures, a community-nested system would be practically impossible to implement. Given these challenges, Carlsson and Berkes (2005) third typology framing a state-nested system, could be useful, provided central government empowers local government and local government follows the process of devolution of resource governance to the destination communities of concern (Berkes, 2010).

It is of note that the current research emphasises the influence of social capital on tourism resource governance within a sustainable livelihood framework, rather than concentrating on Indigenous tourism where Indigenous cultures and practices are the primary foci of development. However, Indigenous communities provide an important frame within the current research context and have led to the separation of cultural capital in the CCSLF. Subsequently, the findings of this study modify the proposed CCSLF of Rahman et al. (2020) by separating *cultural* capital from *social* capital. Such a separation draws attention to the contextual dimensions in tourism or destination development research. Future studies could focus on indigeneity as a distinctive feature and be correspondingly broadened to employ Indigenous methods.

A significant implication of this research is that the interrogation of the CCSLF within this CHT context reveals a pathway towards improved sustainable livelihood outcomes for destination communities. Among the various capitals presented in Sustainable Livelihood contexts, the significance of social and cultural capitals as a key enabling dimension is inescapable. In essence, cultural capital can be seen to filter and redirect bridging and bonding mechanisms at a local-, or micro-, level for improved destination community outcomes via the community resource governance co-management. This study and local understandings appeal as a compelling starting point for practitioners. Specifically, the findings are particularly useful for policymakers in Bangladesh, especially at this time when the local governments are considering developing tourism plans for coordinated efforts towards tourism development in CHT region. Notwithstanding the idiosyncrasies of the study area and its unique context, which might limit its generalisability, the co-management debates and resultant models are worthy of further testing and debate in the context of other developing and developed resource management systems.

Notes

1. Or '*Parbatya Chattagram*' in Bengali, an Indo-Aryan language.
2. The word 'Bengali' here indicates people from the broader community in Bangladesh, who are living predominantly throughout the country.
3. The word 'Indigenous' here indicates peoples from the eleven ethnic groups who reside in the CHT.
4. A Upazila is a 'sub-district' administrative region of local government in Bangladesh.

5. A Bengali word indicating 'central point' or 'headquarter'.
6. 'Adi' here, is interpreted to indicate Bengali residents in the region since the British colonial regime.

Acknowledgements

This paper reports the findings from the lead author's Doctoral dissertation. Doctoral study of the lead author was funded by 'New Zealand Aid' and awarded as 'NZ Commonwealth Scholarship 2014'. The authors would like to thank New Zealand Government and Lincoln University (Postgraduate School) for allocating fund to complete this study.

Disclosure statement

No potential conflict of interest was reported by the authors.

ORCID

Muhammad Shoeb-Ur-Rahman http://orcid.org/0000-0003-2667-4794
David Simmons http://orcid.org/0000-0003-4488-1877

References

Adnan, S. (2004). *Migration land alienation and ethnic conflict: Causes of poverty in the Chittagong hill tracts of Bangladesh*. Research & Advisory Services.

Andriotis, K., Stylidis, D., & Weidenfeld, A. (Eds.). (2019). *Tourism policy and planning implementation*. Routledge. https://doi.org/10.4324/9781315162928

Ashley, C. (2000). *The impacts of tourism on rural livelihoods: Namibia's experience*. ODI Working Paper No. 128. Overseas Development Institute (ODI).

Bangladesh Bureau of Statistics. (2013a). *District Statistics 2011: Bandarban*. Statistics and Informatics Division, Ministry of Planning, Government of the People's Republic of Bangladesh.

Bangladesh Bureau of Statistics. (2013b). *District Statistics 2011: Khagrachari*. Statistics and Informatics Division, Ministry of Planning, Government of the People's Republic of Bangladesh.

Bangladesh Bureau of Statistics. (2013c). *District Statistics 2011: Rangamati*. Statistics and Informatics Division, Ministry of Planning, Government of the People's Republic of Bangladesh.

Barnes-Mauthe, M., Arita, S., Allen, S. D., Gray, S. A., & Leung, P. (2013). The influence of ethnic diversity on social network structure in a common-pool resource system: Implications for collaborative management. *Ecology and Society*, *18*(1), 23. https://doi.org/10.5751/ES-05295-180123

Berkes, F. (2010). Devolution of environment and resources governance: Trends and future. *Environmental Conservation*, *37*(4), 489–500. https://doi.org/10.1017/S037689291000072X

Bodin, Ö., & Prell, C. (Eds.). (2011). *Social networks and natural resource management: Uncovering the social fabric of environmental governance*. Cambridge University Press. https://doi.org/10.1017/CBO9780511894985

Bosworth, G., & Turner, R. (2018). Interrogating the meaning of a rural business through a rural capitals framework. *Journal of Rural Studies*, *60*, 1–10. https://doi.org/10.1016/j.jrurstud.2018.02.002

Bourdieu, P. (1986). Forms of capital. In J. Richardson (Ed.), *Handbook of theory and research for the sociology of education* (pp. 241–258). Greenwood.

Çakir, O., Evern, S., Tören, E., & Kozak, N. (2018). Utilizing the sustainable livelihoods approach to evaluate tourism development from the rural host communities'point of view: The case of Cappadocia (Turkey). *GeoJournal of Tourism and Geosites*, *21*(1), 7–25.

Carlsson, L., & Berkes, F. (2005). Co-management: Concepts and methodological implications. *Journal of Environmental Management*, *75*(1), 65–76. https://doi.org/10.1016/j.jenvman.2004.11.008

Carr, A., Ruhanen, L., & Whitford, M. (2016). Indigenous peoples and tourism: The challenges and opportunities for sustainable tourism. *Journal of Sustainable Tourism*, *24*(8–9), 1067–1079. https://doi.org/10.1080/09669582.2016.1206112

Castiglione, D., Van Deth, J. W., & Wolleb, G. (Eds.). (2008). *The handbook of social capital*. Oxford University Press Inc.

Castle, E. N. (1998). A conceptual framework for the study of rural places. *American Journal of Agricultural Economics*, *80*(3), 621–631. https://doi.org/10.2307/1244569

Castro, A. P., & Nielsen, E. (2001). Indigenous people and co-management: Implications for conflict management. *Environmental Science & Policy*, *4*(4–5), 229–239. https://doi.org/10.1016/S1462-9011(01)00022-3

Chakma, M. K., & Chakma, S. (2016). Adivasi oddhushito onchole parjatan o unnoyon: Adibasider ongshidaritto o sorkarer vumika [Tourism and development in indigenous inhabited region: Indigenous partnership and the role of government]. In A. Chakma (Ed.), *Khoi Sha: Boishuk-Sangrai-Bizu-Bishu-Bihu Collections'16* (pp. 28–44).Jum Aesthetic Council (JAC).

Chowdhury, K. (2008). Politics of identities and resources in Chittagong Hill Tracts, Bangladesh: Ethnonationalism and/or indigenous identity. *Asian Journal of Social Science*, *36*(1), 57–78. https://doi.org/10.1163/156853108X267567

Cohen, N., & Arieli, T. (2011). Field research in conflict environments: Methodological challenges and snowball sampling. *Journal of Peace Research*, *48*(4), 423–435. https://doi.org/10.1177/0022343311405698

Dalziel, P., Saunders, C., Fyfe, R., & Newton, B. (2009). *Sustainable development and cultural capital* (Vol. 5). Official Statistics Research Series, Statistics New Zealand.

DFID. (1999). *Sustainable livelihoods guidance sheets*. Department for International Development (DFID).

Dhamai, B. M. (2014). An overview of indigenous peoples in Bangladesh. In M. S. Chowdhury (Ed.), *Survival under threat: Human rights situation of indigenous peoples in Bangladesh* (pp. 10–26). Asia Indigenous Peoples Pact (AIPP).

Esfehani, M. H., & Walters, T. (2018). Lost in translation? Cross-language thematic analysis in tourism and hospitality research. *International Journal of Contemporary Hospitality Management*, *30*(11), 3158–3174. https://doi.org/10.1108/IJCHM-10-2017-0701

Fennell, D., Plummer, R., & Marschke, M. (2008). Is adaptive co-management ethical? *Journal of Environmental Management*, *88*(1), 62–75. https://doi.org/10.1016/j.jenvman.2007.01.020

Flora, C. B., Flora, J., & Fey, S. (2004). *Rural communities: Legacy and change* (2nd ed.). Westview Press.

Gray, R. A. (1994). Genocide in the Chittagong hill tracts of Bangladesh. *Reference Services Review*, *22*(4), 59–79. https://doi.org/10.1108/eb049231

Huda, K. S. (2008). *Co-management of protected areas in Bangladesh: A strategy for establishing an institutional framework*. Nishorgo Support Project, Bangladesh Forest Department (BFD), Ministry of Environment and Forest, Government of Bangladesh.

Islam, M. W., Ruhanen, L., & Ritchie, B. W. (2018). Adaptive co-management: A novel approach to tourism destination governance? *Journal of Hospitality and Tourism Management*, *37*, 97–106. https://doi.org/10.1016/j.jhtm.2017.10.009

Jennings, G. (2010). *Tourism research* (2nd ed.). John Wiley and sons.

Johnson, R. B. (1997). Examining the validity structure of qualitative research. *Education*, *118*(2), 282–292.

Joo, D., Woosnam, K. M., Strzelecka, M., & Boley, B. B. (2020). Knowledge, empowerment, and action: Testing the empowerment theory in a tourism context. *Journal of Sustainable Tourism*, *28*(1), 69–85. https://doi.org/10.1080/09669582.2019.1675673

Karlsson, S. (2005). The social and the cultural capital of a place and their influence on the production of tourism – A theoretical reflection based on an illustrative case study. *Scandinavian Journal of Hospitality and Tourism*, *5*(2), 102–115. https://doi.org/10.1080/15022250510014400

Laplaza, A., Tanaya, I. G. L. P. & Suwardji, (2017). Adaptive comanagement in developing world contexts: A systematic review of adaptive comanagement in Nusa Tenggara Barat, Indonesia. *Climate Risk Management*, *17*, 64–77.

Macbeth, J., Carson, D., & Northcote, J. (2004). Social capital, tourism and regional development: SPCC as a basis for innovation and sustainability. *Current Issues in Tourism*, *7*(6), 502–522. https://doi.org/10.1080/1368350050408668200

Marvasti, A. (2014). Analysing observations. In U. Flick (Ed.), *The sage handbook of qualitative data analysis* (pp. 354–366). Sage.

Mavridis, D. (2015). Ethnic diversity and social capital in Indonesia. *World Development*, *67*, 376–395. https://doi.org/10.1016/j.worlddev.2014.10.028

Miles, M. B., & Huberman, A. M. (1994). *Qualitative data analysis: An expanded sourcebook* (2nd ed.). Sage.

Ministry of Chittagong Hill Tracts Affairs. (2020, July 12). *CHT at a glance*. http://mochta.portal.gov.bd/site/biography/ce3e959e-e77d-4f17-b610-185a29a8cf00

Mitchell, R. K., Agle, B. R., & Wood, D. J. (1997). Toward a theory of stakeholder identification and salience: Defining the principle of who and what really counts. *Academy of Management Review*, *22*(4), 853–886. https://doi.org/10.5465/amr.1997.9711022105

Moscardo, G. (2011). Exploring social representations of tourism planning: Issues for governance. *Journal of Sustainable Tourism*, *19*(4–5), 423–436. https://doi.org/10.1080/09669582.2011.558625

Moscardo, G. (2014). Tourism and community leadership in rural regions: Linking mobility, entrepreneurship, tourism development and community well-being. *Tourism Planning & Development*, *11*(3), 354–370. https://doi.org/10.1080/21568316.2014.890129

Musavengane, R., & Kloppers, R. (2020). Social capital: An investment towards community resilience in the collaborative natural resources management of community-based tourism schemes. *Tourism Management Perspectives, 34*, 100654. https://doi.org/10.1016/j.tmp.2020.100654

Nunkoo, R. (2017). Governance and sustainable tourism: What is the role of trust, power and social capital. ? *Journal of Destination Marketing & Management, 6*(4), 277–285. https://doi.org/10.1016/j.jdmm.2017.10.003

Nyaupane, G. P., Poudel, S., & York, A. (2020). Governance of protected areas: An institutional analysis of conservation, community livelihood, and tourism outcomes. *Journal of Sustainable Tourism*, https://doi.org/10.1080/09669582.2020.1858089.

OECD. (2001). *The well-being of nations: The role of human and social capital.* OECD.

Ooi, N., Laing, J., & Mair, J. (2015). Social capital as a heuristic device to explore sociocultural sustainability: A case study of mountain resort tourism in the community of Steamboat Springs, Colorado, USA. *Journal of Sustainable Tourism, 23*(3), 417–436. https://doi.org/10.1080/09669582.2014.957211

Ostrom, E. (2005). *Understanding institutional diversity.* Princeton University Press.

Patton, M. Q. (2002). *Qualitative research and evaluation methods* (3rd ed.). Sage Publications.

Plummer, R., & FitzGibbon, J. (2004). Some observations on the terminology in co-operative environmental management. *Journal of Environmental Management, 70*(1), 63–72. https://doi.org/10.1016/j.jenvman.2003.10.005

Plummer, R., & FitzGibbon, J. (2006). People matter: The importance of social capital in the co-management of natural resources. *Natural Resources Forum, 30*(1), 51–62. https://doi.org/10.1111/j.1477-8947.2006.00157.x

Plummer, R., Kulczycki, C., & Stacey, C. (2006). How are we working together? A framework to assess collaborative arrangements in nature-based tourism. *Current Issues in Tourism, 9*(6), 499–515. https://doi.org/10.2167/cit284.0

Pomeroy, R. S., & Berkes, F. (1997). Two to tango: The role of government in fisheries co-management. *Marine Policy, 21*(5), 465–480. https://doi.org/10.1016/S0308-597X(97)00017-1

Putnam, R. D. (2000). *Bowling alone: The collapse and revival of American community.* Simon and Schuster.

Rahman, M. S-. U-., Simmons, D. G., Shone, M., & Ratna, N. (2020). Co-management of capitals for community well-being and sustainable tourism development: A conceptual framework. *Tourism Planning & Development, 17*(2), 225–236. https://doi.org/10.1080/21568316.2019.1600161

Rahman, M. S-. U-. (2019). *Co-management for Tourism Development and Community Wellbeing: The Case of Hill Tracts, Bangladesh* [Unpublished PhD thesis]. Lincoln University.

Royhan, S. A. (2016). *The Chittagong hill tracts.* Kalikolom Prokashona.

Sachs, J. D. (2012). From millennium development goals to sustainable development goals. *The Lancet, 379*(9832), 2206–2211. https://doi.org/10.1016/S0140-6736(12)60685-0

Saldaña, J. (2016). *The coding manual for qualitative researchers.* SAGE.

Sarantakos, S. (2012). *Social research* (4th ed.). Red Globe Press.

Scheyvens, R. (1999). Ecotourism and the empowerment of local communities. *Tourism Management, 20*(2), 245–249. https://doi.org/10.1016/S0261-5177(98)00069-7

Schmidt, J., & Uriely, N. (2018). Tourism development and the empowerment of local communities: The case of Mitzpe Ramon, a peripheral town in the Israeli Negev Desert. *Journal of Sustainable Tourism, 27*(6), 805–825. https://doi.org/10.1080/09669582.2018.1515952

Scoones, I. (1998). *Sustainable rural livelihoods: A framework for analysis* (No. IDS Working Paper 72). Institute of Development Studies. https://opendocs.ids.ac.uk/opendocs/bitstream/handle/123456789/3390/Wp72.pdf?seque

Sessin-Dilascio, K., Prager, K., Irvine, K. N., & de Almeida Sinisgalli, P. A. (2015). The dynamics of co-management and social capital in protected area management—The Cardoso island state park in Brazil. *World Development, 67*, 475–489. https://doi.org/10.1016/j.worlddev.2014.11.004

Shamsuddoha, M., Alamgir, M., & Nasir, T. (2011). Cultural tourism: Bangladesh tribal areas perspective. *Revista de Turism-Studii si Cercetari in Turism, 12*, 28–32.

Shelly, M. R. (Ed.) (1992). *The Chittagong hill tracts of Bangladesh: The untold story.* Centre for Development Research, Bangladesh.

Shen, F., Hughey, K. F. D., & Simmons, D. G. (2008). Connecting the sustainable livelihoods approach and tourism: A review of the literature. *Journal of Hospitality and Tourism Management, 15*(01), 19–31. https://doi.org/10.1375/jhtm.15.1.19

Simmons, D. G. (1994). Community participation in tourism planning. *Tourism Management, 15*(2), 98–108. https://doi.org/10.1016/0261-5177(94)90003-5

Stone, M. T., & Nyaupane, G. P. (2016). Protected areas, tourism and community livelihoods linkages: A comprehensive analysis approach. *Journal of Sustainable Tourism, 24*(5), 673–693. https://doi.org/10.1080/09669582.2015.1072207

Stone, M. T., & Nyaupane, G. P. (2017). Ecotourism influence on community needs and the functions of protected areas: A systems thinking approach. *Journal of Ecotourism, 16*(3), 222–246. https://doi.org/10.1080/14724049.2016.1221959

Stone, M. T., & Nyaupane, G. P. (2018). Protected areas, wildlife-based community tourism and community livelihoods dynamics: Spiraling up and down of community capitals. *Journal of Sustainable Tourism*, *26*(2), 307–324. https://doi.org/10.1080/09669582.2017.1349774

Tao, T. C. H., & Wall, G. (2009). Tourism as a sustainable livelihood strategy. *Tourism Management*, *30*(1), 90–98. https://doi.org/10.1016/j.tourman.2008.03.009

Timothy, D. J. (2007). Empowerment and stakeholder participation in tourism destination communities. In A. Church & T. Coles (Eds.), *Tourism, power and space* (pp. 199–216). Routledge.

Titeca, K., & Vervisch, T. (2008). The dynamics of social capital and community associations in Uganda: Linking capital and its consequences. *World Development*, *36*(11), 2205–2222. https://doi.org/10.1016/j.worlddev.2007.10.021

Uddin, M. N., Hossain, M. M., Chen, Y., Siriwong, W., & Boonyanuphap, J. (2019). Stakeholders' perception on indigenous community-based management of village common forests in Chittagong hill tracts. *Forest Policy and Economics*, *100*, 102–112. https://doi.org/10.1016/j.forpol.2018.12.005

UNICEF Bangladesh. (2019). *Many tracts one community: UNICEF's work in the Chittagong Hill Tracts*. https://www.unicef.org/bangladesh/sites/unicef.org.bangladesh/files/2019-09/CHT-report-LR-August20-website.pdf

Wall, G. (1999). Partnerships involving indigenous peoples in the management of heritage sites. In M. Robinson & P. Boniface (Eds.), *Tourism and cultural conflicts* (pp. 269–286). CABI.

Wang, Y., & Bramwell, B. (2012). Heritage protection and tourism development priorities in Hangzhou, China: A political economy and governance perspective. *Tourism Management*, *33*(4), 988–998. https://doi.org/10.1016/j.tourman.2011.10.010

Warner, M. (2001). Building social capital: The role of local government. *The Journal of Socio-Economics*, *30*(2), 187–192. https://doi.org/10.1016/S1053-5357(00)00105-0

Woolcock, M. (2000). Why should we care about social capital? *Canberra Bulletin of Public Administration*, *98*, 17–19.

Yin, R. (2014). *Case study research: Design and methods* (5th ed.). Sage.

Zurba, M., Ross, H., Izurieta, A., Rist, P., Bock, E., & Berkes, F. (2012). Building co-management as a process: Problem solving through partnerships in aboriginal country, Australia. *Environmental Management*, *49*(6), 1130–1142. https://doi.org/10.1007/s00267-012-9845-2

Governance of protected areas: an institutional analysis of conservation, community livelihood, and tourism outcomes

Gyan P. Nyaupane, Surya Poudel and Abigail York

ABSTRACT

Protected areas are managed for multiple, often competing, goals including biodiversity conservation, community livelihoods, and tourism. However, there is a lack of clear evidence on which governance approaches are more effective in governing protected areas to achieve these goals. This paper sheds light on the effectiveness of different protected area governance approaches using Ostrom's Institutional Analysis and Development (IAD) framework. The research was conducted in three protected areas of Nepal: Chitwan National Park, the Annapurna Conservation Area, and the Kanchenjunga Conservation Area, which represent state-community, NGO, and community-managed protected areas, respectively. Data were collected through interviews, supplemented by published and unpublished documents, and field observations. Results revealed that the outcomes vary across the protected areas primarily because of their governance approaches and local contexts. Due to rigid monocentric and simple polycentric approaches, the PAs failed to adapt to the changing social-ecological systems. Based on the findings, a multi-layered polycentric adaptive co-management approach has been proposed. The study also developed a multi-goal IAD model to assess PA governance. Additional theoretical, methodological, and managerial implications with respect to protected areas governance are also discussed.

Introduction

The modern day protected area (PA) management paradigm began with the establishment of Yellowstone National Park in the United States in 1872. Over time, this paradigm has seen expansion and evolution across the globe. There were only 154 sites extended over 1,13,635 km^2 in 1911, which increased to 20,054 sites covering 28,49,048 km^2 in 1971, and 258,608 sites with an area of 4,76,65,242 km^2 in 2020 (IUCN, 2020; UNEP, 2016). Despite the growth, PAs around the world face enormous challenges, such as human encroachment, poaching, land development, inadequate resources for management, and institutional problems including inappropriate policies, ownership, and governance (Hockings, 2003). Extensive literature published on PA planning and management shows both successful and failed cases of individual PAs (Kingsford et al.,

2011; Parker & Thapa, 2011; Sebele, 2010; Tumusiime & Vedeld, 2012). Their success and failure, however, cannot be measured in isolation, as the interactions between various actors, their power relationships, and wider policies have a bigger influence on how PAs function; hence, governance is a key factor. However, there is a dearth of research oriented toward assessing the performance of contemporary PA governance and developing new governance approaches to improve the outcomes (Hayes, 2006).

PAs vary in terms of ownership of the land, management authority, functions, and the use of resources. They are governed and managed by many actors including state, private individuals and companies, local communities, NGOs, and environmental organizations (Dudley, 2008). The PA functions and the use of resources vary from strict protection to multiple-use of natural resources, including scientific research, recreation and tourism, and community livelihoods (Gardner et al., 2013; Hassanali, 2013; Pfueller et al., 2011; Tumusiime & Vedeld, 2012; van Wilgen & Biggs, 2011).

PA goals including protecting biodiversity conservation, improving community livelihoods, and managing sustainable tourism may be incompatible with each other leading managers the difficult task of balancing of human and economic development and conservation of biological diversity priorities (Borrini-Feyerabend & Hill, 2015). Governance is fundamental to setting priorities and obtaining the public and political support for protection, development, and survival of PA systems (Lockwood, 2010), and it is about who holds the power, authority, and responsibility on defining and implementing goals and objectives (Borrini-Feyerabend & Hill, 2015). For the purpose of this study, governance is operationalized as processes or entities through which the institutions are established and implemented (McGinnis, 2011) with institutions defined as the rules, norms, and shared strategies not organizations (Ostrom, 2010b). In addition to rules, we also focus on management and the implementation of rules set by the organizations (Sen, 2001).

Due to the complex nature of PAs with interdependencies across multiple levels and scales (Gruby & Basurto, 2013; van Wilgen & Biggs, 2011), management of PAs should incorporate their multi-scaled nature (Gruby & Basurto, 2013; Lockwood, 2010; van Wilgen & Biggs, 2011), as less than 4% of all reported PAs are co-managed by more than one entity (UNEP-WCMC & IUCN, 2018). Using an Institutional Analysis and Development (IAD) framework approach (Ostrom, 2005) that explicitly incorporates multiple levels and scales, we explore the biophysical, socioeconomic, and institutional context, and evaluate the PA outcomes in a case study of three PAs in Nepal: Chitwan National Park, the Annapurna Conservation Area, and the Kanchenjunga Conservation Area. These PAs provide an opportunity to explore the role of institutions, particularly the nature of polycentric governance structures including co-management by the government and local communities, management by a non-governmental organization with community input, and local community management.

Our contributions include the development of an IAD model, with associated evaluative criteria, to explicitly explore the success of biodiversity conservation, sustainable tourism, and community livelihood goals. The empirical results demonstrate that polycentric governance can fall short without clear mechanisms for adaptive management and community-based and non-government approaches are inadequate to address all three goals because of intransigence of existing power relationships and lack of governance capacity. Building upon these results we develop a new conceptual framework of polycentric governance of protected areas that positions approaches based on their livelihood and protection focus, level of governance, and adaptive management. Our empirical test of protected area governance in Nepal advances institutional theory through union of insights from both adaptive co-management and polycentric governance and highlights the need for explicit channels for feedbacks between entities to facilitate adaptive co-management, accountability of powerful actors, and role for diverse types of entities particularly when seeking to advance multiple, sometimes competing, goals for protected areas.

Literature review

Devising governance models of PAs to achieve anticipated outcomes is a global challenge (Bown et al., 2013). The initial intent to establish PAs was to save the natural landscape from privatization and exploitative use of resources (Nash, 2014). The concept of PAs persisted for almost a century in the western world through the fence-and-fine approach, also known as the Yellowstone model (Kitamura & Clapp, 2013). Although the governance of PAs is evolving, about 90% of all reported PAs are still managed by state agencies, 6% are managed by private parties, and less than 1% are managed by communities (UNEP-WCMC & IUCN, 2018). This type of approach in which a single entity (states, private parties, or communities) assumes the authority, responsibility, and accountability for managing PAs is known as the monocentric approach (Dudley, 2008). Gardner et al. (2013) noted that the current PA management approaches are not sufficiently flexible, dynamic, or iterative despite the increasing consensus among scholars and practitioners for adaptive management. In addition, the experience of almost 150 years in PA management, which has been treated as the blueprint approach, resists change and adaptation (Ostrom, 2010a; Ostrom & Cox, 2010). The fact is that PAs are simultaneously used by various stakeholder groups, and the interests of these groups may be similar, different, or independent (Kitamura & Clapp, 2013; Pfueller et al., 2011). In addition, PA issues, goals, and priorities are evolving with the changing social-ecological systems. However, the current approach of uniformity in PA governance does not take into account the diversity of needs among stakeholders and dynamic social-ecological processes; thus, PA governance should leverage complementary environmental governance insights, especially the large literature on governance of common pool resources.

Common pool resources

Common pool resources include resources such as forests, pasturelands, irrigation systems, and fisheries, which are difficult to exclude others from using but the consumption of resources by one person diminishes the availability to others (Ostrom, 2005). The empirical research shows that the type of rules and governance affects the outcomes of common pool resources (Ostrom et al., 2006). Tang's (2006) evaluation of the effectiveness of farmer-owned and government-owned irrigation systems shows that the former system performs better. There are three main underlying reasons for positive outcomes. First, rules of appropriations are more likely to be followed when the rules are collectively developed by the communities themselves than those imposed and handed down from the distant government agencies. Second, local conditions are better understood by the locals and they are more likely to solve problems being adaptive and flexible, whereas the government and distant bureaucrats are less knowledgeable about the local context to resolve the issues, and finally, social capitals built within communities are more effective to develop and enforce institutional rules (Sullivan et al., 2017; Tang, 2006). Another empirical study shows that community forest governance outcomes, such as reduction of resource degradation and conservation, depend on institutional design including effective monitoring of rules and sanctioning of violators who break rules (Agarwal, 2006). Relevant for PAs, the common pool research literature demonstrates the need to connect institutions to outcomes (panaceas do not work) and there is a wealth of diverse institutional structures developed or co-developed by resource users that can be more successful at balancing diverse priorities than monocentric, state institutions (Ostrom et al., 2007).

Polycentricity

Moving beyond a dichotomy of government versus community-based management, polycentric refers to multiple centers of semiautonomous decision making which are independent of each other (Carlisle & Gruby, 2019; McGinnis, 2011). A polycentric system is characterized by a governing system in which multi-sector institutions (public, private, community) belonging to different levels (such as local, provincial, national) collaborate for different functions (appropriation, provision, monitoring, enforcement, conflict resolutions) (McGinnis, 2011; McGinnis & Ostrom, 2012). It has several independent but nested decision-making centers that operate and function coherently under an overarching system. The interplay of actors across vertical and horizontal levels is deemed necessary in polycentricism (Tumusiime & Vedeld, 2012). Scholars argue that excessive emphasis on a single form of governance (i.e. state, market, or communal) could fail to recognize local contexts (Ostrom & Cox, 2010). There are many situations involving small and medium-sized, common-pool resources that are best suited for decentralized bottom-up institutional arrangements, whereas for large-scale global commons, institutional arrangement can (and perhaps should) be more complex (Nagendra & Ostrom, 2012; Pennington, 2013).

In the context of PA governance, polycentric governance is useful because it enhances multiple organization's effectiveness by creating a clear boundary of responsibilities and authorities in decision making (Carlisle & Gruby, 2019; Nagendra & Ostrom, 2012; Ostrom, 2002). In addition, it allows for several independent trial-and-error exercises in policy-making that not only foster the adaptability of a system but also reduce the chance of whole system failure (Ostrom, 1990). Polycentric governance system is based on the complex-adaptive systems theory that exhibits greater adaptive capacity (Carlisle & Gruby, 2019). The compatibility of polycentrism with the emergence of an adaptive co-management paradigm further enhances its applicability in the management of common-pool resources (Ostrom & Cox, 2010; Plummer & Fennell, 2009).

Adaptive co-management

Adaptive co-management is also an emerging governance approach for managing common-pool resources that blends the learning function of adaptive management and linking function of co-management (Kingsford et al., 2011; Plummer et al., 2012; Schultz et al., 2011). Adaptive co-management is a flexible, collaborative, and learning-based governance system that relies on teamwork among several stakeholders operating at multiple levels (Bixler, 2014). Adaptive co-management includes iterative decision making processes wherein management strategies are changed or adjusted with the availability of new information (van Wilgen & Biggs, 2011). Nagendra and Ostrom (2012) noted that a polycentric, multilevel, and decentralized arrangement emboldens adaptive co-management by allowing trial-and-error learning. Both polycentric governance and adaptive co-management are guided by the principles of involvement of diverse stakeholders, sharing of power and responsibilities, and linkages between two or more spatial scales (Bixler, 2014). The adaptive governance of common-pool resources requires management strategies to be flexible enough to respond to changing social, economic, and ecological environments, but how it fits with traditional forms of PA governance warrants further research.

Research framework

Organizations such as the World Commission on Protected Areas (WCPA) have developed frameworks to assess the effectiveness of PA management (Hockings, 2003). Although these frameworks have alluded to complex managerial issues, there has been very limited effort to analyze institutions and governance of PAs. We adopt and adapt the Institutional Analysis and Development (IAD) framework, originally developed by Nobel Laureate Elinor Ostrom and associates; IAD enables exploration of complex governance questions, such as how institutions affect decision-making of various actors at different scales, how institutions change over time, and how various institutions or elements of institutions may affect each other shaping outcomes (McGinnis, 2011). Instead of focusing on operations and management of organizations, institutional analysis examines who makes rules and policies, how rules are made, how rules address

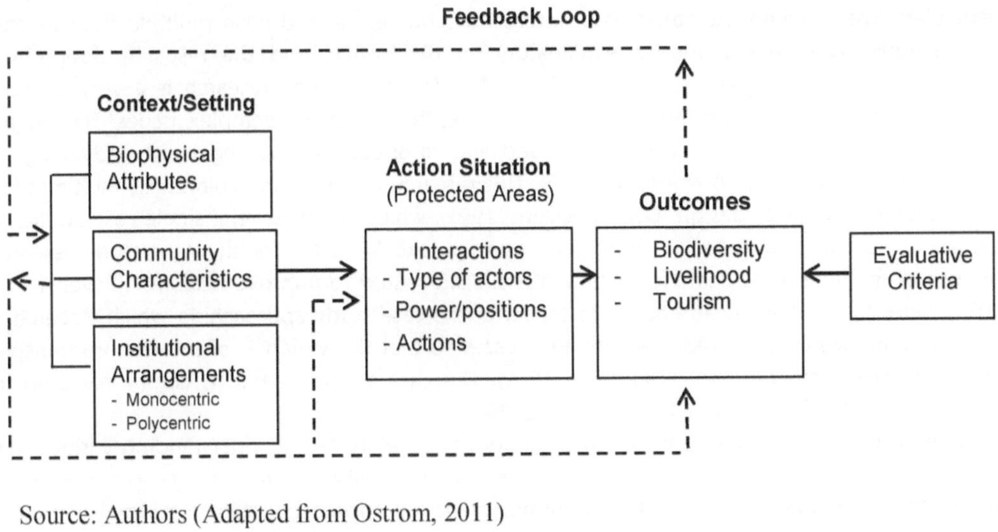

Figure 1. Multi-goal Institutional Analysis and Development (IAD) Model for Protected Area Governance Evaluation. Source: Authors (Adapted from Ostrom, 2011).

the issues, how rules are used to allocate and distribute resources, how rules are enforced, and how rules help solve problems (Imperial & Yandle, 2005). This is a simple yet useful framework to identify, categorize, and organize a range of complex explanatory factors and variables within a foundational structure of logical relationships (McGinnis, 2011).

Building on this foundation, we developed a model based on the IAD framework to evaluate the governance of protected areas (Figure 1). The IAD framework has three major components: inputs and action situations, both of which lead to outcomes that can be evaluated. Inputs include the contextual factors (such as natural conditions, communities, and institutions) that set the context within which an action situation is situated (Ostrom, 2010a). The action situation is where various types of actors interact and policy choices are made. Outcomes are shaped by the inputs, action situations, and their interactions.

In our model, protected areas are defined as the action situations where interactions among different types of actors occur. These interactions are affected by three variables describing context or setting: biophysical attributes, community characteristics, and institutional arrangements. The protected area outcomes in our model include biodiversity conservation, community livelihoods, and sustainable tourism (Nyaupane & Poudel, 2011). Evaluation of these outcomes helps determine the relative success of different governance approaches used in the management of protected areas. Eventually, the outcomes and the process help improve the institutional arrangements.

Our framing integrates three, often competing goals, sustainable tourism, biodiversity conservation, and community livelihoods. It also builds explicitly integrated adaptative management through the feedback loops. We highlight how and whether PAs are using monocentric or polycentric governance approaches and the types of entities involved in decision-making (government, non-government, community). Our approach enables examination of PAs success in achieving sustainable tourism, biodiversity conservation, and community livelihood goals with diverse institutional structures and management approaches.

Study design and data collection

This research employed a case study method to collect data incorporating both quantitative and qualitative evidence, such as documents, archives, interviews, and observations. This method is particularly useful when the variables of interest are numerous and have multiple data sources. Case studies may be exploratory, explanatory, or descriptive, and the research design may involve single- or multiple-case studies (Yin, 2014). The case study research is a very effective tool when only a few cases are available for in-depth analysis of complex issues. Having too many cases, however, leads to a shallow understanding of context and negates the advantage of case study research (Dyer & Nobeoka, 2000). This research adopted an exploratory, multimethod, and multiple-case study design (Dyer & Wilkins, 1991), which allows several single-case studies to be simultaneously conducted in more than one context. Similarly, for the case study research, the use of multiple researchers for data analysis is suggested to improve reliability (Ravenswood, 2011). Selection of cases is an important aspect of the case study approach, in which researchers select one or two cases to represent multiple cases so that the variables of interest are transparently observable to compare (Eisenhardt, 1989). The case study method is commonly used for analyzing the IAD framework (Ostrom et al., 2006).

Nepal's protected area system offers an important case study for understanding various governance approaches, as the country is experimenting with a wide range of governance approaches from strict protection to community-managed approaches. Nepal, a Himalayan country in South Asia, has diverse topographic conditions ranging from lowland plain to the tallest mountain in the world, Mt Everest, which makes the country very rich in biodiversity. Nepal's protected area system consists of 12 national parks, one wildlife reserve, one hunting reserve, six conservation areas, and 13 buffer zones, covering about 24% of the land area of the country (Department of National Parks and Wildlife Conservation [DNPWC], 2018). The protected area governance in Nepal is primarily guided by three approaches. First, protected areas that are co-managed by both the government and local communities, in which core areas of national parks, conservation areas, and wildlife reserves are managed by the government, whereas the buffer zones are managed by the communities. Second, protected areas that are managed by a national NGO, the National Trust for Nature Conservation (DNPWC, 2018; National Trust for Nature Conservation [NTNC], 2018; WWF, 2018) and currently there are three conservation areas managed by the NGO. Third, protected areas that are managed by local communities. The Kanchenjunga Conservation Area is the only protected area under this third category, which was handed over to the local community by the Government of Nepal in 2006.

Using the purposive sampling strategy, three protected areas were chosen as case studies to represent the three approaches, respectively: Chitwan National Park, the Annapurna Conservation Area, and the Kanchenjunga Conservation Area (Figure 2). These areas were also selected to represent the range of protected areas in Nepal in terms of physiographic zones (lowland Terai, mid-hills, and high mountains, respectively).

Data collection

One of the key strategies for successful data collection in case study research is to use a variety of data sources, which facilitates triangulation to enhance the validity, credibility and trustworthiness of the information (Eisenhardt, 1989). This study combines several methods for data collection, including published and unpublished documents, interviews, and observations. Published sources include books, journal papers, newspaper articles, periodic plans, laws, and policies. For additional information, the authors also searched organizational documents, including annual progress reports, periodic financial reports, meeting minutes, and internal databases. Additionally, 55 face-to-face interviews (15 short interviews and 40 in-depth interviews) were conducted with park managers (11), nongovernmental and community-based organizations (8), community leaders/residents (20), and tourism businesses owners and employees, tour operators, and tour guides (12). A purposeful and snowball sampling technique was employed to recruit participants from different stakeholder groups. The initial list of participants through purposeful

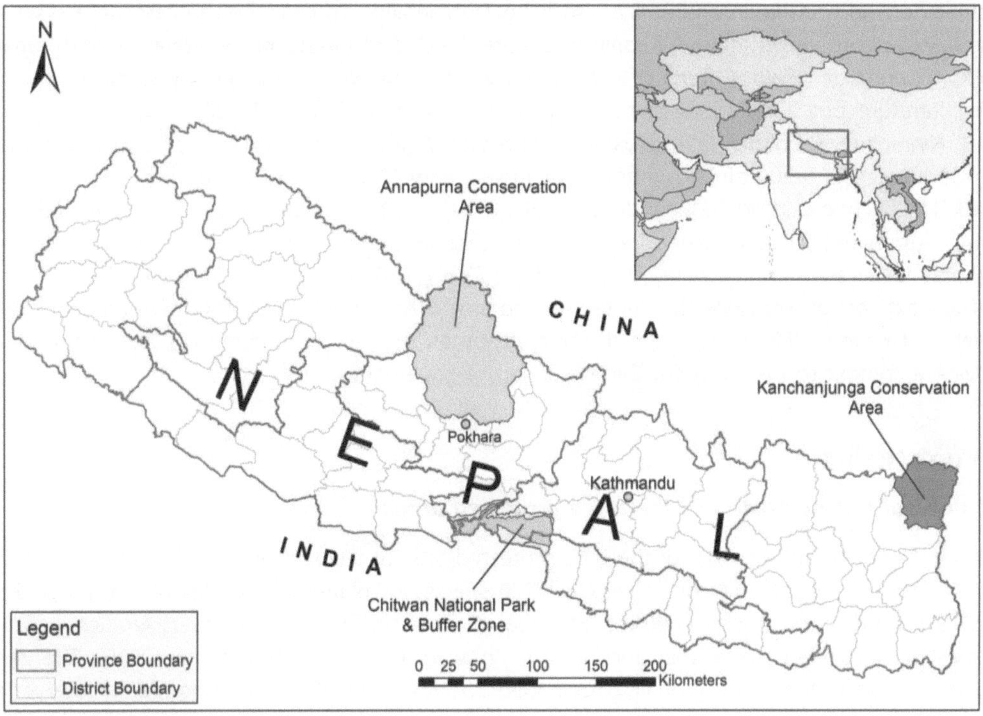

Source: Authors

Figure 2. Map of Study Sites.
Source: Authors.

sampling included 18 participants, followed by an additional 37 participants through snowball sampling. The total number of interviewees was guided by the fulfillment of theoretical saturation (Padgett, 2008). The authors made several field trips and office visits to conduct both short and in-depth interviews with relevant stakeholders at local, regional, and central levels. Short interviews lasted 10-20 minutes and were particularly useful for identifying participants for in-depth interviews, locating secondary data sources, and triangulating the secondary data. In-depth interviews were longer (30-60 minutes) and were useful as a complement to the secondary data. Interview guides were prepared in advance to maintain consistency across the interviewees and protected areas. Interviews with local residents were conducted with the support of locally hired research assistants. Notes were taken for both types of interviews, and the in-depth interviews were also audio-recorded with the permission of interviewees. The researchers took notes to record observations during their field trips. Most of the interviews were conducted in Nepali and later translated into English during transcribing. The first and second authors have substantial experience in working with various PA management institutions, both governmental and nongovernmental, and researching PA systems in Nepal, which was critical in identifying key interviewees, interview and document translation, accessing and analyzing documents, and data triangulation.

The interviews, secondary sources, and observations produced a massive amount of data, which were organized into a Microsoft Excel spreadsheet according to the variables in the model developed to evaluate protected area governance approaches (see Figure 1). Using the technique developed by Bowen (2009), our goal was to triangulate the data and come up with a confluence of evidence. This process helps investigators to be objective and reduce bias inherent to a single method or single source of qualitative data analysis. One of the co-authors who was not involved in data collection directly, but fluent in both English and Nepali verified the accuracy of the translation and analysis to increase reliability. To evaluate the cases in a systematic way, a-priori specifications of constructs were developed based on the literature and frameworks (Eisenhardt, 1989). Criteria of evaluation of outcomes were also developed based on the prior literature but adapted to fit the purpose of this study (McGinnis, 2011; Nyaupane et al., 2018; Nyaupane & Poudel, 2011). Using Institutional Analysis evaluation criteria, the outcomes were evaluated in a 20-point scale (four categories): poor (0-5), fair (6-10), good (11-15), and very good (16-20). Two authors independently evaluated the outcomes after reading the detailed outcomes summarized in an Excel spreadsheet. In case of discrepancy in the assigned value, the authors met face-to-face, exchanged each other's views, and agreed to a common value.

The next section provides the context of the study. As suggested by the IAD framework, we developed a model (Figure 1), in which three attributes: biophysical, community, and institutional provide a context to the study for the evaluation of governance.

Context setting

Biophysical characteristics of the three protected areas

Chitwan National Park consists of the core protected area of 932 km^2 and the buffer zone, which covers 750 km^2 around the core area. Over 570 species of plants, 70 species of mammals, 546 species of birds, 120 species of fishes, 150 species of butterflies, 47 species of reptiles, and 55 species of amphibians have been recorded in Chitwan (Chitwan National Park, 2019). The park provides habitats for many globally threatened species of wildlife, including the greater one-horned Asian rhinoceros, royal Bengal tiger, and gharial crocodile. Chitwan contains the second largest population of the greater one-horned rhinoceros in the world. The park was designated as a UNESCO World Heritage Natural Site in 1984.

The Annapurna Conservation Area, the largest protected area of Nepal that covers 7629 km^2, is known for its diverse geography, as the altitude ranges between 1,000 m and 8,091 m (Mt. Annapurna, the tenth highest peak in the world). The Annapurna region is rich in floral and faunal diversity, including 1,226 flowering plant species, 105 species of mammals, 518 species of birds, 40 species of reptiles, and 23 species of amphibians (NTNC, 2018). It is home to several rare and endangered wild animals, including the snow leopard, musk deer, Tibetan argali, Tibetan fox, Tibetan wolf, Tibetan wild ass, and Himalayan brown bear.

The Kanchenjunga Conservation Area, extending over 2035 km^2 in the Himalayas of eastern Nepal, ranges from 1,200 m to 8,586 m (Mount Kanchenjunga, the third highest peak in the world). The area is home to 844 species of plants (including 24 species of rhododendron and 48 varieties of orchids), over 30 species of mammals, 10 species of reptiles, and 158 species of birds (Kanchenjunga Conservation Area Management Council, 2013b; WWF, 2018). Endangered species recorded in the Kanchenjunga area include the snow leopard, red panda, Himalayan black bear, musk deer, and Tibetan wolf.

Community characteristics of the three protected areas

The buffer zone of Chitwan National Park is the most populated among the three areas, with a population of 273,977 people (Chitwan National Park, 2019). This population is so diverse that virtually every major religious, ethnic, cultural, and linguistic group of Nepal is represented. The area is very accessible and well connected via road and air networks with major cities of Nepal, including Kathmandu and Pokhara. The economy of the area is diverse, with farming, animal husbandry, trade, service, and tourism being the major engines of the local economy. Its agriculture sector is more modernized and profitable compared to those of other parts of Nepal. Many people own micro-enterprises, such as retail, grocery, and hardware stores, hotels, and restaurants. Jobs in the government, nonprofit, and private sectors (e.g. guide, cook, and server) are growing rapidly.

The Annapurna Conservation Area is inhabited by about 120,000 people (NTNC, 2018). This population is diverse in terms of race (Tibeto-Burman and Indo-Aryan), religion (Hindu, Buddhist, and pre-Buddhist), and ethnicity (over 10 ethnic groups). The Annapurna region is relatively more accessible compared to other mountain-protected areas of Nepal. Several points in this conservation area are connected by all-weather or seasonal roads to Pokhara, a major city and tourism hub of Nepal. People in the Annapurna region are historically dependent on agriculture, with livestock as a secondary livelihood option. Their agricultural practices are traditional and barely meet the subsistence needs of the local population. Other sources of income are service (foreign and domestic), trade (with Tibet and other parts of Nepal), and tourism. People are traditionally dependent on forest resources for timber, fuelwood, fodder, grazing, and non-timber forest products. The demand for timber and fuelwood is especially high in the areas receiving a large number of tourists.

The Kanchenjunga Conservation Area is sparsely populated and has approximately 6,500 inhabitants (World Wildlife Fund (WWF), 2018). People of more than 10 ethnicities live in the area; the Sherpa, Limbu, Rai, and Bhote are the major ethnic groups. People practice Hinduism, Buddhism, and Animistic religious traditions. The region is very remote and lacks basic infrastructures. It has no connection to road networks, and the nearest road head is about a one-day-walk (12 hours) away. Agriculture is the main source of livelihood for the majority of households. Agricultural practices are traditional and barely meet subsistence needs. Shifting cultivation, also termed as slash-and-burn agriculture, has been practiced for generations. The cultivation of herbal plants, specifically chiraito and cardamom, has appeared as a major income source in recent years. Other possible livelihoods include pastoralism, non-timber forest products, and tourism. Forests are sources of timber for house construction, firewood for cooking and heating, and fodder for livestock.

Institutional arrangements of the three protected areas

Nepal's National Park and Wildlife Conservation Act of 1973 (Government of Nepal, 2018) provides a legal basis for the establishment and governance of protected areas at the national level. The National Park and Wildlife Conservation Act requires the Department of National Parks and Wildlife Conservation to manage national parks. The management of conservation areas may be delegated to any institution established to conserve nature and natural resources. Institutional arrangements of these three protected areas are summarized in Table 1.

For Chitwan National Park, the DNPWC handles the financial, legal, and technical matters related to forests, wildlife, and the natural environment. The buffer zone of Chitwan National Park, however, is managed by local people, embracing the Integrated Conservation and Development Project (ICDP) approach. This approach recognizes some traditional rights of local people and considers public participation as an indispensable tool to achieve conservation goals. The Chitwan National Park manages the core area of the national park with the help of the Nepalese Army. The Army is responsible for park protection, including controlling poaching activities, illegal fishing and extraction of resources, and livestock grazing. The park office carries out works related to buffer zone management, including the provision of technical support to establish community-based organizations. Records show that one buffer zone management committee, 21 buffer zone user committees, 1781 buffer zone user groups (856 women only, 852 men only, and 73 mixed groups), and 61 buffer zone community forests have been formed in the buffer zone (Chitwan National Park, 2019).

Management of the Annapurna Conservation Area is entrusted to the NTNC, an autonomous, not-for-profit organization. The NTNC has been managing this conservation area with the help of

Table 1. Institutional arrangements of the three protected areas.

Protected areas	Chitwan National Park	Annapurna Conservation Area	Kanchenjunga Conservation Area
Act/Regulations	NPWC Act, 1973 Buffer Zone: 4th Amendment in 1993	NPWC Act, 1973 Conservation Area: 3rd Amendment in 1989	NPWC Act, 1973 Kanchenjunga Conservation Regulation, 2008
IUCN category	II (core area) and VI (Buffer Zone)	VI	VI
Management	DNPWC and Buffer Zone Council	NGO (NTNC)	KCAMC
Establishment	National Park 1973 Buffer zone 1996	1986	1998
Partners	Nepalese Army, CBOs, NTNC and WWF Nepal	GON/MFE/DNPWC and CBOs	GON/MFE/DNPWC, and WWF Nepal
CBO structure	Three tiers: protected area, management unit, and settlement	One tier: village development committee	Three tiers: conservation area, management unit, and settlement
Number of CBOs	1781	1000	110
Management Committees	21 Buffer Zone User Committee	57 Conservation Area Management Committee	1 Conservation Area Management Council

CBO- Community-based organization, DNPWC- Department of National Parks and Wildlife Conservation, GON- Government of Nepal, KCAMC- Kanchenjunga Conservation Area Management Council, MFSC- Ministry of Forest and Environment, NPWC-National Parks and Wildlife Conservation, NTNC- National Trust for Nature Conservation,.
Source: Authors document analysis.

the Government of Nepal and local people. The roles of the government are to approve laws and policies, monitor and evaluate the progress of the project, and issue executive orders for sustainable management of the conservation area (Government of Nepal, 1997; 2018). The Department of National Parks and Wildlife Conservation maintains a liaison office to act as the bridge between the Government of Nepal and the NTNC. DNPWC is also responsible for law enforcement to control wildlife-related crimes. The Annapurna Conservation Area is the first protected area in Nepal without military protection.

The NTNC is globally credited for pioneering the ICDP model through the Annapurna Conservation Area Project (ACAP) (NTNC, 2009). This project's ICDP model considers local communities as both principal actors and prime beneficiaries of the conservation and development programs. In more than 33 years of implementation, the project has formed 1,000 community-based organizations, including 57 conservation area management committees, 135 forest management subcommittees, 46 tourism management subcommittees, 75 green force clubs, 303 women groups, and 58 savings and credit groups (Annapurna Conservation Area Project Monitoring & Evaluation Team, 2012; NTNC, 2018).

Management of the Kanchenjunga Conservation Area was initially a collaboration between the DNPWC, the WWF Nepal, and local communities. The DNPWC, with the technical and financial support of WWF Nepal, launched the Kanchenjunga Conservation Area Project in March 1998. This project also embraced the ICDP approach and focused on the formation and strengthening of local institutions. As a result, the Kanchenjunga Conservation Area Management Council was formed in 2003, representing various self-organized conservation area user committees, user groups, and mother groups. The Government of Nepal handed over the management responsibilities of the conservation area to the Kanchenjunga Conservation Area Management Council in 2006 (Kanchenjunga Conservation Area Management Council, 2013b). This decision was a major milestone in the history of PA governance in Nepal, as the Kanchenjunga Conservation Area became Nepal's first community-managed protected area. Since the handover, the DNPWC has provided necessary technical, legal (including law enforcement), and financial support. WWF Nepal works as an intermediary in garnering technical and financial assistance of donors to the management council. Within the last twenty years of community management, significant

Table 2. Summary of evaluation of the outcomes of protected areas.

Outcomes	Chitwan	Annapurna	Kanchenjunga
Biodiversity conservation			
Community forests	Good	Fair	Good
Forest protection	Good	Fair	Fair
Wildlife research and monitoring	Good	Fair	Fair
Wildlife protection	Good	Poor	Poor
Human-wildlife conflict mitigation	Poor	Fair	Fair
Community livelihoods			
Access to resources	Good	Fair	Good
Capacity building	Fair	Good	Fair
Economic opportunities	Fair	Fair	Poor
Community development	Fair	Good	Fair
Sustainable tourism			
Visitation and growth	Good	Good	Poor
Tourism marketing	Fair	Fair	Poor
Facilities and services	Good	Good	Poor
Overall tourism impacts	Fair	Fair	Poor

Note: Poor (0–5), fair (6–10), good (11–15), and very good (16–20).

progress has been made in forming one council, 129 community-based organizations, including seven conservation area user committees, 27 community forests user groups, 42 user groups, 35 mothers' groups, four snow leopard conservation committees, eight community-based anti-poaching units, and six community-based fire management groups (Kanchenjunga Conservation Area Management Council, 2013a; WWF, 2018).

Among the three PAs, communities have a relatively higher level of agency to make decisions in Kanchenjunga than Chitwan and Annapurna. Both Chitwan National Park and the Annapurna Conservation Area have management committees authorized by the regulations to make decisions on provisioning, appropriation, developing operation plans, monitoring, and enforcement. However, the Kanchenjunga Conservation Area has an apex body, the conservation area management council, which has a higher level of decision-making power than the committees in Chitwan and Annapurna. The role of the state is the highest in governing Chitwan National Park's core area.

Evaluation of outcomes- conservation, livelihood and tourism

Outcomes for the protected areas are evaluated with respect to three major criteria: biodiversity conservation (5 sub-criteria), community livelihoods (4 sub-criteria), and sustainable tourism (4 sub-criteria) (See Table 2).

Biodiversity conservation

Biodiversity conservation is one of the major accomplishments of Chitwan National Park, but there has been less conservation success in the Annapurna Conservation Area and Kanchenjunga Conservation Area. Community forestry programs have been implemented in Chitwan and Kanchenjunga, but no such provision exists in the Annapurna. There are 61 community forest user groups managing 10,888 ha. of forests in the buffer zone of Chitwan (Chitwan National Park, 2015), and 27 community forest user groups manage 73,327 ha. of forest and pasturelands in the Kanchenjunga area (Kanchenjunga Conservation Area Management Council, 2013b; WWF, 2018). Illicit collection of forest products, forest fires, and encroachment pose varying degrees of threat to forests in all three protected areas. Illegal extraction and smuggling of timber and fuelwood is a common ubiquitous challenge among all three cases. In the Annapurna and Kanchenjunga regions, non-timber forest products suffer problems of improper harvesting, illegal collection, and smuggling; forest fires are also recurring problems. The encroachment of land to resettle so-called "landless" people under the protection of political parties is yet to be regulated in all areas. Furthermore, hotel construction in Annapurna, infrastructure projects in Chitwan, and slash-and-burn agriculture and non-timber forest product farming in Kanchenjunga also encourage forest encroachment.

Several studies on wildlife, including population status, species ecology, and habitat condition, have been conducted in these protected areas. In the Annapurna region studies document increase in forest cover and overall bio-diversity richness. In Chitwan National Park, there have been several longitudinal studies and population censuses of certain endangered species. The studies show considerable increases in wildlife populations, including the one-horned rhinoceros (310 in 1978, 544 in 2000, 372 in 2005, 605 in 2015), Bengal tiger (60 in 2000, 93 in 2018), and gharial crocodile (53 in 1970, 119 in 2015), mostly reared in the Crocodile Breeding Center and released into the rivers (Chitwan National Park, 2015, 2019; DNPWC, 2018). In the Kanchenjunga Conservation Area, research shows a sizable population of the snow leopard (19-29 adults), blue sheep (1404 individuals), and musk deer (310 individuals) (Kanchenjunga Conservation Area Management Council, 2013a, 2013b). The population of snow leopards increased by 28% in four years (2009 to 2013), and blue sheep by 40% in eight years (2007 to 2015) (WWF, 2018).

Wildlife protection efforts are found to be more challenging, but successful in Chitwan compared to Annapurna and Kanchenjunga. With the help of the Nepalese Army, Chitwan National Park has been successful in law enforcement for wildlife protection. Data show that 148 rhinos were killed in the ten-year period between 2001-2010, but casualties dropped to zero in 2011, 2013, and 2015 (Chitwan National Park, 2013; 2015), as a result of strong anti-poaching initiatives and awareness programs. In the interviews we conducted, park officials in all three areas repeatedly acknowledged the support of local communities in conservation initiatives. Illegal hunting/poaching practices by local people have been significantly reduced, and community-based organizations have been proactive to control wildlife-related crimes. Due to the high economic value of wildlife parts and escalating international demand, anti-poaching operations need concerted and continuous efforts. However, the Annapurna Conservation Area and Kanchenjunga Conservation Area do not have a strong mechanism to deal with organized wildlife crimes, and their community-based organizations are not effective in controlling high-level wildlife crimes. Hence, high-level wildlife-related crimes are seen as a big challenge in community-based conservation approaches, where neither military not armed forest/wildlife guards are deployed. Interviewed park officials from all three PAs repeatedly mentioned sustainability of funding as the major challenge to conducting anti-poaching operations.

The negative interactions between humans and wildlife as a result of attacks on humans and property damage by wildlife and subsequent retaliation by humans are generally referred to as "human-wildlife conflict" (Lamichhane et al., 2018). The incidents of "human-wildlife conflict" caused by crop damage, livestock depredation, and human casualties are more frequent in recent years in all three protected areas. Local residents seem to be very concerned about crop damage by pest animals, such as rhinos, elephants, bears, wild boar, deer, and monkeys. The loss of livestock due to predators such as the snow leopard, common leopard, and tiger is also significant. Human casualties by wildlife is a major issue in Chitwan, where the incidents of human injury and death associated with wildlife attacks are disturbing, with 329 attacks by wildlife between 2003 and 2013 (Silwal et al., 2017). Most of these attacks were carried out by rhinos (38%), followed by tigers (21%), sloth bears (18%), elephants (9%), and wild boar (8%) (Silwal et al., 2017). In the fiscal year 2018-2019 alone, 12 people were killed and 32 were seriously injured in Chitwan (Chitwan National Park, 2019). These human-wildlife conflict incidents frequently sour the relationship between the protected areas and the local people, and they increase retaliatory killing of wildlife.

Both the secondary data and interviews reveal that various preventive and compensatory mechanisms have been put into practice to reduce human wildlife-conflicts. Preventive measures include fencing of crops, guarding of livestock, and construction of predator-proof corrals in the

Annapurna and Kanchenjunga regions, and construction of electric fences, watchtowers, and game-proof (mesh wire) fences in Chitwan. Chitwan National Park provides cash compensation to victims of human-wildlife conflict, based on government-approved wildlife damage relief guidelines. In Kanchenjunga, all four snow leopard conservation subcommittees run community-based livestock insurance schemes, and all seven conservation area user committees have established endowment funds to provide compensation for wildlife-caused financial loss. However, the budget allocated for compensation is limited in all three protected areas, so the partnership with local communities and other stakeholders helped initiate insurance schemes. There are other forms of compensation to support families. Chitwan National Park has initiated a free scholarship program for the children of those who have lost their lives because of wildlife attacks. In the fiscal year 2017/18, 16 students were provided a total amount of NRs 150,000 (about US$1,500) toward their tuitions and educational costs (Chitwan National Park, 2019).

Community livelihoods

The interviewees reported that the three protected areas have brought many benefits to local people, including access to natural resources, capacity building, income and employment, saving and credit facilities, and infrastructure development to local communities. Residents are allowed to collect forest products from the community forests in the Kanchenjunga Conservation Area and the buffer zone of Chitwan National Park. In the Annapurna Conservation Area, the forest management subcommittees allocate forest products to the residents. However, these subcommittees are not given the level of autonomy enjoyed by the community forests user groups regarding self-organization, forest management, forest products distribution, and use of community funds.

The protected areas have also offered many capacity-building opportunities to local people through leadership development, empowerment, and skill development. For example, various levels of community-based organizations provide leadership opportunities to local people. Special programs have been conducted to empower socioeconomically disadvantaged groups, such as the poor and women. These programs provide stipends for students (both girls and boys) from poor and marginalized families. For example, in the Annapurna area, scholarships and stipends were provided to more than 200 students in the fiscal year 2017/2018, some of them aimed toward socio-economically marginalized students (NTNC, 2018). Various skill development training such as beekeeping, cash crop farming, horticulture, sewing, weaving, carpet cutting, tour guiding, and cooking/baking have been offered to small farmers and economically disadvantaged groups to enhance their livelihood.

At the household level, the protected areas have brought several economic benefits, including income diversification, enterprise development, and access to saving and credit. Local people now derive income from cash crops, vegetable farming, fruit orchards, livestock, and non-timber forest products. For example, the Annapurna Conservation Area Project in particular is credited with the introduction of tea and coffee farming and the improvement of animal husbandry practices (NTNC, 2018). In the buffer zone of Chitwan National Park, four veterinary centers are established by buffer zone user committees, and the buffer zone program supports the purchasing of drugs and vaccines. The Kanchenjunga Conservation Area promotes green enterprises based on non-timber forest products.

Development of community infrastructure has been a focus area since the establishment of the Annapurna Conservation Area and Kanchenjunga Conservation Area, and the declaration of the buffer zone in Chitwan National Park. Infrastructure projects contribute to transportation (road, trails, and bridges), education (school buildings), health (drinking water systems, toilets, and waste management), culture (temples and monasteries), and energy (biogas, solar, and hydropower). Alternative energy has been ubiquitously adopted as a measure of forest protection. The Annapurna Conservation Area Project was the first project in Nepal to promote micro-hydro projects; the project has helped to construct 28 micro hydro-power projects that generate 468.3 KW electricity and benefit 2,228 households. In the Chitwan buffer zone, special emphasis has been given to biogas unit installation at the household level using human and animal waste. These biogas units reduce fuelwood need by up to 80% and help sanitation if connected to toilet waste. In the Kanchenjunga region, installation of solar panels and micro-hydro plants substitute the use of kerosene for producing light. The Annapurna Conservation Area and Kanchenjunga Conservation Area have distributed several hundred units of energy-saving devices, such as improved cooking stoves, low wattage cookers, back boilers, and solar water heaters.

Sustainable tourism

Chitwan National Park is the most visited protected area in Nepal. The number of tourists visiting the area has constantly increased, from 57,876 tourists in the FY 2003-2004 to 187,109 tourists in the FY 2018-2019 (Chitwan National Park, 2015; 2019). The Annapurna region is the most popular trekking destination in Nepal. This area received 172,730 international tourists in the FY 2017-2018, which is more than triple of its 44,969 visitors in the FY 2003-2004 (DNPWC, 2015; 2018). Tourism is limited in the Kanchenjunga; the area received 418 tourists in the FY 2003-2004 and only 821 tourists in the FY 2017-2018 (DNPWC, 2018). This low visitation is due to the area's remoteness, lack of infrastructure, and closed to free and independent tourists.

Marketing for tourism is poor in the Kanchenjunga Conservation Area, while the Annapurna Conservation Area and Chitwan National Park have been promoted as trekking tourism and wildlife tourism destinations, respectively, both nationally and internationally. Tourism facilities and services are highly developed in Chitwan and Annapurna, and poorly developed in the Kanchenjunga region. The buffer zone of Chitwan National Park contains well-developed tourism superstructures, including more than 90 professionally managed hotels. In the Annapurna region, there are more than 500 village inns/homestays, which are predominantly owned and operated by locals. Tourism establishments in the Kanchenjunga region include only 46 village inns/tea shops and 53 campsites.

Tourism is the main source of park revenue in Annapurna and Chitwan. In the FY 2016-2017, the Annapurna Conservation Area Project collected 2,150,000 United States dollars (USD); about 95% was contributed by tourism. Chitwan National Park in the FY 2017-2018 collected 2,188,109 USD, which came from tourist entry fees (83.7%), other tourism-related income (8.2%), and non-tourism sources (8.1%). The tourism revenue in the Kanchenjunga Conservation Area was only 8,850 USD in the FY 2017-2018 (DNPWC, 2018). The Government of Nepal allows conservation areas (including Annapurna and Kanchenjunga) to retain 100% of revenue, and national parks (including Chitwan National Park) to retain only 30-50% of revenue for buffer zone communities (Government of Nepal, 1996; 1997; 2008). However, the communities are allowed to keep 100% of the revenue generated in Chitwan's buffer zone. The impacts of tourism vary across the protected areas. Based on the type of tourism enterprises, the economic benefit of tourism is high in Chitwan and Annapurna but negligible in the Kanchenjunga region (low volume). Economic benefits are, however, asymmetrically distributed in Annapurna (benefits received only by the few families who own tourism businesses along the trekking routes) and Chitwan (tourism is concentrated in Sauraha, a gateway community). The social impacts (e.g. crowding) and environmental impacts (e.g. littering) of tourism are visible in Sauraha of Chitwan National Park and Ghandruk and Ghorepani of the Annapurna Conservation Area.

Discussion

Overall, outcomes related to biodiversity conservation, community livelihoods, and sustainable tourism are not uniform across Chitwan National Park, the Annapurna Conservation Area, and the Kanchenjunga Conservation Area (Table 2). The achievements of Chitwan National Park are good with respect to biodiversity conservation (except human-wildlife conflict), and fair or good regarding community livelihoods and sustainable tourism. The Annapurna Conservation Area has attained fair in four sub-categories and poor in wildlife protection in terms of biodiversity conservation. Their success in implementing livelihood and tourism development programs in the Annapurna Conservation Area is mixed (fair or good). Though the condition of tourism development is poor, conservation outcomes in the Kanchenjunga Area are mixed (one good, three fair, and one poor). Livelihood outcomes are also mixed (one good, two fair, and one poor) in the Kanchenjunga Conservation Area. The outcomes vary primarily because of their governance approaches and local contexts, as indicated in the IAD model. Overall, more desirable outcomes occur in state-community co-managed PA than NGO or community alone managed protected areas. Livelihood and tourism outcomes could be better achieved through NGO and community-managed PAs, but the protection of endangered species requires greater involvement of the state. The results indicate that no single governance approach should be regarded as the silver bullet to solve challenges associated with all protected areas. Thus, it is not imperative to discuss whether protected areas should be put under one entity, such as government, private, and community governance (Dudley, 2008). Similarly, it would be naïve to prescribe a governance approach according to the IUCN category of protected areas without first considering the biophysical, socioeconomic, and institutional contexts.

None of the three PAs are traditional state-led monocentric governance structures, yet existing institutional governance arrangements, a major component of IAD framework, are unnecessarily rigid, focusing on the marketing of limited successes rather than learning and improvement that is the most essential function of adaptive co-management (Plummer et al., 2012). For example, the Annapurna Conservation Area is a proud, self-proclaimed, and first-of-its-kind NGO-managed protected area in Nepal, which is reluctant to change to incorporate current political reform and involve the state, province, and local government, particularly in protecting wildlife and empowering local communities. In Chitwan National Park, no evaluation of the effectiveness of the Nepalese Army in park protection has been conducted, given that a large amount of money is being spent on their salary. Community participation is sought in conservation, but the presence of hundreds of armed guards might not be compatible with participatory conservation approaches. Additionally, the large footprints of the army inside the park, their demands of natural resources including water, firewood and timber, and waste disposal can have huge impacts on the ecosystem. The park was initially established as a Royal Hunting Reserve and it was the interest of the King and Royal families to keep the Army in the park, but now the political system has changed, as the country's royal dynasty has been abolished. In the case of the Kanchenjunga Conservation Area, although the area is handed over to the local community, the community's income generation opportunities, primarily from tourism, are limited due to the low number of tourists, which also limits the revenue needed for infrastructure and community development work. In sum adaptive co-management fails in these NGO, community-led, and polycentric systems due to intransigence of existing management priorities and the difficulty of advocating for change given existing power relationships.

Therefore, the multiple-functions, large-scale, and cross-scale stakeholder nature of PA systems require more than a monocentric community-based approach or simple polycentric approach. In these three PAs, the state, nongovernmental, community-based organizations, and local people are all involved at different levels in managing protected areas. Intergovernmental agencies, such as the Convention of Biological Diversity (CBD), the United National Framework Convention on Climate Change (UNFCCC), the United Nations Educational, Scientific and Cultural Organization (UNESCO), are also actively engaged in PA management. Leveraging the diversity of actors and further developing the polycentric nature of these arrangements with explicit inclusion of adaptive co-management feedbacks and pathways could provide flexibility to navigate change, yet as already discussed simply adding polycentric structures without the inclusion of adaptive co-management

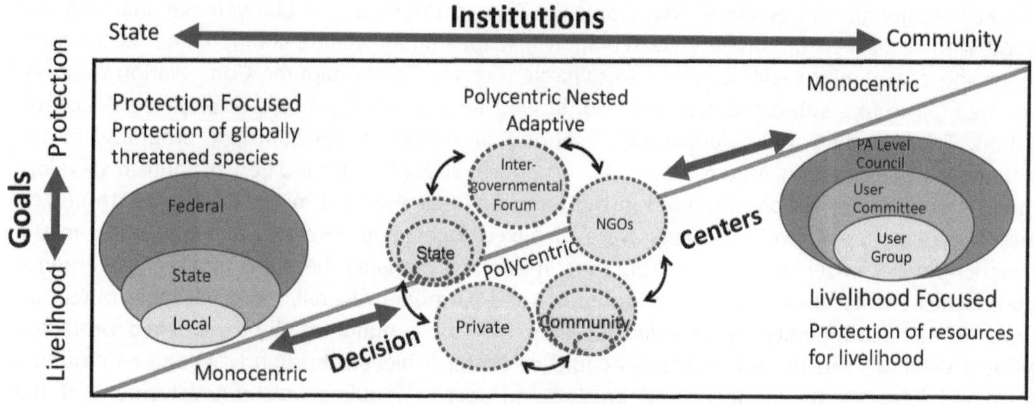

Source: Authors

Figure 3. Polycentric, adaptative co-management governance model.
Source: Authors.

pathways may lead to, or even increase, the inability of PA governing entities to deal with changing local needs, shifting political priorities, and new environmental challenges.

In order to elucidate the complexity of PA governance, we propose a new polycentric, adaptative co-management governance model (Figure 3). The model is built upon the previous literature on common-pool resources (McGinnis, 2011; Nagendra & Ostrom, 2012; Ostrom, 1990, 2002, 2005, Ostrom, 2010a; Ostrom et al., 2006) and the findings of this study. Given the multiple goals of protected areas and involvement of multiple actors working both vertically and horizontally, PA governance should go beyond a monocentric (single entity) or polycentric (multiple entities) approach to embrace a multi-layered (nested) polycentric adaptive co-management approach (Nagendra & Ostrom, 2012). The proposed model has three dimensions- goals, institutions, and decision centers. The goals range from protection focused to livelihood focused, and the institutions include state to community. The decision center dimension includes monocentric to polycentric governance. PA managed by a single entity, such as state, community, or NGO, is a monocentric approach, whereas PA co-managed by multiple entities collaboratively is polycentric. The monocentric approach has different layers, but they are vertical, also known as the nested model. For example, the state can be comprised of federal, state, and local governments. Similarly, the community can have multiple vertical layers, such as a higher-level council at the PA level, user committee (mid-level), and user group (community-level with few households).

The polycentric ideal-type PA governance structure also provides adaptive co-management pathways where diverse entities across governance levels provide feedback (center of the figure). As demonstrated in the study, a shift toward community-led PA governance, NGO led, or polycentric does not necessarily enable adaptive co-management. These alternatives to state-led monocentric governance balance some competing PA conservation, livelihood, and sustainable tourism goals, although there is evidence that none are able to balance all three simultaneously. In prior work, attention has primarily focused on governance structure, and a shift away from state-led PA governance, as a means to balance competing goals. The structure is not enough to overcome power relations or deal with novel threats; we must also focus on process and specifically the adaptive co-management pathways that allow feedbacks across entities and levels.

Conclusion

Establishing good governance across various ownership and responsibility arrangement is critically important for the effectiveness of PA systems (Lockwood, 2010). This study examined the types of governance most effective in managing protected areas in the given social-ecological systems developing a model that incorporates polycentricity and adaptive co-management in institutional design. A successful PA governance should involve state, communities, non-governmental organizations, intergovernmental forums, and private sectors acting together at different scales and levels with pathways to adapt to new challenges and means to adjust management priorities. The polycentric, NGO, and community-managed PA's in our study demonstrate the influence of power relations at play outside of state managed PAs. Thus, a thorough reform is needed to change the mindset of policymakers and park officials for them to consider locals and other stakeholders as equal partners in protected area conservation, as well as include less privileged community groups and priorities in community-led efforts.

This paper makes significant theoretical and practical contributions to PA governance. Theoretically, the study makes the case that multi-layered polycentric adaptive co-management approaches with explicit feedback mechanisms are more effective than rigid monocentric or simple polycentric approaches. Various organizations, including communities, governments, private parties, NGOs, and inter-governmental forums should be engaged in developing and implementing institutions for different functions, such as making rules, operation, monitoring, enforcement, conflict resolution, provision, marketing, fundraising, and appropriation. Emphasis should be placed on encouraging learning and adaptation across the stakeholders active at multiple scales and levels.

The multi-goal IAD model with evaluation criteria provides a means for future empirical research on PAs and other types of common-pool resources to evaluate diverse, sometimes competing, goals. The study contributes to the protected and sustainable tourism literature with a tailored Institutional Analysis and Development framework to analyze how local settings, particularly biophysical, socioeconomic and institutional context, impact outcomes. Moreover, this study established Institutional Analysis evaluation criteria, goals, sub-criteria and scales, which can be applied in other studies.

Researchers and practitioners of PAs should go beyond a quest for an institutional panacea for governance. Instead, it is imperative to know how to maximize the resilience and adaptability of PA systems. PAs may vary in terms of institutional structure and process, including the power distribution among stakeholders and provision of community-based organizations. The theoretical development of a polycentric adaptive co-management conceptual framework integrates adaptive co-management and polycentric literatures within the larger institutional literature providing a means to explore governance structures from local to state, protection or community livelihood orientation, and assess adaptation feedbacks. Theoretically, the study reveals a need for greater attention in institutional scholarship on protected areas on power relations and feedback mechanisms between decision-makers, as there is inherent intransigence in many, if not all, governance structures.

Acknowledgments

The authors would like to thank Mr Rabindra Maharjan for his help with the map.

Disclosure statement

No potential conflict of interest was reported by the authors.

ORCID

Gyan P. Nyaupane http://orcid.org/0000-0002-0814-8507

References

Agarwal, A. (2006). Rules, rule making, and rule breaking: Examining the fit between rule systems and resource use. In E. Ostrom, R. Gardner, & J. Walker, (Eds.), *Rules, games, & common-pool resources.* (pp. 267–282). The University of Michigan Press.

Annapurna Conservation Area Project Monitoring and Evaluation Team. (2012). *Monitoring and evaluation report of Annapurna Conservation Area Project*. Ministry of Forests and Soil Conservation.

Bixler, R. P. (2014). From community forest management to polycentric governance: Assessing evidence from the bottom up. *Society & Natural Resources, 27*(2), 155–169. https://doi.org/10.1080/08941920.2013.840021

Borrini-Feyerabend, G., & Hill, R. (2015). Governance for the conservation of nature. In G. L. Worboys, M. Lockwood, A. Kothari, S. Feary & I. Pulsford (Eds.), *Protected area governance and management* (pp. 26–71). ANU Press.

Bowen, G. A. (2009). Document analysis as a qualitative research method. *Qualitative Research Journal, 9*(2), 27–40. https://doi.org/10.3316/QRJ0902027

Bown, N. K., Gray, T. S., & Stead, S. M. (2013). Co-management and adaptive co-management: Two modes of governance in a Honduran marine protected area. *Marine Policy, 39*, 128–134. https://doi.org/10.1016/j.marpol.2012.09.005

Carlisle, K., & Gruby, R. L. (2019). Polycentric systems of governance: A theoretical model for the commons. *Policy Studies Journal, 47*(4), 927–952. https://doi.org/10.1111/psj.12212

Chitwan National Park (2013). *Annual report (16 July 2012 to 15 July 2013).*

Chitwan National Park (2015). *Annual report (16 July 2014 to 15 July 2015).*

Chitwan National Park (2019). *Annual report (17 July 2018 to 16 July 2019).*

DNPWC (2015). *Annual report (17 July 2014 to 16 July 2015).*

DNPWC (2018). *Annual report (17 July 2017-16 July 2018).*

Dudley, N. (2008). *Guidelines for applying protected area management categories.* IUCN.

Dyer, J. H., & Nobeoka, K. (2000). Creating and managing a high-performance knowledge-sharing network: The Toyota case. *Strategic Management Journal, 21*(3), 345–367. https://doi.org/10.1002/(SICI)1097-0266(200003)21:3<345::AID-SMJ96>3.0.CO;2-N

Dyer, W. G., & Wilkins, A. L. (1991). Better stories, not better constructs, to generate better theory: A rejoinder to Eisenhardt. *Academy of Management Review, 16*(3), 613–619. https://doi.org/10.5465/amr.1991.4279492

Eisenhardt, K. M. (1989). Building theories from case study research. *Academy of Management Review, 14*(4), 532–550. https://doi.org/10.5465/amr.1989.4308385

Gardner, C. J., Nicoll, M. E., Mbohoahy, T., Oleson, K. L. L., Ratsifandrihamanana, A. N., Ratsirarson, J., René de Roland, L.-A., Virah-Sawmy, M., Zafindrasilivonona, B., & Davies, Z. G. (2013). Protected areas for conservation and poverty alleviation: Experiences from Madagascar. *Journal of Applied Ecology, 50*(6), 1289–1294. https://doi.org/10.1111/1365-2664.12164

Government of Nepal (1996). *Buffer Zone Management Regulation, 1996.*

Government of Nepal (1997). *Conservation Area Management Regulation, 1997.*

Government of Nepal (2008). *Kanchanjunga Conservation Area Regulation, 2008.*

Government of Nepal. (2018). *National Park and Wildlife Conservation Act, 1973.*

Gruby, R. L., & Basurto, X. (2013). Multi-level governance for large marine commons: Politics and polycentricity in Palau's protected area network. *Environmental Science & Policy, 33*, 260–272. https://doi.org/10.1016/j.envsci.2013.06.006

Hassanali, K. (2013). Towards sustainable tourism: The need to integrate conservation and development using the Buccoo Reef Marine Park, Tobago, West Indies. *Natural Resources Forum, 37*(2), 90–102. https://doi.org/10.1111/1477-8947.12004

Hayes, T. M. (2006). Parks, people, and forest protection: An institutional assessment of the effectiveness of protected areas. *World Development, 34*(12), 2064–2075. https://doi.org/10.1016/j.worlddev.2006.03.002

Hockings, M. (2003). Systems for assessing the effectiveness of management in protected areas. *BioScience, 53*(9), 823–832. https://doi.org/10.1641/0006-3568(2003)053[0823:SFATEO]2.0.CO;2

Imperial, M. T., & Yandle, T. (2005). Taking institutions seriously: using the IAD framework to analyze fisheries policy. *Society & Natural Resources, 18*(6), 493–509. https://doi.org/10.1080/08941920590947922

IUCN (2020). Protected Planet Digital Report. https://livereport.protectedplanet.net/chapter-2

Kanchenjunga Conservation Area Management Council (2013a). *Annual progress report (16 July 2012 to 15 July 2013).*

Kanchenjunga Conservation Area Management Council (2013b). *Kanchenjunga Conservation Area Management Plan (2013-2018).*

Kingsford, R. T., Biggs, H. C., & Pollard, S. R. (2011). Strategic adaptive management in freshwater protected areas and their rivers. *Biological Conservation, 144*(4), 1194–1203. https://doi.org/10.1016/j.biocon.2010.09.022

Kitamura, K., & Clapp, R. A. (2013). Common property protected areas: Community control in forest conservation. *Land Use Policy, 34*, 204–212. https://doi.org/10.1016/j.landusepol.2013.03.008

Lamichhane, B. R., Persoon, G. A., Leirs, H., Poudel, S., Subedi, N., Pokheral, C. P., Bhattarai, S., Thapaliya, B. P., & de longh, H. H. (2018). Spatio-temporal patterns of attacks on human and economic losses from wildlife in Chitwan National Park. *Plos One, 13*(4), e0195373. https://doi.org/10.1371/journal.pone.0195373

Lockwood, M. (2010). Good governance for terrestrial protected areas: A framework, principles and performance outcomes. *Journal of Environmental Management, 91*(3), 754–766. https://doi.org/10.1016/j.jenvman.2009.10.005

McGinnis, M. D. (2011). An introduction to IAD and the language of the Ostrom Workshop: A simple guide to a comlex framework. *Policy Studies Journal, 39*(1), 169–183. https://doi.org/10.1111/j.1541-0072.2010.00401.x

McGinnis, M. D., & Ostrom, E. (2012). Reflections on Vincent Ostrom, public administration, and polycentricity. *Public Administration Review, 72*(1), 15–25. https://doi.org/10.1111/j.1540-6210.2011.02488.x

Nagendra, H., & Ostrom, E. (2012). Polycentric governance of multifunctional forested landscapes. *International Journal of the Commons, 6*(2), 104–133. https://doi.org/10.18352/ijc.321

Nash, R. (2014). *Wilderness and the American mind* (5th ed.). Yale University Press.

National Trust for Nature Conservation (NTNC). (2009). *Management plan of Annapurna Conservation Area (2009-2012).*

NTNC. (2018). *Annual Report 2018.*

Nyaupane, G. P., & Poudel, S. (2011). Linkages among biodiversity, livelihood, and tourism. *Annals of Tourism Research, 38*(4), 1344–1366. https://doi.org/10.1016/j.annals.2011.03.006

Nyaupane, G. P., Poudel, S., & Timothy, D. J. (2018). Assessing the sustainability of tourism systems: A social–ecological approach. *Tourism Review International, 22*(1), 49–66. https://doi.org/10.3727/154427218X15202734130440

Ostrom, E., Gardner, R., & Walker, J. (Eds.) (2006). *Rules, games, & common-pool resources.* The University of Michigan Press.

Ostrom, E. (1990). *Governing the commons: The evolution of institutions for collective action.* Cambridge University Press.

Ostrom, E. (2002). Common-pool resources and institutions: Toward a revised theory. *Handbook of Agricultural Economics, 2*, 1315–1339.

Ostrom, E. (2005). Doing institutional analysis digging deeper than markets and hierarchies. In C. Menard & M. M. Shirley (Eds.). *Handbook of new institutional economics* (pp. 819–848). Springer.

Ostrom, E. (2010a). Beyond markets and states: Polycentric governance of complex economic systems. *American Economic Review, 100*(3), 641–672. https://doi.org/10.1257/aer.100.3.641

Ostrom, E. (2010b). Institutional analysis and development: Elements of the framework in historical perspective. *Historical Developments and Theoretical Approaches in Sociology, 2*, 261–288.

Ostrom, E. (2011). Background on the institutional analysis and development framework. *Policy Studies Journal, 39*(1), 7–27. https://doi.org/10.1111/j.1541-0072.2010.00394.x

Ostrom, E., & Cox, M. (2010). Moving beyond panaceas: A multi-tiered diagnostic approach for social-ecological analysis. *Environmental Conservation, 37*(4), 451–463. https://doi.org/10.1017/S0376892910000834

Ostrom, E., Janssen, M. A., & Anderies, J. M. (2007). Going beyond panaceas. *Proceedings of the National Academy of Sciences of the United States of America, 104*(39), 15176–15178. https://doi.org/10.1073/pnas.0701886104

Padgett, D. K. (2008). *Qualitative methods in social work research* (2nd ed.). Sage.

Parker, P., & Thapa, B. (2011). Distribution of benefits based on household participation roles in decentralized conservation within Kanchenjunga Conservation Area Project. *Environment, Development and Sustainability, 13*(5), 879–899. https://doi.org/10.1007/s10668-011-9296-6

Pennington, M. (2013). Elinor Ostrom and the robust political economy of common-pool resources. *Journal of Institutional Economics, 9*(4), 449–468. https://doi.org/10.1017/S1744137413000258

Pfueller, S. L., Lee, D., & Laing, J. (2011). Tourism partnerships in protected areas: Exploring contributions to sustainability. *Environmental Management, 48*(4), 734–749. https://doi.org/10.1007/s00267-011-9728-y

Plummer, R., Crona, B., Armitage, D. R., Olsson, P., Tengö, M., & Yudina, O. (2012). Adaptive comanagement: A systematic review and analysis. *Ecology and Society, 17*(3), 11. https://doi.org/10.5751/ES-04952-170311

Plummer, R., & Fennell, D. A. (2009). Managing protected areas for sustainable tourism: Prospects for adaptive co-management. *Journal of Sustainable Tourism, 17*(2), 149–168. https://doi.org/10.1080/09669580802359301

Ravenswood, K. (2011). Eisenhardt's impact on theory in case study research. *Journal of Business Research, 64*(7), 680–686. https://doi.org/10.1016/j.jbusres.2010.08.014

Schultz, L., Duit, A., & Folke, C. (2011). Participation, adaptive co-management, and management performance in the world network of biosphere reserves. *World Development, 39*(4), 662–671. https://doi.org/10.1016/j.worlddev.2010.09.014

Sebele, L. S. (2010). Community-based tourism ventures, benefits and challenges: Khama Rhino Sanctuary Trust, Central District. *Tourism* Management, *31*(1), 136–146. https://doi.org/10.1016/j.tourman.2009.01.005

Sen, S. (2001). Involving stakeholders in aquaculture policy-making, planning and management. *Aquaculture in the Third Millennium*, 83–94.

Silwal, T., Kolejka, J., Bhatta, B. P., Rayamajhi, S., Sharma, R. P., & Poudel, B. S. (2017). When, where and whom: assessing wildlife attacks on people in Chitwan National Park, Nepal. *Oryx, 51*(2), 370–377. https://doi.org/10.1017/S0030605315001489

Sullivan, A., York, A., White, D., Hall, S., & Yabiku, S. (2017). De jure versus de facto institutions: trust, information, and collective efforts to manage the invasive mile-a-minute weed (Mikania micrantha). *International Journal of the Commons*, *11*(1), 171–199. https://doi.org/10.18352/ijc.676

Tang, S. Y. (2006). Institutions and performance in irrigation system. In E. Ostrom, R. Gardner, & J. Walker (Eds.), *Rules, games, & common-pool resources* (pp 225–246). The University of Michigan Press.

Tumusiime, D. M., & Vedeld, P. (2012). False promise or false premise? *Conservation and Society*, *10*(1), 15. https://doi.org/10.4103/0972-4923.92189

UNEP (2016). Protected Planet Report 2016. Retrieved June 25, 2020, from https://www.protectedplanet.net/

UNEP-WCMC & IUCN (2018). *Protected Planet: The World Database on Protected Areas (WDPA), July 2018 version.* .

van Wilgen, B. W., & Biggs, H. C. (2011). A critical assessment of adaptive ecosystem management in a large savanna protected area in South Africa. *Biological Conservation*, *144*(4), 1179–1187. https://doi.org/10.1016/j.biocon.2010.05.006

World Wildlife Fund (WWF) (2018). *WWF Report- Kanchanjunga Conservation Area Project 1998-2017.*https://d2ouvy59p0dg6k.cloudfront.net/downloads/kcap_retro_report_2018.pdf

Yin, R. K. (2014). *Case study research: Design and methods* (5th ed.). Sage Publications.

Index

Page numbers in **bold** refer to tables and those in *italic* refer to figures.